CNC Machining Certification Exam Guide

CNC Machining Certification Exam Guide

Setup, Operation and Programming

Ken Evans

Industrial Press

Industrial Press, Inc.

32 Haviland Street, Suite 3
South Norwalk, Connecticut 06854
Phone: 203-956-5593
Toll-Free in USA: 888-528-7852
Fax: 203-354-9391
Email: info@industrialpress.com

Author: Ken Evans
Title: CNC Machining Certification Exam Guide
   Setup, Operation and Programming
Library of Congress Control Number is on file with the Library of Congress.

© by Industrial Press, Inc.
All rights reserved. Published in 2020.
Printed in the United States of America.

ISBN (print):   978-0-8311-3637-6
ISBN (ePUB):   978-0-8311-9501-4
ISBN (eMOBI): 978-0-8311-9502-1
ISBN (ePDF):   978-0-8311-9500-7

Publisher and Editorial Director: Judy Bass
Copy Editor: Elise Davies
Compositor: Paradigm Data Services (P) Ltd., Chandigarh

industrialpress.com
ebooks.industrialpress.com

10 9 8 7 6 5 4 3 2 1

Contents

Contents

Acknowledgments

First, I give thanks to God, for blessing me with the opportunity, knowledge, and ability to share in this work. Thanks to all machining instructors who help change the lives of their students by teaching them skills that will provide for a bountiful future. Special thanks to the publisher, Industrial Press, Inc. and specifically to: Owner/CEO Alex Luchars; Managing Editor Laura Brengleman; Editorial Director/Publisher Judy Bass; Typesetting and Design Devanand Madhukar; Project Manager Patricia Wallenburg; and Elise Davies, Copy Editor. Thanks also to Daniel E. Puncochar for his work on the second edition (original text based on ASME Y 14.5 1994) of *Interpretation of Geometric Dimensioning and Tolerancing* (based on ASME Y14.5-2009, 3rd edition). Thanks also to all of the following contributors:

- ATAGO USA, Inc.
  Miyuki Clauer, Marketing Supervisor for Brix Scale and Refractometer images
- Autodesk, Inc.
  Al Whatmough, Sr. Product Manager, for granting the use of Autodesk Fusion 360 and Autodesk Inventor Professional CAD/CAM software for design graphics, program creation, and screen shots used throughout the text; and Tim Paul, Manager – Manufacturing and Business Strategy, for his support.
- Buckeye Fire Equipment Company
  Kevin Bower, President for fire extinguisher-related images and text.
- CAMplete Solutions, Inc.
  Jeff Fritsch, VP of Sales and Marketing for temporary license and use of CAMplete Trupath simulation software and screenshot graphic images from within.
- Command Tooling, LLC
  Bounti Hanedae, CFO for Haas BMT style lathe turret tooling models and images.
- DATRON Dynamics, Inc.
  Steve Carter, Brand Manager for 15-station ATC (automatic tool changer) with tool-length sensor on DATRON M8Cube high-speed milling machine image.
- Haas Automation, Inc.
  Scott Rathburn, Marketing Product Manager/Senior Editor for Haas CNC machine images used throughout the text.
- Harvey Performance Company, LLC
  Tim Lima, Senior Content Marketing Specialist for text excerpts and graphics from the Helical Machining Guidebook and the 2018 Product Catalog.
- Kurt Manufacturing Inc., Industrial Products Division
  Steve Kane, Sales and Marketing Manager for vise 3D models and images of work-holding accessories downloaded from the Technical Library, Kurt Images, Gallery at https://www.kurtworkholding.com.
- The L.S. Starrett Company
  Krys Swan, Senior Graphic Designer for excerpts from Bulletin 1211, Tools and Rules, including related graphic images.
- National Fire Protection Association (NFPA) for images and text on the NFPA/OSHA Comparison of NFPA and HazCom Labels 2012 Labels "Quick Card."
- Occupational Safety and Health Administration (OSHA) for image and text on the Hazard Communication Standard Labels "Quick Card" and the Hazard Communication Safety Data Sheets "Quick Card."

- Sandvik Coromant
  Angela Roxas, PR and Sponsorships for 3D models and images of inserts, cutting tools, holders, and assemblies from their CoroPlus Tool Guide application and the CoroPlus ToolLibrary.

- Vericut
  Meridith Choy, Marketing Specialist for temporary software license and screenshot images from the software.

Preface

Congratulations on your decision to seek skills and earn certification in CNC Machine Operation, Setup, and Programming. The career of machining has been around for a very long time. This career is constantly changing with new technology influencing manufacturing methods at a pace that is fascinating. Hang on—you are in for a ride! Have an open mind to discovery, never be afraid to change, and keep learning. With these new skills you will not be out-of-work unless you wish to be.

Instructors, teach your students. Let them do. The best way to learn is by doing. If they fail or miss the mark, encourage and be patient. The old saying, "This is the way we have always done it" is a death sentence to new ideas and productivity. Never accept the "status quo," keep learning.

Manufacturing is in crisis with a shortage of skilled workers needed to fill open positions. In times of crisis, people pull together to get the task done and rebuild. Sometimes, new and more productive methods are the result. Humans are amazingly resilient and able to overcome unthinkable odds. You don't have to look back very far in history to see examples. Never be afraid of change, embrace it!

The most important thing to remember is to never compromise when it comes to safety, quality, and work ethic. Be tenacious in your endeavors and don't be hard on yourself when things go awry. Keep trying, keep learning.

May God bless all who read this text with extraordinary skill and a positive career experience for a better world. Life is good, my friends! Live it!

1

CNC Machining Fundamentals

Machine Safety

Objectives

1. Find and understand machine and chemical hazard labeling.
2. Define safety-related acronyms.
3. Properly demonstrate an understanding of fire extinguisher applications and use.
4. Identify appropriate workplace clothing and personal protective equipment (PPE).
5. Be able to find and use safety data sheets.

When working with any type of machinery, *you* are the only one who can protect yourself from injury. This is only possible if you understand the risks associated with machining and apply safe work practices that are outlined here and in every machine operation/maintenance manual provided by the manufacturer. By following these best practices, you and those you work with will be protected from injury on the job. Also, under the Occupational Safety and Health Act of 1970 (OSH Act), employers are responsible for providing a safe and healthy workplace. All provided safety-related training should be documented and updated at least annually.

Some requirements related to working safely on machines are

- Wear appropriate clothing—no loose-fitting sleeves or neck ties, limit skin exposure (hot metal chips can find their way into places and burn you where they land).
- No jewelry or lanyards—like rings, dangling chains or watches. Lanyards are common for employee badging in today's workplace but pose a risk around machines. For safety, either remove lanyards during machining activities or, at minimum, use only the breakaway type.

- Keep hair tied back.
- Avoid distractions—like horseplay or loud music.

You should not operate any machine without first understanding basic safety and operation procedures necessary to control and stop the machine in emergency situations and when necessary. Computer Numerical Control (CNC) machines with enclosures are equipped with safety devices (door interlocks) that prevent machine operation with the doors open to protect personnel from injury. Operators should never modify or disable these safety devices.

The machine enclosure guarding and especially the windows are not a guarantee of protection and caution should be taken at all times during machining. Machine tools are required to display safety warning decals near high-risk areas of the machine, such as on electrical cabinet doors, where the potential for electrical shock is present. Electrical cabinet doors should only be opened by qualified maintenance personnel. Read and heed these labels on all machines and equipment. Some examples and brief explanations are given in Figures 1-1, 1-2 and 1-3.

NOTE: *Never alter or remove any safety decal or symbol.*

Each hazard is defined and explained on the general safety decal at the front of the machine. Review and understand each safety warning, and familiarize yourself with the symbols.

F1.1:   Standard Warning Layout. [1] Warning Symbol, [2] Severity and Word Message, [3] Action Symbol. [A] Hazard Description, [B] Consequence of Ignoring the Warning, [C] Action to Prevent Injury.

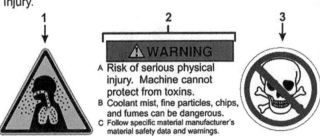

Figure 1-1.   Warning label layout, *courtesy Haas CNC Automation*

Figure 1-2. Machine danger, warning and notice labels, *courtesy Haas CNC Automation*

Figure 1-3. Machine warning labels, *courtesy Haas CNC Automation*

Personal Protective Equipment (PPE)

Eye Protection

As a rule, you should always wear appropriate eye protection when working with machines and tools. Even with safety glasses it is still possible that debris or a liquid splash can get into your eyes. If this happens, DO NOT rub the eye(s) or use any object in an attempt to remove the item. Sometimes the natural flow of tears will dislodge the item. In every case, you should know the location of the eye wash station in your workplace and use it to flush out the offending object or substance. Remember to allow the water to flow into your eye(s) for several minutes.

Hearing Protection

The machining work environment can be quite loud at times, so it is important to wear proper hearing protection. The OSH Act states that when the noise levels are at or above 85 decibels (dB), protection measures should be taken. How do you know when the level is at 85 dB? A general rule is that if you are standing within three feet of a person and you have to raise your voice to be heard, then you should be wearing hearing protection. A common situation is the noise caused when using compressed air to clean parts after cutting. The most common types of hearing protection found in machine shops are plugs and ear muffs. Both are effective, so which one you use depends mostly on preference.

Hand Protection

Machinists perform much of their work directly with their hands and the fact is 13 percent of all industrial injuries are cuts to the fingers and hands. The Bureau of Labor Statistics data report that 70 percent of on-the-job hand injuries happen to workers not wearing gloves. For many years, wearing gloves around machine tools had been considered a risk. Today, there are times when it is acceptable and even advisable to wear gloves and is considered safe practice. Some multipurpose gloves made with advanced materials (polyurethane-coated, cut-resistant gloves) enable the user to perform different tasks to avoid cuts and chemical exposure while maintaining grip and dexterity. For example, multipurpose and leather gloves are suitable for materials handling, Nitrile gloves perform well for light chemical applications, rubber may be required for more hazardous chemicals. Do not wear gloves if the danger of entanglement exists. To avoid entanglement risks, always keep hands out of the path of moving parts during machining operations and perform all setup procedures and loading or unloading of workpieces with the spindle stopped.

Footwear

Appropriate protective footwear is required in every shop environment. At minimum, shoes covering the toes should be worn, but steel or composite toe are most effective at preventing injury. Shop floors are often slippery due to coolant and oil residue in the area, so anti-slip soles are highly recommended. Many employers subsidize the purchase of work boots, shoes and other PPE.

Some of the most common causes for injuries in the workplace are slips, trips and falls. Always be aware of your surroundings and pay special attention to floor hazards like cords, unmarked obstacles, uneven or wet surfaces, steps, drop-offs and traffic pathways. Take special care when accessing raised work areas like portable stairs and scaffolding. Pay attention when fork-lifts are being used in the work area and walk at a safe distance from the forks.

Dust Masks

In some cases, machining can create airborne particles that are small enough for the operator to inhale. This is especially true when machining foams, wood products, carbon-fiber or composites.

A particulate dust-mask should be worn. In rare circumstances, where hazardous materials are being machined, respirators may be required. Your employer is required to train and equip you with protective equipment if this is the case.

Contact with Blood

The most common injuries that occur when using CNC machines are minor cuts and lacerations caused by mishandling of sharp cutting tools or the burr edges of parts after machining. When these incidents happen, it is important to protect yourself and others from coming in contact with blood. Blood-borne pathogens are pathogenic microorganisms that can cause serious disease.

Seek first aid immediately and report the accident. Do not continue working or touching equipment until the bleeding has stopped. Apply pressure to the wounded area with a clean towel. Wash the wound thoroughly with soap and water and apply an appropriate bandage or dressing as needed. If you are helping someone else with an injury, be sure to put on non-latex or nitrile gloves before administering first aid to the victim.

Note: It is important to decontaminate the area or item where blood is present by cleaning with the proper disinfectant in the blood-borne pathogen kit in your first aid cabinet and properly disposing of cleaning towels (your employer may assign a janitorial crew for this step). For more detailed information, visit https://www.osha.gov/pls/oshaweb/owadisp.show_document?p_id=10051&p_table=STANDARDS.

Lifting Techniques

In the shop there are often times when the part, material or work-holding system weighs more than fifty pounds. Back injuries are a common result of improper lifting in the workplace. In all cases, proper lifting techniques should be followed. When lifting mechanisms such as portable lifts, scissor lifts, hoists and cranes are available they should be used (after being properly trained) rather than lifting by hand. Overhead cranes require special training certification (OSHA CC – Cranes and Derricks in Construction, as specified in 29 CFR 1926.1427) for safe use, including lifting accessories such as chains, cables, hooks and straps. At least two people, the operator and a spotter, must be present when using a crane. The crane operator must wear a safety vest and both the operator and spotter must wear hard hats.

When mechanical lifting methods are not available and the item is less than 50 pounds, lift by bending your knees, lift with your legs— not your back. If the item you are lifting is long or awkward, enlist a helper (see Figure 1-4).

Figure 1-4. Two-person lift

Although the OSH Act does not specify an exact weight limit a person should be required to lift, the National Institute for Occupational Safety and Health (NIOSH) has developed a mathematical model. "Applications Manual for the Revised NIOSH Lifting Equation" (on the web at http://www.cdc.gov/niosh/docs/94-110/) predicts the risk of injury based on the weight being lifted and other criteria as a voluntary guideline. Generally speaking, lifting an item weighing over 50 pounds should not be attempted without some form of assistance. Protect yourself, get help.

Fire Prevention and Suppression

There is always a potential for fire when working in a manufacturing environment. CNC machining itself is not usually flammable; however, sometimes materials being machined may pose a threat. Magnesium and titanium machining can create fine particles that are very flammable. It is more likely some of the chemicals used for lubrication and cleaning might cause a fire. Understand the potential risks. Read all chemical safety data sheets (SDS), follow directions related to flammability and adhere to safe working habits at all times. Discard oily rags in approved safety cans.

Building fire codes require sprinkler systems and fire extinguishers based on square footage of each room. Know the location of the nearest fire extinguisher in your area. The Occupational Safety and Health Administration (OSHA) standards require that your employer have a fire prevention plan and the minimum elements of a fire prevention plan must include (per training requirements in OSHA Standard 1910.39):

- A list of all major fire hazards, proper handling and storage procedures for hazardous materials, potential ignition sources and their control, and the type of fire protection equipment necessary to control each major hazard.
- Procedures to control accumulations of flammable and combustible waste materials.
- Procedures for regular maintenance of safeguards installed on heat-producing equipment to prevent the accidental ignition of combustible materials.
- The name or job title of employees responsible for maintaining equipment to prevent or control sources of ignition or fires.
- The name or job title of employees responsible for the control of fuel source hazards.

OSHA Fire Protection and Prevention Standard 1926.152 states that flammable liquids must be stored in an approved container or storage cabinet. For more information, go to https://www.osha.gov/pls/oshaweb/owadisp.show_document?p_id=10673&p_table=STANDARDS.

Extinguisher Selection Guide

To provide the best protection of life and property from fire, proper extinguisher selection is critical. The classification and rating of an extinguisher are vital pieces of information for making this selection.

The National Fire Protection Association (NFPA) has established the requirements for the number, size, placement, inspection, maintenance, and testing of portable fire extinguishers. These requirements are contained in NFPA 10, "Standard for Portable Fire Extinguishers."

Within this standard it is stated that the selection of fire extinguishers for a given situation is determined by the following:

■ The character and size of the fires anticipated to be encountered
■ The construction and occupancy of the property to be protected
■ The ambient temperature of the area where the extinguisher will be located
■ Other factors that may dictate the selection of a particular type of extinguisher

Fire Classifications

Generally, fire extinguishers are selected based on the hazard(s) they are intended to address. Classifications have been established to categorize these hazards as shown in Figure 1-5.

It is important to be aware of the locations of fire extinguishers in your facility and to understand the function for each of the components as shown in Figure 1-6.

Before trying to fight a fire:

1. Know where all exits are.
2. Alert others that there is a fire in the building.

Class A: Fires involving ordinary combustible materials such as wood, cloth, paper, rubber and many plastics.

Class B: Fires involving flammable liquids, combustible liquids, petroleum grease, tars, oils, oil-based paints, solvents, lacquers, alcohols, and flammable gases.

Class C: Fires that involve energized electrical equipment.

Class D: Fires involving combustible metals such as magnesium, sodium, potassium and titanium.

Class K: Fires involving combustible cooking media such as vegetable oils, animal oils and fats.

Figure 1-5. Fire classification types, *courtesy Buckeye Fire Equipment*

Fire Extinguisher Components

Figure 1-6. Fire extinguisher components, *courtesy Buckeye Fire Equipment*

3. Have someone call the fire department.
4. Assess the magnitude of the fire and, if it is too large, exit the area immediately.
5. Determine if there may be toxic smoke; if in doubt, exit the area immediately.
6. If the fire is small and contained, retrieve the fire extinguisher.

Follow the PASS method: Pull the pin. While holding the extinguisher with the nozzle pointing toward the fire, Aim low and at the base of the fire. Squeeze the lever slowly and evenly and Sweep the nozzle from side to side repeatedly until the fire is out.

The NFPA developed what is called the 704 Diamond and is the standard labeling system for hazardous chemicals that identifies the health hazards, flammability, reactivity and other special hazards intended primarily for emergency response personnel (Figure 1-7).

Hazardous Materials Identification System (HMIS)

In the United States, OSHA, through the revised Hazard Communication System (HCS), requires suppliers and employers to follow labeling standards to identify health hazard levels from 1 to 4 (1–minimal; 2–slight; 3–moderate; 4–serious and severe) for all chemical materials found in the workplace in order to communicate the chemical hazard (see Figure 1-8). This labeling requirement

Figure 1-7. NFPA 704 and HazCom labels, *courtesy NFPA*

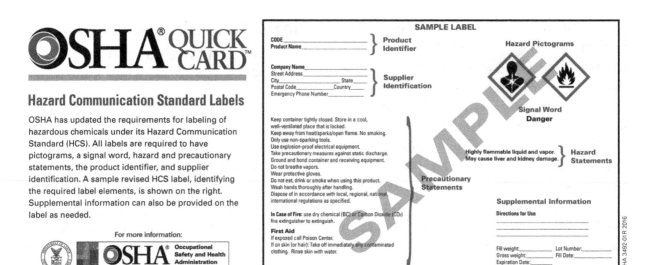

Figure 1-8. Hazard communication standard labels quick card, *courtesy OSHA*

includes all containers, including small bottles used for distribution at the point of use.

Safety Data Sheets

According to the HCS, employees have the right to know the hazards related to the chemicals used in their workplace. You have the right to a safe workplace. The OSH Act was specifically passed to prevent workers from being killed or otherwise harmed at work, and it requires employers to provide their employees with working conditions that are free of known dangers. The OSH Act resulted in the creation of the OSHA, the administration which sets and enforces protective workplace safety and health standards. OSHA also provides information, training and assistance to employers and workers (OSHA 3021-11R 2016, Workers' Rights). Training conducted in compliance with 1926.59,

the Hazard Communication Standard, will help employees to be more knowledgeable about the chemicals they work with and familiarize them with SDS (OSHA 1910.39) (Figure 1-9). (Safety Data Sheets [SDS] is the current terminology used).

Summary

There is nothing more important than your safety and health when it comes to working in the machining environment. There are risks that must be recognized and avoided. Some very basic techniques and tools have been presented here to help you avoid injury. Knowledge about your workplace and the associated risks of working with machining equipment and the use of common sense are your best protection. Protect yourself and always THINK SAFETY FIRST!

OSHA® QUICK CARD™

Hazard Communication Safety Data Sheets

The Hazard Communication Standard (HCS) requires chemical manufacturers, distributors, or importers to provide Safety Data Sheets (SDSs) (formerly known as Material Safety Data Sheets or MSDSs) to communicate the hazards of hazardous chemical products. The HCS requires new SDSs to be in a uniform format, and include the section numbers, the headings, and associated information under the headings below:

Section 1, Identification includes product identifier; manufacturer or distributor name, address, phone number; emergency phone number; recommended use; restrictions on use.

Section 2, Hazard(s) identification includes all hazards regarding the chemical; required label elements.

Section 3, Composition/information on ingredients includes information on chemical ingredients; trade secret claims.

Section 4, First-aid measures includes important symptoms/effects, acute, delayed; required treatment.

Section 5, Fire-fighting measures lists suitable extinguishing techniques, equipment; chemical hazards from fire.

Section 6, Accidental release measures lists emergency procedures; protective equipment; proper methods of containment and cleanup.

Section 7, Handling and storage lists precautions for safe handling and storage, including incompatibilities.

(Continued on other side)

OSHA® QUICK CARD™

Hazard Communication Safety Data Sheets

Section 8, Exposure controls/personal protection lists OSHA's Permissible Exposure Limits (PELs); ACGIH Threshold Limit Values (TLVs); and any other exposure limit used or recommended by the chemical manufacturer, importer, or employer preparing the SDS where available as well as appropriate engineering controls; personal protective equipment (PPE).

Section 9, Physical and chemical properties lists the chemical's characteristics.

Section 10, Stability and reactivity lists chemical stability and possibility of hazardous reactions.

Section 11, Toxicological information includes routes of exposure; related symptoms, acute and chronic effects; numerical measures of toxicity.

Section 12, Ecological information*
Section 13, Disposal considerations*
Section 14, Transport information*
Section 15, Regulatory information*

Section 16, Other information, includes the date of preparation or last revision.

*Note: Since other Agencies regulate this information, OSHA will not be enforcing Sections 12 through 15 (29 CFR 1910.1200(g)(2)).

Employers must ensure that SDSs are readily accessible to employees.
See Appendix D of 29 CFR 1910.1200 for a detailed description of SDS contents.

Figure 1-9. Hazard communiction safety data sheets, *courtesy OSHA*

Machining Safety, Study Questions

1. Some of the most common injuries that occur when working with CNC machines are:
   a. From slips, trips and falls
   b. Cuts and abrasions
   c. From using improper lifting techniques
   d. All of the above

2. Hearing protection is required when shop noise levels exceed:
   a. No specific level; always wear hearing protection
   b. 85 dB
   c. 65 dB
   d. A point where talking at a normal level cannot be heard when standing three feet apart
   e. b or d

3. Safety Data Sheets are:
   a. Chemical manufacturer- and distributor-required information about hazards related to their use
   b. Used to communicate dangers related to handling chemicals
   c. Only needed for flammable materials
   d. Both a and b

4. The acronym HMIS stands for?
   a. Horizontal Machining Information Standard
   b. Heavy Materials Instruction Sheet
   c. Hazardous Materials Information System
   d. Hazardous Materials Identification System

5. Personal protective equipment includes which of the following?
   a. Hearing protection (plugs or muffs)
   b. Safety glasses
   c. Appropriate clothing sometimes including steel- or composite-toe shoes
   d. All of the above

6. When lifting an object that weighs more than 50 pounds you should:
   a. Use the two-person lifting method
   b. Use a lifting device when available
   c. Either a or b
   d. An individual can be required to lift up to 75 pounds by oneself

7. What does the fire extinguisher-related acronym PASS stand for?
   a. Pull pin, Aim, Squeeze and Sweep
   b. Pull pin, Assess the situation, Send for help, Shower the fire
   c. Point, At, and Stand at a Safe distance
   d. None of the above

8. Which of the following is the organization that sets and enforces protective workplace safety and health standards?
   a. USDA
   b. NFPA
   c. OSHA
   d. ORCA

9. When an accident occurs that includes the loss of blood, why is special care and protection needed when cleaning the work area after first aid is administered?

   a. As a courtesy to other workers
   b. Because blood-borne pathogens are pathogenic microorganisms that can cause serious disease
   c. To avoid frightening other workers
   d. The machine coolant will dilute any remnant blood and render it harmless

10. The use of overhead cranes and forklifts to lift heavy objects requires:

    a. Special attention to weight distribution
    b. Special training and certification
    c. The operator to wear a hard hat
    d. All of the above

11. What type of fire extinguisher is used for fighting a burning metals fire?

    a. A
    b. B
    c. C
    d. D

Types of CNC Machines

Objectives

1. List at least three common CNC machine types.
2. Demonstrate understanding of the right-hand rule.
3. Identify the axes present on a three-axis vertical CNC mill.
4. Identify the axes present on a two-axis CNC lathe.
5. Make a machine selection based on part geometry.

CNC machine tools are steadily advancing and have all but eliminated manual machining in a production environment. Manual lathes and milling machines still exist, but they are primarily used for secondary operations or simple one-off parts and tooling. On CNC machines, program data can be entered through manual data input at the control panel keyboard to create programs that are downloaded via ethernet connection from a remote source like a personal computer (PC) network, a USB drive or, in rare cases, wi-fi. Today, many CNC machines offer conversational or shop floor programming right at the control and some even incorporate computer-aided manufacturing (CAM) software within the control. All operations on CNC machines are carried out automatically once programmed. The CNC machine will only do what it is programmed to do and requires human involvement for cutting tool work-holding setups, loading and unloading the workpiece (sometimes robotic) and entering tool and work offsets into registers on the control.

A major objective for using CNC machines is to increase productivity and improve quality by consistently controlling the machining operation. It is necessary for CNC programmers to have a thorough understanding of the CNC machines they are responsible for programming. Knowledge of the exact capabilities of the machine and its components, as well as the tooling involved, is imperative when working with CNC.

Some machines that are CNC-controlled today include milling machines, lathes, routers, wire electronic discharge machines, laser, grinders, turret punches, plasma cutters, water jets, multitask mill turn centers, robotic pick-and-place loading devices, robotic machining systems and many more. There are also several design variations of machining and turning centers. Some of the five-axis machining centers (see Figures 1-10 and 1-11) use a rotary axis on a trunnion table and some turning centers have live tooling, secondary spindles and multiple turrets. All operations on these machines can be carried out automatically, but human involvement is necessary for programming, setting up, loading and unloading the workpiece, and entering the amounts of dimensional offsets into registers on the control.

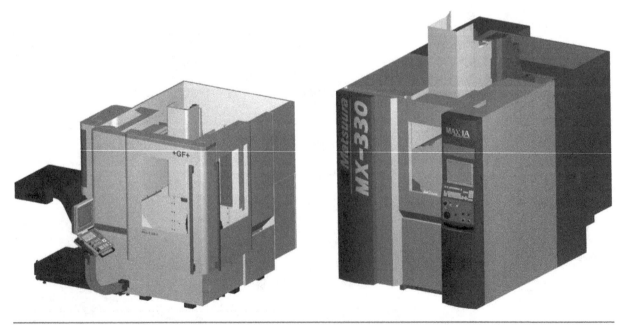

Figure 1-10.   Five-axis machining centers, *courtesy CAMPlete Solutions Inc.*

Figure 1-11. Five-axis machining centers, *courtesy CGTech Vericut*

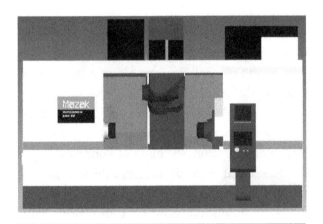

Figure 1-12. Multi-axis mill turn center, *courtesy CGTech Vericut*

In this section, we will focus on three-axis CNC vertical mills and two-axis CNC lathes and turning centers (Figures 1-12, 1-13 and 1-14). These types of machines comprise the foundation of CNC learning and are where most new machinists start their career as operators.

CNC Machine Configurations

All CNC machines are equipped with the basic traveling components, which move in relation to one another and in perpendicular directions.

Figure 1-13. Three-axis CNC mills, *courtesy Haas CNC Automation*

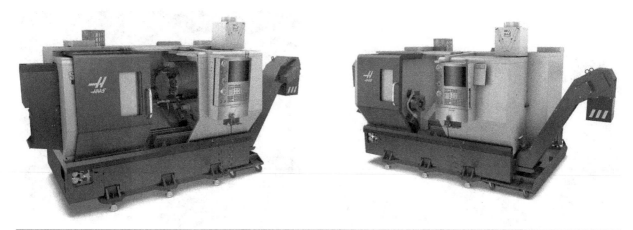

Figure 1-14. Two-axis CNC lathes, *courtesy Haas CNC Automation*

Vertical mills and machining centers are equipped with a traversing worktable or column, which travels along two axes, and a spindle with a driven tool that travels along a third axis (Figure 1-15). All of the axes of these machines are oriented in an orthogonal coordinate system (each axis is perpendicular to the other), as in the Cartesian coordinate system. On vertical mills, the X-axis is left and right movement in either direction, the Y-axis is movement away from and toward the operator in both directions and the Z-axis is

movement of the spindle toward or away from the machine worktable. The *right-hand rule* is often used to describe the axis orientation for vertical mills.

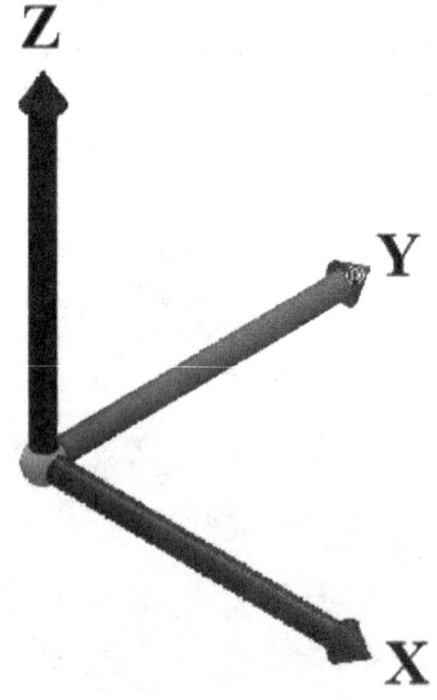

In discussing the X, Y and Z axes, the right-hand rule establishes the orientation and the description of tool motions along a positive or negative direction for each axis. This rule is recognized worldwide and is the standard for which axis identification was established. For the vertical representation, the palm of your right hand is laid out flat in front, face up, the thumb will point in the positive X direction. The forefinger will be pointing the positive Y direction. Now fold over the little finger and the ring finger and allow the middle finger to point up. This forms the third axis, Z, and points in the Z positive direction. The point where all three of these axes intersect is called the origin or zero point. When looking at any vertical milling machine, you can apply this rule.

CNC Lathe Axis Designation

Two-axis CNC lathes and turning centers (Figure 1-16) are equipped with a turret and tool carrier, which travels along two axes. The Z-axis is from right to left with the face of the spindle nose representing zero location of the coordinate system and the X-axis is from front to back, with the

Figure 1-15. Vertical mill axis designation, the right-hand rule

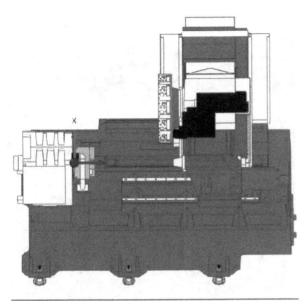
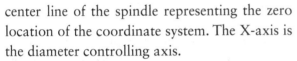

Figure 1-16. CNC lathe, axis designation, *courtesy Haas CNC Automation*

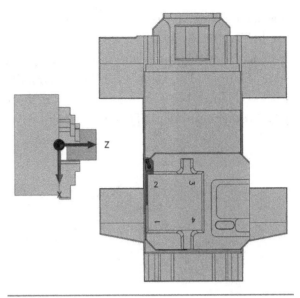

Figure 1-17. CNC tool post lathe, axis designation, *courtesy Haas CNC Automation*

center line of the spindle representing the zero location of the coordinate system. The X-axis is the diameter controlling axis.

With two-axis CNC lathes, tools are mounted in a *tool turret* that can hold eight or more tools at once. The turret normally has every other tool location (pot) dedicated to outside diameter (OD) tools or inside diameter (ID) tools. When tool changes are commanded in the program, the tool turret unclamps, rotates to the desired position and then reclamps. Caution must be exercised when mounting tools in consecutive pot locations to avoid interference with other tools, the part or work-holding system.

There are also two-axis lathes with similar configurations to manual lathes where tools are mounted to the tool post using a wedge lever locking system (Figure 1-17). In this case, the face of the cutting tool or insert is visible from above. These types of tool posts make it possible to mount tools for OD and facing as well as ID boring. On this type of machine, drilling is completed manually by advancing the tailstock into position, locking it and turning the hand crank to advance the drill into the work. At tool change commands, the machine will prompt the

operator to uninstall the current tool and install the next tool. This type of machine is an effective transitional step for those just moving away from manual turning to CNC, and many technical schools start their machinist trainees on these machines.

It is important to note that, in the following drawings of turning centers, the cutting tool and turret is located on the positive side of spindle center line. This is a common design of modern CNC turning centers. For visualization purposes, in this book, the cutting tool will be shown upright as would be true for tool post lathes. In reality for turning centers, the tool is mounted with the insert facing down and the spindle is rotating clockwise for cutting.

Note: The direction of spindle rotation in turning—clockwise (CW) or counterclockwise (CCW)—is determined by looking from the headstock toward the tailstock and tool orientation.

CNC Mill Tool Changer Types

CNC machining centers use an efficient tool change system that incorporates an automatic tool changer (ATC) in the form of an umbrella, magazine or rack style. With the umbrella, tools are held vertically and have dedicated locations for each tool (Figure 1-18); with the magazine style (Figure 1-19), tools are held horizontally. Racks (Figure 1-20) are commonly used on CNC routers. Another ATC type (not shown) is

Figure 1-20. Fifteen-station ATC with tool-length sensor on DATRON M8Cube high-speed milling machine, *courtesy DATRON Dynamics*

the chain style, where tools can be removed and replaced more quickly and incorporate random assignment to tool locations and track tool pot location via Remote Frequency Identification (RFID) chip. The number of tools vary with each machine and options are available for expansion.

Machine Selection

Determining which machine is needed to make a part is driven by the part design. Typically, cylindrical shaped parts are turned on a lathe, whereas parts with rectangular or prismatic shapes are machined in a mill. Some part configurations may require both types of machines for specific operations. This situation is commonly handled by multiple operations including use of both types of machines, but today multi-task mill–turn machines (Figure 1-12) are making it possible to complete all operations in one setup on one machine. As product designs are developed, the manufacturing method should be considered. This is called *design for manufacture*. For example, CNC milling machines cannot ordinarily make square holes and two-axis CNC lathes

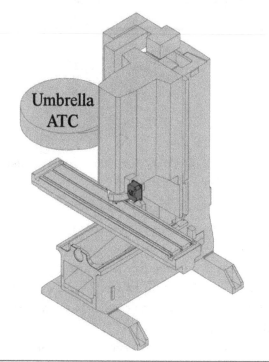

Figure 1-18. CNC mill umbrella ATC, *courtesy Haas CNC Automation*

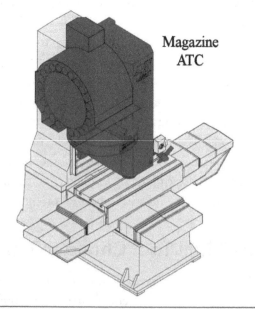

Figure 1-19. CNC mill magazine ATC, *courtesy Haas CNC Automation*

cannot cut rectangular part shapes. The machine used must have the physical ability to perform the machining. If the planned machining cut requires ten horsepower from the spindle motor, a machine with only 7.5 horse power will not be an efficient choice. It is important to work within the capabilities of the machine tool. When planning for machining, always remember that the stability, rigidity and repeatability of the machine are of paramount importance for success. The determination of the machine type needed to produce the part is, in fact, one of many steps involved in *process planning for CNC*.

Summary

There are many types of CNC machines that enable machining of a multitude of part geometry requirements. Because of this, the use of manual machines is limited mostly to secondary operations and simple one-off tooling and is seldom used for production today. Modern manufacturing operations may include automation of part load and unload using robots and it is not unusual for some CNC machines to run unattended, or "lights out," for long periods of time.

Types of CNC Machines, Study Questions

1. Which is not one of the three common CNC machine types?
   a. CNC mill
   b. CNC lathe
   c. Engine lathe
   d. Multi-tasking machines (Mill-turn)

2. The right-hand rule:
   a. Indicates positive axis directions for vertical CNC mills
   b. Indicates positive axis directions for horizontal CNC mills
   c. Indicates positive axis directions for CNC lathes
   d. Indicates negative axis directions for vertical CNC mills

3. CNC mills include the following linear axes:
   a. X and Z
   b. A, B and C
   c. X, Y and Z
   d. U, V and W

4. CNC lathes include the following linear axes:
   a. X and Z
   b. A, B and C
   c. X, Y and Z
   d. U, V and W

5. Proper machine selection is:
   a. Dependent upon part design
   b. One of the many steps in process planning for CNC
   c. Should be considered in the design for manufacture process
   d. All of the above

Machine Maintenance and Workspace Efficiency

Objectives

1. Identify daily maintenance procedures.
2. List the pillars of the 5S system.
3. List coolant types used in CNC machining.
4. Read a refractometer.
5. Understand lockout/tagout procedures.

Maintenance

To get the optimum performance and productivity from all CNC equipment, it is very important to recognize the need for proper maintenance and general upkeep of these machines.

Lubrication Reservoirs

Before performing any work on turning or machining centers, first verify that all lubrication reservoirs are properly filled with the correct oils. Because machine slideways need constant lubrication during operation, automatic oiler systems inject an appropriate amount of oil at intervals determined by the builder. The recommended lubrication oils are listed in the operation/maintenance manuals provided by the manufacturer. Check the machine enclosure at the back of the machine for a placard with a diagram of the machine and numbered locations for lubrication and the required oil types. Modern CNC machines incorporate sensors and monitoring software within the controller that display warning icons or alarms and may not even allow operation of the machine when lubrication oil levels are too low or the issue is not corrected.

Pneumatic Air Pressure

Check that pneumatic (air) pressures are at the recommended level and regulated properly. If the pressure is too low, some machine functions will not operate until the pressure is restored to normal. The standard air pressure setting (USA) is listed in pounds per square inch (psi), and a pressure regulator is commonly located at the rear of the machine near the air inlet.

Machine Guards and Chip Removal

CNC machines of today are equipped with Plexiglas guards that envelope the entire worktable area. These guards are intended to keep the surrounding floor space clean and the operator safe from flying debris.

Note: Plexiglas window machine guards are not a guarantee against the potential of a flying object leaving the work envelope and injuring any person nearby. There are also guards to protect the machine ways from damage by the metal chips. The *ways* are fixed mechanical features (like tracks) that the machine table and spindle column follow in order to create axis travel.

Machines used in high production environments incorporate a chip conveyor, which carries the chips to a drum on the floor on either side of the machine for easy removal. Even with this feature, there is still a need for chip cleanup inside the working envelope at least once a day. It is very important to thoroughly clean the machine and remove metal chips that are present. You can clean the ways and the working envelope without damaging the machine by using coolant to wash the machine table and the guards free of chips. Another effective cleaning method is to use a wet/dry vacuum to pick up the chips. It is NOT recommended that you use compressed air to blow away the chips from the ways. It is, however, appropriate to use compressed air to remove chips and coolant from the workpiece itself or work-holding fixtures such as a vise. The exterior of the machine usually will need only wiping down with a clean rag.

Waste Disposal

Metal shavings, often called chips or swarf, are the byproduct of machining and can accumulate very quickly and must be removed. Metal chips are razor sharp and sometimes very hot; use care during chip removal and when handling them. Always use the proper tools: brushes, chip hooks and t-slot cleaners. Coolant wash-down of the machining envelope should be performed when available. Avoid using compressed air in excess. Compressed air is acceptable to use around clamping tools (such as vise, chuck jaws and tool post) but never around the machine slideways. The machine *slideways* are the rails that the X, Y and Z axes travel along.

Recycling of metal chips is the environmentally friendly thing to do. Check for local outlets that will accept metal in this form. Do not mix different materials or recyclers will not accept them. In some high-volume shops, metal shavings are processed through special equipment to compress them into small slugs, which makes them more convenient to handle and transport and thus, more easily sold.

Broken carbide tools and cutting tool inserts can be recycled. Check with the tool manufacturer you use for directions on how to recycle their products.

Machining coolant will eventually need to be replaced as chemical properties dilute and proper disposal is required. (Check with your local liquid waste disposal facility for instructions.)

Tramp oil is machine lubricating oil that finds its way into the coolant tank, which dilutes and contaminates the coolant mix. It should be removed from coolant tanks with oil skimmers and properly disposed of. Check for waste oil recycling programs in your area that will pick up and process this type of waste for a fee. Some new machine tools are incorporating sealed lubrication systems to help alleviate this problem.

Workplace Efficiency

Just as important as careful maintenance is the cleanliness of the actual worktable, clamping systems, individual tools and the work area. Be sure to clean off any metal chips and remove any nicks or burrs on the clamping or mating surfaces to mitigate runout and planar alignment issues.

The 5S System

The *5S system* is part of Toyota's Production System "lean manufacturing" methodology whose purpose is to reduce waste within a manufacturing facility and to promote continuous improvement (Kaizen). It's a visual process of organizing and labeling task-related products in close proximity to workers so less time, space and materials are wasted. The philosophy is "A place for everything and everything in its place." When the system is properly implemented, product quality and productivity increase. The five pillars of the 5S system are:

1. *Sort.* Remove items from the workstation that are not task critical, including waste or excess tooling and materials.
2. *Set in order.* Organize the workstation so that all tools and raw materials are conveniently located to be within easy reach and make sure tools are returned to a designated, marked space immediately after use Figures 1-22–1-23. This eliminates wasted motion caused when searching for tools or equipment.
3. *Shine.* Keep the work area clean and organized and maintain all tools and equipment in working order. Report broken tooling and replace as soon as possible. Report equipment maintenance concerns to your supervisor. Clean the machine and the surrounding area at the end of the shift or as needed. This is a duty, not an option.
4. *Standardize.* Make sure the workstation you have worked so hard to organize stays that way by developing a consistent habit of cleanliness and order by making this standard operating procedure. Visual aids, like shadow boards, help identify common tool locations. Make sure your coworkers understand that maintaining a productive workstation is a team effort.
5. *Sustain.* Practice all of the 5S steps listed here and implement them into daily procedures. Be open to change and adjust procedures when needed as part of the process of continuous improvement.

Figure 1-21. CNC tool organization, tool cart

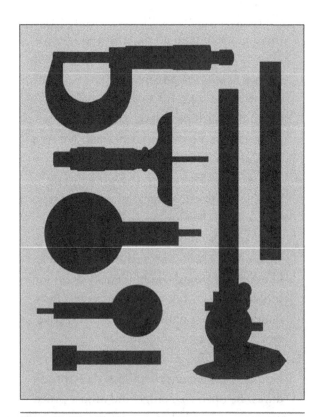

Figure 1-22. Tool box organization, precision tool cutout

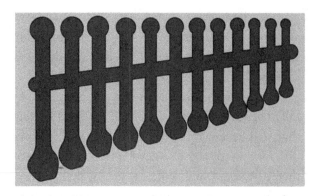

Figure 1-23. Tool box organization, wrench cutout

Control of Shop Tools

Suggestions for the best possible control of shop tools:

- Provide instructions for tool use requirements.
- Identify item inventory at each location.
- Group tooling of items by specific task per drawer.
- Avoid duplicate items.
- Items used most frequently should be easily accessible.
- Allow for future expansion of tool inventory.
- Place smaller items to the front of the drawer.
- Use tool box drawer cutouts made from foam (Figures 1-22 and 1-23).
- Space items an appropriate distance between and add slots to allow for finger access.

Total Productive Maintenance

Another component of lean management is TPM, which aims to maximize the effectiveness of the entire production system. The system goal is to create an accident-free, zero defect and zero breakdown production system with attention paid to maximization of work time throughout the life cycle of the process. Properly scheduled and performed maintenance makes the production equipment more reliable and minimizes the chance of breakdown, which has a stabilizing effect on production scheduling. Check with your company and follow their Total Productive

Maintenance (TPM) program and always refer to the operation/maintenance manuals specific to the machine tool for recommended maintenance activities.

Coolant Reservoir

The coolant tank level should be checked and adjusted as needed, prior to each use. A site glass is sometimes mounted on the coolant tank for easy viewing. The new Haas machines have a coolant level indicator in the upper-right corner of the operation mode screen. When filling the tank, use an acceptable water-soluble coolant mix, synthetic coolant or cutting oil.

Periodically, the coolant tank should be cleaned and refilled. The coolant pH level should be checked routinely with a refractometer; the mixture should be adjusted in order to prevent bacterial growth. Synthetic coolants may help the coolant system stay clean longer.

Machining Coolant

The metal cutting process is one that creates friction between the cutting tool and the workpiece. A cutting fluid or coolant is necessary to lubricate and remove heat and chips from the tool and workpiece during cutting. Water alone is not sufficient because it only cools and does not lubricate; it will also cause rust to develop on the machine ways and table. Water-miscible coolants are used at a concentrate mixture between 5 and 15 percent. The coolant manufacturer will provide proper mixing ratio requirements and pH-level parameters. It is important to follow manufacturer mixing directions and, if possible, incorporation of a mixing device with metering can ensure correct concentration mixture. If mixing manually, remember: "OIL" (oil in last). Fill a container with water, then add the coolant concentrate while stirring continuously. A refractometer is needed to properly check the percent of the mixture. This mineral oil-based soluble oil

and water mixture creates an excellent coolant for most light metal-cutting operations.

Refractometer Readings

Open the refractometer plastic cover and place two drops of coolant mixture on the lens (Figure 1-24). Close the cover and allow the mixture to spread. Hold the eyepiece up to your eye and point toward a light source to observe the scale reading (Figure 1-25). When finished, clean all surfaces with a clean rag and store for next use. Sample checking is simplified with a digital handheld pocket refractometer as shown in Figure 1-26.

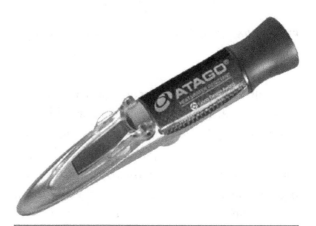

Figure 1-24. Refractometer, *courtesy ATAGO USA*

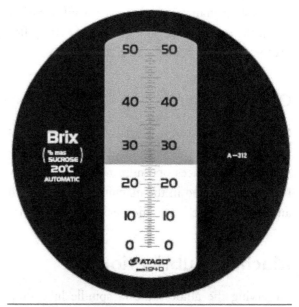

Figure 1-25. Refractometer mixture reading, *courtesy ATAGO USA*

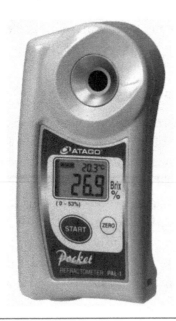

Figure 1-26. Digital pocket refractometer, *courtesy ATAGO USA*

Synthetic coolants offer advanced technology as well. Finally, the flow of coolant should be as strong as possible and be directed at the cutting edge to accomplish its purpose. Some machine tools are equipped with thru-spindle/tool and high-pressure coolant that really aid in cutting zone cooling and removal of chips. Programmers and machine operators should research available resources, such as *Machinery's Handbook* and coolant manufacturer data, for information about the proper selection and use of cutting fluids for specific types of materials.

Oil as a Coolant

In cases where harder materials like stainless steel and high-alloy steels are machined, the use of a cutting oil produces optimum results. When cutting oil is required in the machine, it is dedicated and seldom switched to another form of coolant.

Machine Lubrication

Observing the guideway and spindle lube fluid levels and filling with the correct lubricant (check

machine maintenance manuals for factory recommendations) is part of machine readiness checking. Follow the manufacturer interval recommendations for adding grease at grease points. Many machines are being built today that use sealed spindles and sealed linear guideways, nearly eliminating the need for oil lubrication.

Pneumatic Systems

It is common for CNC machine tools to use pneumatic systems for automatic tool change functions such as an air blast through the spindle during the tool change to keep metal chips from entering the empty spindle taper. Minimum and maximum air pressure machine requirements are published in the operation/maintenance manuals provided with the machine tool. An air pressure regulator is normally located near the air inlet for the machine. Observing and adjusting if necessary, this pressure setting is part of daily machine readiness checking. CNC machine tools require clean, dry air to function properly. Most shops purchase air compressors with air dryers to eliminate the condensation of water in the air lines. Lastly, each machine usually includes an air nozzle for cleaning clamping systems or parts. These air nozzles should be regulated to OSHA standards of 30 psi and should be used sparingly and safely while wearing hearing and eye protection.

Lockout/Tagout

During service, maintenance or disassembly of any machinery, power to the machine tool must be removed—locked out and tagged out—in order to prevent injury from the potential of automatic startup of the equipment, unexpected energization or the release of energy in the equipment. NEVER enter into the machining work envelope or service electrical components without lockout/tagout (LOTO) in place. The LOTO procedure is performed by affixing appropriate lockout or tagout devices on the power-disconnect switch in order to ensure the energy source cannot be restored to the equipment without unlocking it. LOTO kits can include red color-coded padlocks that include "DANGER! LOCKED OUT–DO NOT REMOVE" warning labels; red color-coded hasps used to attach the padlock to the power disconnect; and warning labels that read "DANGER–DO NOT OPERATE!" that are to be attached to the locked system with plastic cable ties. For more information, see https://www.osha.gov/SLTC/controlhazardousenergy/.

Note: This type of work should only be performed by properly trained service engineers or maintenance personnel. Consult with your supervisor to communicate any needs for maintenance and repairs.

Summary

Daily maintenance activities include

- Verify that all lubrication reservoirs are filled.
- Verify air pressure level by examining the regulator on the machine.
- Check that the chip bin/pan is empty.
- Check coolant level and mixture are correct; clean or fill as needed.
- Ensure that automatic chip removal (auger) equipment is operational when the machine is cutting metal.

- Be sure that the worktable and all matting surfaces are clean and free from nicks or burrs.
- Ensure that the chuck pressure setting is adequate for clamping the work to be machined.
- Clean up the machine at the end of use with a wet/dry vacuum or wash the machine guards with coolant to remove chips from the working envelope.
- Keep wrenches and tools stored properly and away from the machine's moving parts.
- Ensure fixtures and workpieces are securely clamped before starting the machine.
- Inspect cutting tools for wear or damage prior to use.
- Keep the surrounding area well lit, dry and free from obstructions.

A safe, clean and organized workspace should be the goal of all employees. Routine maintenance activities ensure the most productive machining experience and prevent machine and tooling breakdowns. A set of daily checklist items should be performed before machining begins. Following the 5S system will enhance the work environment and promote teamwork and overall workplace satisfaction. When maintenance activities do require entry into the machining envelope, for any reason, LOTO procedures should be strictly followed. Whenever possible, recycling of spent cutting tools, coolant and metal chips should be done for the sake of the environment.

Machine Maintenance and Workspace Efficiency, Study Questions

1. Machining swarf (chips) should be cleaned from mill tables and spindle tapers to avoid:
   a. Runout or planar alignment on work-holding devices
   b. Damage to the work-holding device
   c. Damage to the spindle taper or mill table
   d. All of the above

2. A machinist can find the proper lubrication type, schedule and recommended maintenance practices by referring to the:
   a. Instructions in the work traveler
   b. *Machinist Handbook*
   c. OSHA manuals
   d. Operation/maintenance manuals for the machine

3. Which of the following is not considered to be an effective machining coolant?
   a. Water-miscible oil
   b. Cutting oil
   c. Water
   d. Synthetic coolant

4. What is the OSHA requirement for compressed air pressure when using air nozzles?
   a. 120 psi
   b. 85 psi
   c. 50 psi
   d. 30 psi

5. Lockout/tagout procedures are required:
   a. During service, maintenance or disassembly activities on a machine tool
   b. When the machine is left unattended
   c. To prevent injury from the potential of automatic startup of the equipment or the unexpected energization or the release of energy in the equipment
   d. Both a and c

6. The purpose of the 5S system is to:
   a. Provide a safe, clean and organized workspace
   b. Promote productivity and teamwork
   c. Keep management happy
   d. Both a and b

7. What is the most environmentally friendly way to dispose of machining waste?
   a. Recycling programs
   b. Discard frequently in regular trash
   c. Melt and recast
   d. None of the above

8. A refractometer is used to measure:
   a. pH-level
   b. Coolant mixture ratio
   c. Water content
   d. Both a and b

9. The acronym TPM stands for:
   a. Tool Performance Machining
   b. Total Productive Maintenance
   c. Kaizen event
   d. Tool Preventative Maintenance

Cutting Tool Selection

Objectives

1. Communicate the fundamentals related to metal cutting tools.
2. Identify technical data resources to aid the tool selection process.
3. Properly select cutting tools specific to CNC milling applications.
4. Properly select cutting tools specific to CNC turning applications.
5. Identify cutting tool holders for both milling and turning applications.
6. Assemble and maintain cutting tools and holders for both CNC milling and CNC turning.

A major objective of CNC machining is to increase productivity and improve quality by consistently controlling the machining operation, and the ultimate goal of a CNC programmer is to create a repeatable process to obtain optimum metal-cutting conditions that decrease cycle time without compromising quality.

Machining of any kind can only be effective if proper cutting tools are selected and used. In this section, we will first address the selection process for milling tools and then do the same for tools required for turning. Over 80 percent of the machining that takes place on CNC milling machines is the making of hole-type features (commonly called hole-making). Tool selection for machining operations is one of the most important steps in process planning for CNC. The scope of this text is not intended to cover every possible situation, but to introduce the necessary information regarding tooling selection and to encourage you to tap all available resources in order to identify and use the appropriate tooling.

Whether milling or turning, these basic steps constitute the selection process:

- Identify the type of machining operation to be performed.
- Identify the workpiece material.
- Identify the machine capacities (limitations).
- Identify the type of cutter that can be used to create the required geometry.
- Calculate the appropriate cutting data for successful efficient cutting (feeds and speeds as well as depth of cut [DOC]).

Cutting Tool Selection for Milling

The designed feature of the part determines the type of milling operation and the tool required. The basic operation types are: face milling, face and shoulder milling, 2D contour, pocket and circular milling (Figures 1-33 to 1-35), as well as slot milling, drilling, tapping and boring. There are many variations of these and more advanced methods as well, but we will focus our attention on these basic types in this text.

Facing Milling

These milling operations are used to remove material from a planar face of the part. End mills can be used to perform this operation but, because of their limited size (requiring multiple step-over passes), are not always the best choice. For best results, the preferred choice is to use a face mill that is larger than the surface being machined and one with indexable inserts positioned at 45 degrees (Figures 1-27 and 1-28).

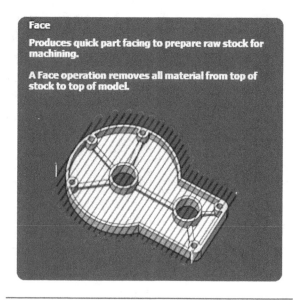

Figure 1-27.   Flat face milling, *courtesy Autodesk Fusion 360*

Figure 1-28.   Flat face milling cutter, *courtesy Sandvik Coromant*

Face and Shoulder Milling

Face and shoulder milling operations are used to remove material from a planar face of the part and can include orthogonal shoulders (Figure 1-29). Again, end mills can be used to perform this operation—especially for finishing. This style of face mill can be used for roughing or finishing. In either face milling case, the insert size is the determining factor for maximum DOC.

Figure 1-29. Face and shoulder milling cutter, *courtesy Sandvik Coromant*

Indexable Inserted Cutting Tools

Insert Attributes/Characteristics

Indexable inserted cutting tool materials consist of: uncoated carbide, coated carbide and ceramic.

Inserted tool coatings that are applied through chemical vapor deposition (CVD), or physical vapor deposition (PVD) are: cubic boron nitride (CBN), polycrystalline diamond (PCD) and diamond.

Insert Selection by Materials Application

Inserts for different workpiece materials fall into six main color-coded groups (Figure 1-30) that are identified in the following International

P	Steel
M	Stainless Steel
K	Cast Iron
N	Aluminum
S	Heat Resistant Super Alloy
H	Hardened Material

Figure 1-30. Insert material applications

Organization for Standardization (ISO) form shown in Chart 1-1: Steel = Blue; Stainless Steel = Yellow; Cast Iron = Red; Aluminum = Green; Heat Resistant Super Alloy = Salmon; and Hardened Material = Blue Gray.

2D Contour, 2D Pocket and Circular Milling

2D Contour, 2D Pocket and Circular Milling are sometimes called *profile milling*. Here we limit our focus to two-dimensional milling. It is common in these operations to use a fluted solid end mill made of high-speed steel (HSS) or solid carbide (Figure 1-31). In some cases, as previously described, smaller diameter face and shoulder milling cutters can also be a high productivity option (Figure 1-32). The number of cutting flutes (cutting edges) depends on the type of operation and the part and cutter material. For roughing, a large amount of materials is being removed so a larger distance between the flutes (lower flute count) is recommended to provide for better chip evacuation. For finishing, a higher flute count is recommended for the best results because smaller amounts of material are being removed on these paths (Figures 1-33 to 1-35).

General Code Key for CoroMill Inserts

R	390	-	11	T3	12	M	-	P	L	W
1	2		3	4	5	6		7	8	9

1 Hand of Insert

R = Right Hand
L = Left Hand

2 Main Code

For Example:
390 = CoroMill© 390

3 Insert Width

For Example:
11 = 11 mm (.433 inch)

4 Insert Thickness, S

For Example:
T3 S = 3.97 (.156")
04 S = 4.76 (.187")
06 S = 6.33 (.250")

5 Corner Radius

For Example:
12 = 1.2 mm (.047 inch)

6 Edge Performance

M = Highest Edge Security
E = Highest Sharpness & Precision
H = High Edge Sharpness & Precision
K = High Cutting Sharpness

7 Main ISO Application Area

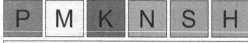

P M K N S H

8 Operation

L = Light Cutting
M = Medium
H = Heavy
T = Turn Milling

9 Wiper

W = Wiper

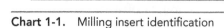

Chart 1-1. Milling insert identification

Figure 1-31. Fluted solid end mill, *courtesy Sandvik Coromant*

Figure 1-32. Carbide inserted end mill, *courtesy Sandvik Coromant*

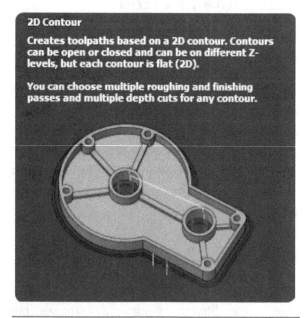

2D Contour

Creates toolpaths based on a 2D contour. Contours can be open or closed and can be on different Z-levels, but each contour is flat (2D).

You can choose multiple roughing and finishing passes and multiple depth cuts for any contour.

Figure 1-33. 2D Contour milling, *courtesy Autodesk Fusion 360*

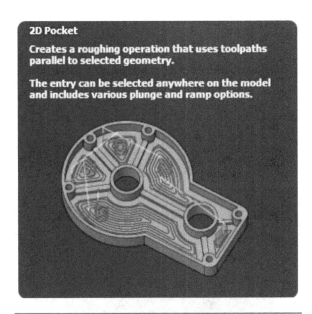

2D Pocket

Creates a roughing operation that uses toolpaths parallel to selected geometry.

The entry can be selected anywhere on the model and includes various plunge and ramp options.

Figure 1-34. 2D Pocket milling, *courtesy Autodesk Fusion 360*

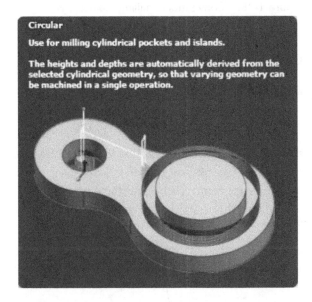

Circular

Use for milling cylindrical pockets and islands.

The heights and depths are automatically derived from the selected cylindrical geometry, so that varying geometry can be machined in a single operation.

Figure 1-35. Circular milling, *courtesy Autodesk Fusion 360*

Slot Milling

End mills are used for all types of slotting. Entry into the cut is critical and can be done by pre-drilling, ramping or helical ramping into the cut depth (avoid straight plunging). The cutter diameter must be smaller than the slot width. For slotting, a four-flute end mill is a good option (Figure 1-36).

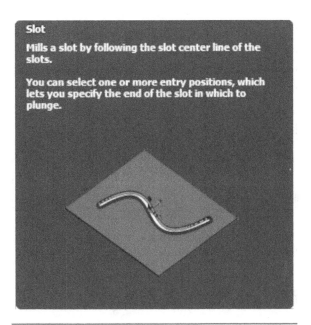

Slot

Mills a slot by following the slot center line of the slots.

You can select one or more entry positions, which lets you specify the end of the slot in which to plunge.

Figure 1-36. Slot milling, *courtesy Autodesk Fusion 360*

Cutter Features and Attributes

Cutting Tool Materials and Coatings

Fluted cutting tools are made up of these materials: high speed steel (HSS), cobalt and carbide. Advances in cutting tool materials and coatings have enabled machining of difficult materials resulting in increased cutting speeds and tool life while decreasing tool wear, providing for higher production throughput. Tool coatings that are applied through chemical vapor deposition (CVD), or physical vapor deposition (PVD) are: titanium nitride (TiN), titanium carbon nitride (TiCN), titanium aluminum nitride (TiAlN), aluminum titanium nitride (AlTiN), aluminum titanium nitride + silicon nitride (nACo), aluminum chrome nitride + silicon nitride (nACRo), zirconium nitride (ZrN), and amorphous diamond. Drilling tool materials include: HSS, cobalt, TIN, and TICN-coated, solid carbide and black oxide.

Cutting Tool Characteristic Attributes

The characteristic attributes that define the fluted cutting tool are: the cutter material and coating, diameter, length of cut (LOC), overall length

(OAL), shank diameter, number of flutes (cutting edges), the helix angle of the flutes, whether the tool allows center cutting (ability to plunge) and the shank style (straight or Weldon). A Weldon-style shank allows for a set screw in the tool holder to lock the tool in place (Figure 1-37).

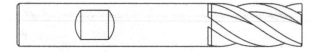

Figure 1-37. Weldon shank end mill, *courtesy Sandvik Coromant*

Milling Methodology

The following text and graphics are used with permission by Harvey Performance Company, from the *Helical Machining Guidebook*.

There are two distinct ways to cut materials when milling—conventional (up) milling and climb (down) milling. The difference between these two techniques is the relationship of the rotation of the cutter to the direction of feed. In conventional milling, the cutter rotates against the direction of the feed while during climb milling, the cutter rotates with the feed. Conventional milling is the traditional approach when cutting because the backlash, the play between the lead screw and the nut in the machine table, is eliminated (Figure 1-38). Climb milling has been recognized as the preferred way to approach a workpiece due to the fact that more and more machines compensate for backlash or have a backlash eliminator (Figure 1-39). Below are some key properties for both conventional and climb milling.

Conventional Milling

- Chip width starts from zero and increases, which causes more heat to diffuse into the workpiece and produces work hardening.
- Tool rubs more at the beginning of the cut, causing faster tool wear and decreasing tool life.

- Chips are carried upward by the tooth and fall in front of cutter, creating a marred finish and re-cutting of chips.

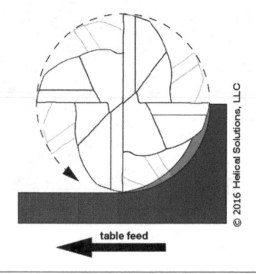

Figure 1-38. Conventional milling, *courtesy Harvey Performance Company*

Climb Milling

- Chip width starts from maximum and decreases so heat generated will more likely transfer to the chip.
- Creates cleaner shear plane, which causes the tool to rub less and increases tool life.
- Chips are removed behind the cutter, which reduces the chance of re-cutting.

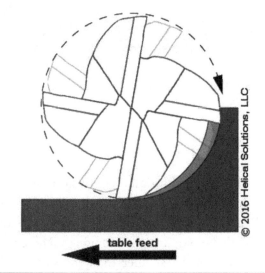

Figure 1-39. Climb milling, *courtesy Harvey Performance Company*

When to Choose Conventional or Climb Milling

Climb milling is generally the best way to machine parts today since it reduces the load from the cutting edge, leaves a better surface finish and improves tool life. During conventional milling, the cutter tends to dig into the workpiece and may cause the part to be cut out of tolerance. Even though climb milling is the preferred way to machine parts, there are times when conventional milling is the recommended choice. Backlash, which is typically found in older and manual machines, is a huge concern with climb milling. If the machine does not counteract backlash, conventional milling should be implemented. Conventional milling is also suggested for use on casting or forgings or when the part is case-hardened since the cut begins under the surface of the material.

Tool Deflection

Rigidity during a milling operation is key for optimal tool performance and desired results. Keeping tool deflection to a minimum will help increase success on a deep reach application (Figure 1-40).

A Deflection Rule of Thumb

Tool overhang length decreases rigidity as a third power (L3), but even more importantly, tool diameter increases rigidity by the fourth power (D4).

Common Techniques to Combat Deflection

- Ensure tools are sharp.
- Increase tool diameter.
- Decrease depth of cut.
- Climb mill in lieu of conventional milling.
- Decrease inches per minute (IPM).

- Use shorter tool and/or employ necked tooling.
- Increase number of flutes.
- Re-evaluate surface feet per minute (SFM) parameter.

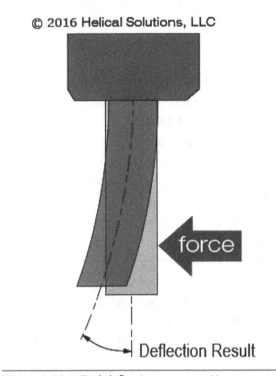

© 2016 Helical Solutions, LLC

Figure 1-40. Tool deflection, *courtesy Harvey Performance Company*

Finishing

Finishing cuts are used to complete the part and achieve the final dimension, tolerances, and surface finish. The goal when finishing a component is to avoid, or at least minimize, the necessity for manual re-touching (Figure 1-41).

Factors that Influence Finish

- Specific material and hardness
- Proper cutting tool speeds and feeds
- Tool holder accuracy
- Proper tool design and deployment
- Tool projection/deflection
- Tool-to-workpiece orientation
- Rigidity of work-holding
- Coolant/lubricity

Tips for Successful Finishing

- Using an increased helix angle will help to improve surface finish.
- 45 degrees or higher for aluminum.
- 38 degrees or higher for hard metal machining
- Increasing the number of flutes will help to improve surface finish.
- 3–4 for aluminum.
- 5–7+ for hard metal.
- Utilize tools with corner radii.
- Tool runout of .0003 inch or less.
- Using precision tool holders that are in good condition, undamaged and run true.
- Climb milling vs. conventional produces a better surface finish.
- Variable pitch tooling helps to reduce chatter and increase part finish.
- Proper radial depth of cut (RDOC) between 2 to 5 percent of tool diameter.
- For long reach walls, consider using "necked down" tools, which allow less deflection, with LOC overlap, a good step blending will occur (see Figure 1-42).
- Extreme contact finishing (> 3 × dia. deep) may require a 50 percent feed rate reduction.

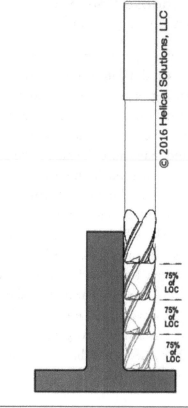

Figure 1-42. Necked-down finishing tools, *courtesy Harvey Performance Company*

Surface finishing is an important step in the operations sequence for the production of any high-quality part. Many times, this requirement is aesthetically driven but other times has to satisfy print specification. See Figure 1-43 for a list of common surface finish nomenclature.

High-Efficiency Milling

High-efficiency milling (HEM) has become a common term in machine shops worldwide, but what does it mean? Simply, HEM is a milling technique for roughing that utilizes the entire flute length, spreading the wear evenly across the cutting length of the tool (Figures 1-44 and 1-45).

How It (HEM) Works

Machining technology has been advancing with the development of faster, more powerful machines.

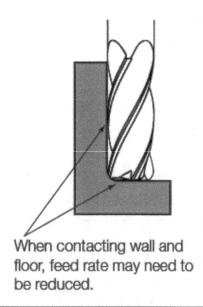

When contacting wall and floor, feed rate may need to be reduced.

Figure 1-41. Finishing, *courtesy Harvey Performance Company*

- R_a = Roughness Average
- R_q = RMS (Root Mean Square) = R_a x 1.1
- R_z = R_a x 3.1

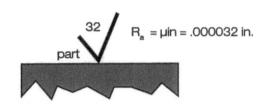

R_a = μin = .000032 in.

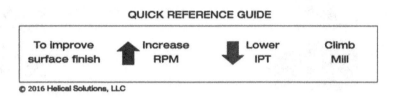

© 2016 Helical Solutions, LLC

Figure 1-43. Surface finish, *courtesy Harvey Performance Company*

In order to keep up, many CAM applications are generating more efficient HEM tool paths. These tool paths adjust parameters to maintain constant tool load throughout the entire roughing operation and allow more aggressive speeds and feeds.

Advantages of HEM

- Increased metal removal rates.
- Reduced cycle times.
- Increased tool life.

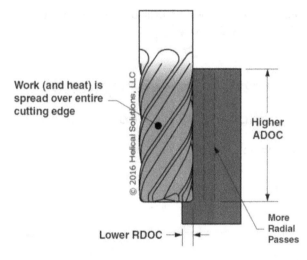

Figure 1-45. High-efficiency milling, *courtesy Harvey Performance Company*

Depth of Cut

Radial Depth of Cut (RDOC)

- The distance a tool is stepping over into the material.
- Stepover, cut width, XY.

Axial Depth of Cut (ADOC)

- The distance a tool is being sent into the cut along its center line.
- Directly related to MRR.

See Figures 1-46 to 1-56.

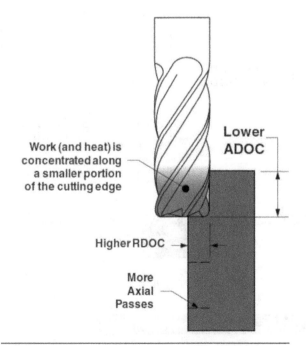

Figure 1-44. Standard milling, *courtesy Harvey Performance Company*

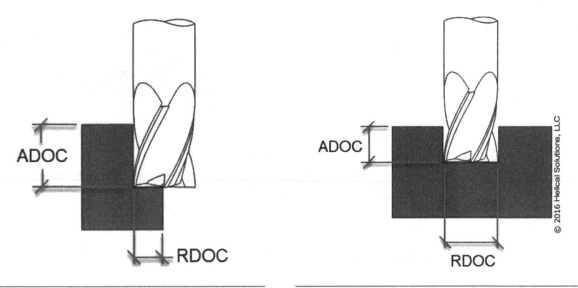

Figure 1-46. Peripheral milling, *courtesy Harvey Performance Company*

Figure 1-47. Slotting, *courtesy Harvey Performance Company*

Depth of Cut – Peripheral

Traditionally, it has been:

- Heavy RDOC
- Light ADOC
- Conservative IPM

New strategies include:

- Light RDOC
- Heavy ADOC
- Increased IPM

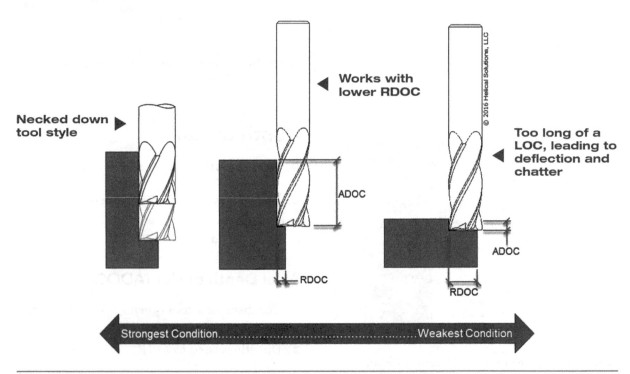

Figure 1-48. Depth-of-cut peripheral, *courtesy Harvey Performance Company*

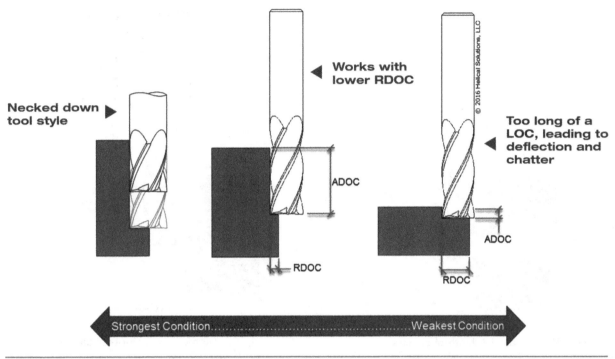

Figure 1-49. Depth-of-cut peripheral conditions, *courtesy Harvey Performance Company*

High-Efficiency RDOC Strategy

- Uniform cuts will increase tool life.
- Climb milling increases tool life.
- Lowered RDOC = 7%–30% × D.

- Increased IPM = chip thinning parameter must include "inside arc" feed reduction.
- Multi-fluted tools can be used.
- Utilize HE type of tool paths for best results.

High-Efficiency ADOC Strategy

- Controlled slice milling.
- Accommodates lower RDOC.
- Up to 2 × dia. ADOC.
- Utilizes majority of LOC.
- Multi-fluted tooling allows larger core dia. = less deflection.
- Tends to stabilize cutter by disbursing load for entire axial LOC length.
- Increased ADOC = Increase in MRR.
- Utilize HE type of tool paths for best results.

Light RDOC/Heavy ADOC Strategy

- Higher ADOC can help stabilize the cutter.
- Uniform cuts accommodate less mechanical stress and increase tool life.
- Better heat/chip management.

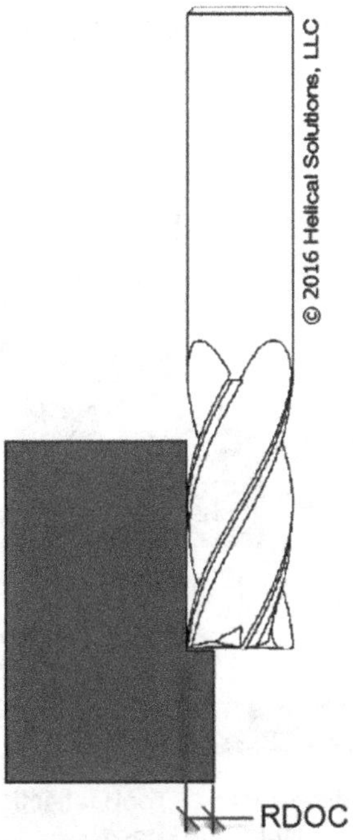

Figure 1-50. High-radial depth-of-cut strategy, *courtesy Harvey Performance Company*

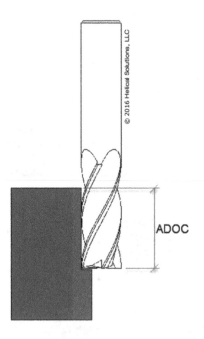

Figure 1-51. High-axial depth-of-cut strategy, *courtesy Harvey Performance Company*

- Increased IPM (due to chip thinning).
- CAM applications with HP tool paths are good solutions for this type of strategy.

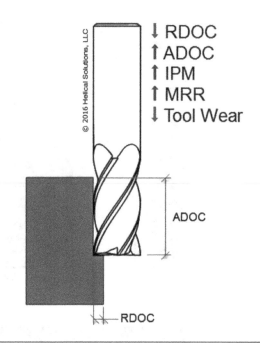

Figure 1-52. High-axial depth-of-cut strategy, *courtesy Harvey Performance Company*

Depth of Cut – Slotting

Controlled Slice Milling – RDOC Strategy

- Uniform cuts will increase tool life.

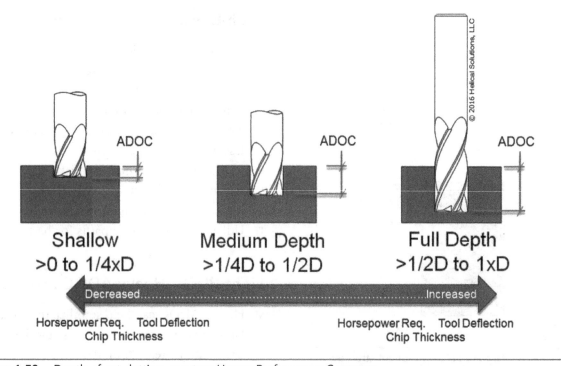

Figure 1-53. Depth-of-cut slotting, *courtesy Harvey Performance Company*

- Lowered RDOC = 10–15 percent of tool diameter.
- Increased IPM = chip thinning parameter must include "inside arc" feed reduction.
- Max cutter diameter 55–65 percent of slot width.
- Multi-fluted tools can be used.
- Must be able to evacuate the chip.

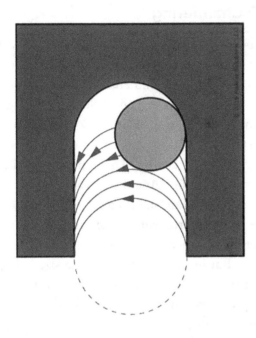

Figure 1-54. Controlled slice milling, *courtesy Harvey Performance Company*

Controlled Slice Milling – ADOC Strategy

- Allows for increased ADOC.
- Up to 2 × dia. ADOC.
- Utilizing entire LOC.
- Multi-fluted tooling allows larger core dia. = less deflection.
- Tends to stabilize cutter by disbursing load for entire axial LOC.
- Increased ADOC = increase in MRR.
- Utilize HE type of tool paths for best results.

End of excerpt from the *Helical Machining Guidebook*.

Hole-Making Operations

Drilling

The most common operation done on CNC milling machines is drilling. Spot or center drilling tools are used to prepare for follow-up drilling operations. Standard jobber length drills are most common and shorter length NC drills specifically designed for CNC applications are used on CNC machines. Drill, spot drill and center drill size charts can be found in the *Machinery's Handbook*.

Figure 1-55. Drilling icon, *courtesy Autodesk Fusion 360*

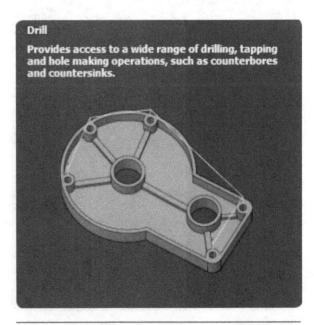

Drill

Provides access to a wide range of drilling, tapping and hole making operations, such as counterbores and countersinks.

Figure 1-56. Drilling description, *courtesy Autodesk Fusion 360*

Figure 1-57. Drilling types, *courtesy Autodesk Fusion 360*

Additional hole-making operations are categorized under the drilling and include: center drilling, spot drilling, reaming, countersinking, and tapping (Figure 1-57). Holes that do not penetrate through the part are considered blind and are defined with a depth requirement that is measured from the apex of the drill point angle at the full diameter of the drill.

Boring

Boring is categorized under milling in the Fusion 360 software, as, oftentimes, milling tools can be used to rough machine prior to programming a single-point boring head to finish the precision hole (Figure 1-58).

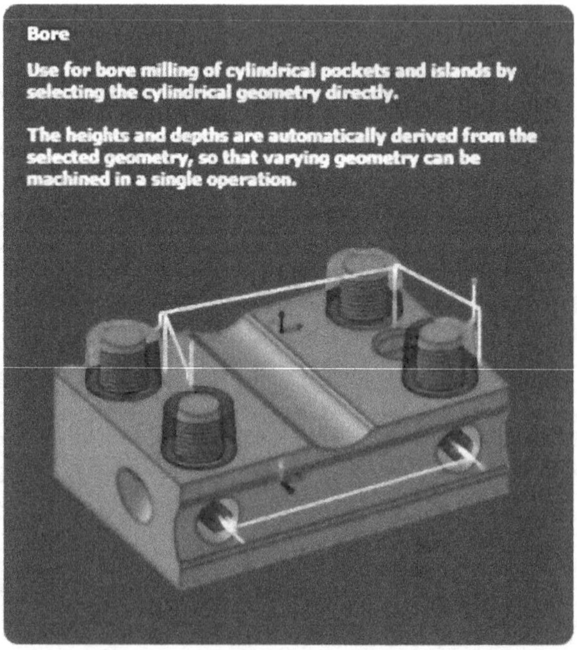

Other Milling Operations

2D Chamfering

Deburring of post machined parts can be virtually eliminated with the use of chamfering operations and tools. Solid chamfer tools are available with straight or helical flutes and range in diameter from ⅛ inch up to ¾ inch and come in standard cutting angles of 60 degrees, 90 degrees or 120 degrees. This type of tool allows for programming of tool paths that follow contours to eliminate sharp edges. Machined holes can be chamfered with center and spot drilling operations when hole size and tool diameter allow. But in cases where the hole size is larger than the tool, a circular interpolation path can be programmed to accomplish the chamfer (Figure 1-59).

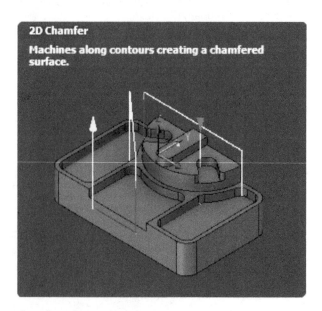

Figure 1-58. Bore description, *courtesy Autodesk Fusion 360*

Figure 1-59. 2D chamfering, *courtesy Autodesk Fusion 360*

Larger diameter indexable inserted tools are available for more difficult to machine materials. Single- and multi-tooth tools are available in angles from 30 degrees up to 75 degrees in 15-degree increments (Figures 1-60 and 1-61).

Figure 1-60. Multi-insert chamfer cutter, *courtesy Sandvik Coromant*

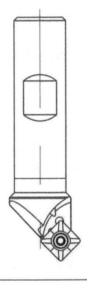

Figure 1-61. Single-insert chamfer cutter, *courtesy Sandvik Coromant*

Engraving

Engraving operations allow very detailed part marking capabilities. The center line of the text profile is programmed and cut to a depth that is defined (Figure 1-62). These tools range in diameter from ⅛ to ¼ inch and angles from 30 degrees to 120 degrees.

Figure 1-62. Engraving, *courtesy Autodesk Fusion 360*

Tool Assembly Considerations for the CNC Mill

Tool-Clamping Methods

Proper selection of cutting tools and work-holding methods are paramount to the success of any machining operation. The scope of this text is not intended to teach all of the necessary information regarding tooling. You must consult the appropriate tooling catalogs, websites, and online resources for selection of tool holders and cutting tools that are relevant to the required operation. Use sound machining principals that require the most rigid setup possible and avoid excessive tool extension. Also, avoid over extending the part in the clamping system. Vibration and tool deflection are your enemies and will cause a poor surface finish and, eventually, tool damage that makes it difficult to maintain dimensional accuracy. Therefore, it is very important to carefully select the most effective tool-clamping method and use proven best practices.

Tool holders for vertical milling machines have industry standardized taper sizes depending on the machine specification being used and are classified as: CAT (Caterpiller "V-flange") or HSK (Hohl Schaft Kegel, which translates to hollow taper shank in English). The tapered portion of the holder is the actual surface that is in contact with the mating taper of the spindle. Big-Plus style tool holders mate the taper and the flange to further increase the contacted surface area. Common sizes for CAT tooling are nos. 30, 40 and 50. Common sizes for HSK tooling are nos. 40, 50, 63, 80 and 100. HSK styles are used on machines with high-spindle speed and high-performance specifications. For this text, the focus will be exclusively on the CAT 40 taper style.

Another feature on the tool holder are the notches or cutouts on the center line of the tool. These cutouts enable axial orientation within the spindle and tool changer. As the holder is inserted into the spindle, the cutouts enable it to be locked into place in exactly the same orientation each and every time it is used. This orientation makes a real difference when trying to perform very precise operations such as boring a diameter. These notches also aid the spindle driving mechanism. Machining centers need the retention knob or pull-stud to pull the tool into the spindle and clamp the holder. This knob is threaded into the small end of the taper, as shown in Figure 1-63.

Note: There are several styles of knobs available. The operator should consult the appropriate manufacturer manual for specifications required in their situation.

Figure 1-63. Retention knob, *courtesy Sandvik Coromant*

Milling Tool Fitting

Make sure the holder insert pocket or bore is clean and free of burrs before installing a new insert or cutter by wiping the surfaces with a clean rag or by applying a light air blast. Tighten the clamping screws to the recommended torque determined by the tool maker. If this info is not available, use good judgement. **Do not over-tighten!** When clamping the tool in the holder, always pay attention to the minimum extension requirement (also called *extension offset height*) required by the feature to be machined and communicated in the setup information. In the case of Weldon shank tools, be sure the setscrew seats into the cutout in the shank. For high volume/accuracy applications, hydraulic shrink fit tool holders may perform best; when high rev/min are required, tool balancing is imperative for best accuracy.

Face Mill Applications

Face milling cutters, such as those shown in Figures 1-27 and 1-28, are mounted to drive arbors such as that shown in Figure 1-64. Inserts are changed out as they become worn, broken or the material application changes.

End Mill Applications

In the case of a simple milling operation of a contour, an ER-style collet (see Figures 1-64 and 1-66) or a Weldon style locking end mill holder for the end mill could be used (see Figures 1-64 and 1-65). The correct choice would depend on the actual features of the part to be machined and its dimensional tolerance. If the amount of metal to be removed is minimal and the tolerance allows, then a collet would probably suffice. But if a considerable amount of metal is to be removed (more than two-thirds of the tool diameter on a single DOC pass), then the Weldon style holder selection would be better. The reason for selecting the Weldon style holder is that under heavy cuts, a collet may not be able to grip the tool tightly enough. This situation could allow the tool to spin within the collet while cutting is in progress, which could damage the collet and possibly damage the part being machined. There is a tendency for the tool to dive into the workpiece when the tool spins within the collet and so damage to the part may occur.

Note: Most HSS end mills have a flat ground on them (called a Weldon shank) to facilitate the use in these types of toolholders. This flat area allows for a setscrew to lock into it, creating a rigid and stable tool-clamping method. This same shank style is used with the holders for indexable inserted end mills also.

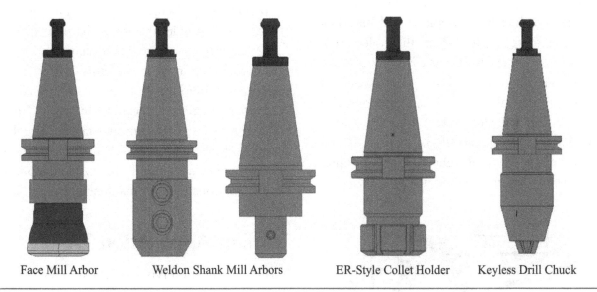

| Face Mill Arbor | Weldon Shank Mill Arbors | ER-Style Collet Holder | Keyless Drill Chuck |

Figure 1-64. CAT V40 tool holder styles, *courtesy Sandvik Coromant*

Types of Tool Holders and Determining Factors

End mill Weldon flat specifications are typically based on an HSS tool standard (NAS 986), and are measured from the shank end. Currently, there is no standardized specification for high performance (HP) solid carbide end mills. The flat ground into the shank precisely forces the tool extension length based on the flute length and the setscrew locks the tool from rotation in the holder.

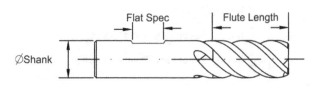

Figure 1-65. Weldon shank specifications, *courtesy Harvey Performance Company*

Length of Cut

The design geometry of the feature determines the tool chosen based on the LOC, width of cut and tool extension required. For example, if the feature requires a shoulder of one-half-inch, the tool used must have at least one-half-inch flute length (unless a necked tool is used). Sometimes this is called *extension offset height*.

Collet Applications

Collets are available in many different sizes and configurations and offer a very quick and accurate way to clamp straight shank tools. The ER collet system is the most commonly used system for milling (Figure 1-66). The range of diameter in sizes are typically available from one-eighth to one-inch, incremented in thirty-seconds of an inch. The shank of the tool is inserted into the collet, *not* the cutting flutes. When possible, the tool should be inserted as far as possible to mitigate tool deflection. When assembling the tool holder, the collet nut should be snapped onto the collet first and then installed in the CAT holder—but only after that the cutting tool is installed.

Note: The collet is marked for a specific size of tool shank. There is some downward size mobility possible but caution should be taken so as not to spring (over-flex) the collet. As a general rule, when downsizing is needed, place the cutting tool to be used into an unassembled collet, grip the collet in your hand and pinch to see if you can still slide or turn the cutting tool. If it clamps snuggly, then it can be used. If you cannot pinch the tool tight enough by hand, then the combination should not be used.

Figure 1-66. ER style collet, *courtesy Sandvik Coromant*

Drilling Applications

The tool-clamping method for drills could be either a collet (previously described) or a drill chuck. A keyed drill chuck is usually selected for heavier metal removal or larger holes, whereas the keyless drill chuck is suitable for small holes (generally ≤ .500 inch). Generally, in the case of larger drills, a collet will be necessary to hold the tool. When holes are to be drilled, remember to center drill or spot drill first, so that the tool

does not wander off target. The center drill may be held in the same manner as a drill.

Cutting Tool Selection for Turning

For turning, selection of the type of tool holder is determined by the finished part geometry and the part material. There are a variety of toolholder styles, as well as indexable insert shapes available to accomplish the desired part shape and size. For more information on the proper selection of inserts and toolholders, refer to the *Machinery's Handbook*, section "Indexable Inserts." Other valuable resources for technical data regarding the selection of inserts and tool-holding are ordering catalogs, online advice and optimization applications from tool and insert manufacturers.

Turning Application Types

Turning application types are listed in the drop-down menu of the Fusion 360 turning icon (Figure 1-67). These are common to all CNC turning.

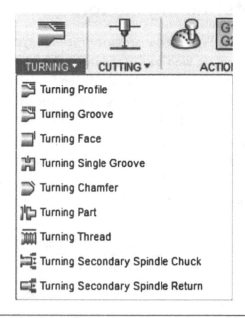

Figure 1-67. Turning icon menu, *courtesy Autodesk Fusion 360*

Turning Profile

Removing excess material from the outer and inner profiles on lathes is the most common operation during machining on a lathe. The roughing passes remove material following the design geometry at a suitable depth-of-cut for the tool used, leaving a small amount of material (1–2.5 times the tool nose radius) for a finishing pass along the profile. As shown in Figure 1-68, the cut direction can be either toward the spindle face (Z-axis) or toward spindle center line (X-axis). The same technique can be used for internal operations by using boring bar tools. Examples of the tool types used are given in Figure 1-69. The insert shape for the roughing tool uses an 80-degree, the finish tool uses either a 35-degree or 55-degree diamond-shaped insert. The tool shank size depends on the machine configuration. Always consult the operation/maintenance manuals to be certain before ordering. For internal profiles, the same insert shapes as previously described can be used for roughing and finishing when mounted on boring bars (not shown). For this type of operation, a predrilled hole is required that is greater than the shank diameter of the boring bar.

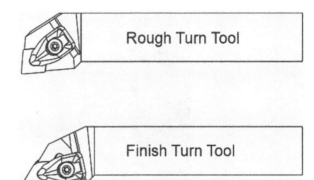

Figure 1-69. Rough and finish turning tools, *courtesy Sandvik Coromant*

Turning Groove

Grooving is commonly used to reach part profile geometry configurations wherever it is impossible to use the roughing and finishing tools described for turning profiles. The path is designed to make multiple plunging cuts along the X-axis toward the spindle center line or along the Z-axis toward the spindle face. The same technique can be used for internal operations by using internal diameter (ID) grooving tools. An example of the tool type is given in Figure 1-71.

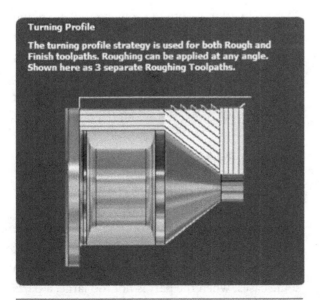

Figure 1-68. Turning profile, *courtesy Autodesk Fusion 360*

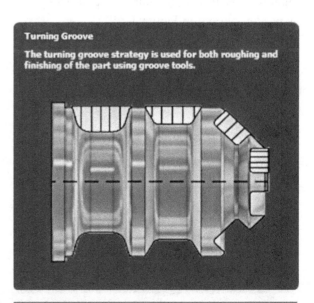

Figure 1-70. Turning groove, *courtesy Autodesk Fusion 360*

Figure 1-71. Grooving tool, *courtesy Sandvik Coromant*

Turning Face

Sometimes the front face of the part needs to be machined in a separate operation. When this operation is needed, the same rough turning tool is used as shown in Figure 1-69.

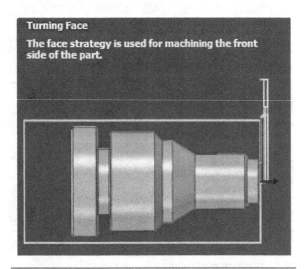

Figure 1-72. Turning face, *courtesy Autodesk Fusion 360*

Turning Single Groove

This type of grooving activity is well suited for machining what is called an *escape groove* for an outside diameter (OD) threading operation that will follow. The same tool as shown for Turning Groove is suitable for this operation as well (Figure 1-71). The only limiting factor may be the groove depth requirement. The tool used must be capable of attaining the required depth.

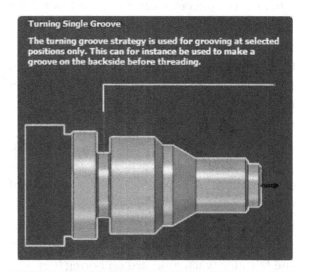

Figure 1-73. Turning single groove, *courtesy Autodesk Fusion 360*

Turning Chamfer

This tool path type uses a grooving tool as shown in Figure 1-74 to chamfer the edges and profile of a groove requiring it. A grooving tool used for this operation must have the ability to cut in along the Z-axis direction also and the groove width must be greater than the insert width (Figure 1-74).

Turning Part

This tool path type (Figure 1-75) is used for *parting-off* and uses a tool as shown in Figure 1-76 to cut off the finished part from the bar stock. The blade extension amount must be slightly greater than the radius of the part being machined. In some cases, when the part is not too large, a parts catcher is automatically positioned by the CNC to keep it from falling into the chip pan.

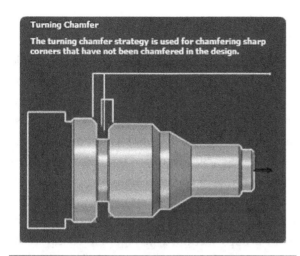

Figure 1-74. Turning chamfer, *courtesy Fusion 360*

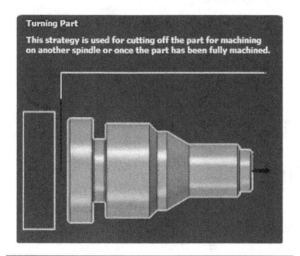

Figure 1-75. Turning part, *courtesy Fusion 360*

Figure 1-76. Part-off tool, *courtesy Sandvik Coromant*

Turning Thread

Single-point threading is performed on the CNC lathe using this tool path type. The insert determines the thread form (commonly 60 degrees). Other thread forms are available. Check cutting tool vendor catalogs and online resources for unique thread form applications (Figures 1-77 and 1-78).

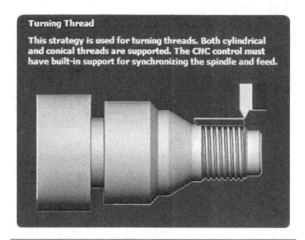

Figure 1-77. Turning thread, *courtesy Fusion 360*

Figure 1-78. O.D. threading tool, *courtesy Sandvik Coromant*

Drilling

Drilling operations performed on a CNC lathe use the same drill cutting information previously given for milling when using drill bits. In the case of turning, the tool is stationary and the part rotates. The toolholder may be a drill chuck mounted in a bushing or, in the case of larger drills, the drill is mounted directly into an exact size bushing and collet holder in the turret pocket. Two-axis lathes are only able to drill on spindle center line.

Turning Cutter Features/ Attributes

Today, modern CNC lathes rarely use high speed cutting tools for turning. Indexable carbide inserts are now the norm. The characteristic attributes that define the insert cutting tool are the insert

material and coating, shape, size, nose radius, thickness, inscribed circle and others. The most common inserts are uncoated cemented carbide and coated. The strength and durability of the insert is based on the insert shape in order from strongest to weakest: Round (strongest and most durable), Square, 80 degrees Diamond, 80 degrees 6-Sided (W), 60 degrees Triangle, 55 degrees Diamond and 35 degrees Diamond (least strength and most fragile). The insert applications for different workpiece materials fall into the same six main groups identified in Figure 1-30 earlier in this section. Charts 1-2a and 1-2b fully describe these attributes.

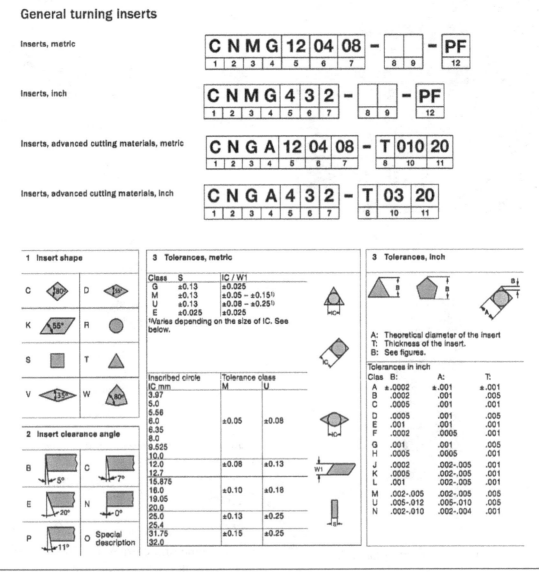

Chart 1-2a.   Insert code key, *courtesy Sandvik Coromant*

General turning inserts

4 Insert type

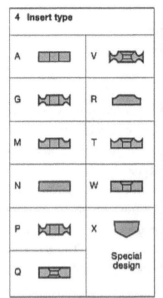

A		V	
G		R	
M		T	
N		W	
P		X	Special design
Q			

5 Insert size

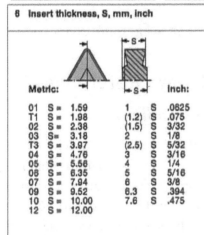

Inscribed circle is indicated in 1/8".

For rectangular and rhombic inserts cutting edge length is indicated in mm.

[1] Metric base design
[2] Inch base design

Cutting edge length, metric		C	D	R	S	T	V	W	K
IC mm	IC inch								
3.18	1/8"					05			
3.97	5/32"					06		02	
5.0				05					
5.56	7/32"			09					
6.0			06						
6.35	1/4"	06	07			11	11	04	
8.0				08					
9.525	3/8"	09	11	09	09	16	16	06	16[1]
10.0	10.0			10					
12.0				12					
12.7	1/2"	12	15	12	12	22	22	08	
13			13				13		
15.875	5/8"	16		15	15	27			
16.0				16					
19.0	3/4"	19		19	19	33			
20.0				20					
25.0				25[1]					
25.4	1"	25		25[2]	25				
31.75	1/4"			31					
32				32					

6 Insert thickness, S, mm, inch

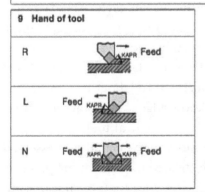

Metric:

01	S =	1.59
T1	S =	1.98
02	S =	2.38
03	S =	3.18
T3	S =	3.97
04	S =	4.76
05	S =	5.56
06	S =	6.35
07	S =	7.94
09	S =	9.52
10	S =	10.00
12	S =	12.00

Inch:

1	S	.0625
(1.2)	S	.075
(1.5)	S	3/32
2	S	1/8
(2.5)	S	5/32
3	S	3/16
4	S	1/4
5	S	5/16
6	S	3/8
6.3	S	.394
7.6	S	.475

7 Nose radius, RE, mm, inch

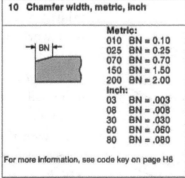

Metric:

00*	=	0
01	=	0.1
02	=	0.2
04	=	0.4
05	=	0.5
08	=	0.8
10	=	1.0
12	=	1.2
15	=	1.5
16	=	1.6
24	=	2.4
32	=	3.2

Inch:

00	=	0
.30	=	.004
.50	=	.008
1	=	.0156
2	=	.031
3	=	.047
4	=	.063
6	=	.094
8	=	.125

Note: See example for approximation of metric nose radius. 16=1.6mm=.063=.0625

8 Cutting edge condition

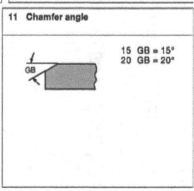

F	Sharp cutting edge
A	ER treated cutting edge (ANSI)
E	ER treated cutting edge
T	Negative land
K	Double negative lands
S	Negative land and ER treated cutting edge

9 Hand of tool

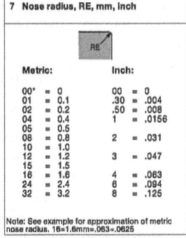

R — KAPR Feed

L — Feed KAPR

N — Feed KAPR / KAPR Feed

10 Chamfer width, metric, inch

Metric:

010	BN = 0.10
025	BN = 0.25
070	BN = 0.70
150	BN = 1.50
200	BN = 2.00

Inch:

03	BN = .003
08	BN = .008
30	BN = .030
60	BN = .060
80	BN = .080

For more information, see code key on page H8

11 Chamfer angle

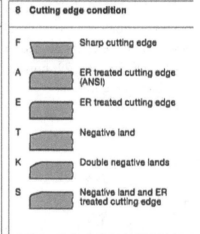

15 GB = 15°
20 GB = 20°

12 Manufacturer's option

The ISO code consists of nine symbols including 8 and 9 which are used only when required. In addition the manufacturer may add further three symbols, e. g.,

- WF = Wiper – finishing
- WMX = Wiper, medium machining
- PF = ISO P – finishing
- PR = ISO P – roughing

7 * Code on round inserts

Code 00 or M0 in position 7 is used on round inserts in the metric code. M0 shows that the diameter of the insert has an even metric dimension. In the inch code for round inserts, position 7 isn't used at all. It's blank.

Chart 1-2b.   Insert code key, *courtesy Sandvik Coromant*

Refer to the tool and insert ordering catalogs and online applications from the tool and insert manufacturers for more information on modular tooling. There are separate charts specific to insert data for grooving, parting and threading listed at http://sandvik.ecbook.se/se/us-en/turning_tools_2017/.

Tool Assembly Considerations for the CNC Lathe

Inserts are mounted into precisely located tool pockets in the tool bar and held in place by torx head screws and, sometimes, top lock clamps. When installing or replacing inserts, take special care to ensure a clean burr-free seat. Carefully examine the shim and check for damage. Check the insert cutting tip for damage or wear and rotate to a new edge, if needed, or replace if all tips are used. Figure 1-79 shows an example of a boring bar configuration.

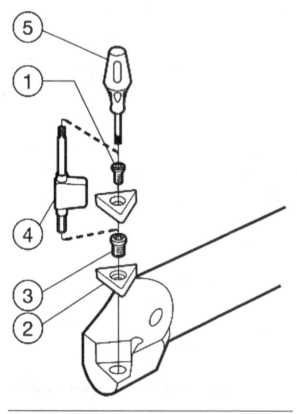

Figure 1-79. OD threading tool, *courtesy Sandvik Coromant*

1. Torx head insert screw.
2. Insert shim.
3. Insert shim clamping screw.
4. Insert and shim clamping tool.
5. Insert clamping tool.

Turning tool bars are mounted in the CNC lathe turret into carriers for external and internal tools. The tool pots alternate with every other tool being opposite in type. Wedge type locks are common for external turning tools. Before inserting a new tool into location, the mating surfaces should be checked for cleanliness and nicks or burrs and corrected, if necessary. Tighten holders to recommended torque specifications. **Do not overtighten!** Pay attention to best practices regarding tool extension and make them as short as possible while allowing ample clearance.

Technical Data References
Machinery's Handbook

Using technical references provided by cutting tool manufacturers is essential to obtaining reliable machining data for successful results. For over one hundred years, the *Machinery's Handbook* ("bible of the metalworking industries") has been the de-facto standard for printed and online reference data for those who work in metalworking, design, engineering and manufacturing facilities, and technical schools and colleges throughout the world. It is universally acknowledged as an authoritative, comprehensive, and practical tool, providing its users with the most fundamental and essential aspects of sophisticated manufacturing practice. It is *the* essential reference for mechanical, manufacturing, and industrial engineers, designers, draftsmen, toolmakers, machinists, engineering and technology students, and the serious home hobbyist. For information specific to all aspects of cutting tool and machining data, refer to the page tabs "Tooling" and "Machining" in the *Machinery's Handbook*.

Machining Advisor Pro (MAP)

Another valuable resource for technical data specific to fluted end mills and high-performance milling data is the Harvey Performance Company. They not only offer cutting tools for purchase but also provide technical references online (specifically, *Machining Guidebook, Quick Reference eBook for CNC Milling Practices and Techniques*, 2016 © Helical Solutions, LLC) and an app for mobile devices called Machining Advisor Pro (MAP), which can be found at http://map .harveyperformance.com/#/login.

CoroPlus® ToolGuide

For data regarding the selection and use of indexable inserts and tool-holding, Sandvik Coromant offers online and offline versions as well as printed catalogs of their CoroPlus® ToolGuide that can be used to identify cutting tools from their catalog with recommended cutting speeds and feeds based on the type of operation being performed. Sandvik Coromant has been referenced often in this text.

Controlling and Measuring Surface Finish

There is a direct effect that feeds and speeds and DOC have on controlling surface finish. There is additional detailed information on this subject at http://new.industrialpress.com/handbook-of -dimensional-measurement-fifth-edition.html.

Summary

In this section, we have introduced the selection process for cutting tools used for CNC milling and turning. The examples given here make up the core of common operations performed at these machines. There are many more possibilities. Cutting tools in and of themselves make up an entire study and thousands of pages of information are available for many other types of cutting tools and insert materials from the online resources referenced here. New developments are always on the horizon so keep an eye out for making chips in more efficient ways.

Cutting Tool Selection, Study Questions

1. Fluted end mills can be used to remove material on part profiles in either feed direction. The preferred method is:

   a. Conventional milling
   b. Climb milling
   c. Helical milling
   d. Ramp milling

2. Once the appropriate feeds and speeds are calculated for a milling operation, what are two more important machining factors that must be considered?

   a. ADOC
   b. RDOC
   c. Both a and b
   d. None of the above

3. What type of an end mill shank uses a setscrew to lock it in place within the toolholder?

   a. Solid carbide
   b. High speed steel
   c. Weldon style
   d. Straight shank

4. Which of these machining applications are not performed on a CNC mill?

   a. Face milling
   b. Drilling
   c. Tapping
   d. Profile turning

5. Machining inserts are sorted by material application. Which letter identifies the aluminum application?

   a. A
   b. K
   c. N
   d. P

6. Turning tool inserts come in different shapes. What shape is a CNMG432 insert?

   a. 80 degrees Diamond
   b. 55 degrees Diamond
   c. Round
   d. Square

7. What is the turning tool insert tool nose radius for the CNMG432 insert?

   a. .031 inch
   b. .015 inch
   c. $3/16$ inch
   d. $1/16$ inch

8. What are the most common insert materials?

   a. Ceramic and coated
   b. Uncoated and coated cemented carbide
   c. Cubic boron nitride and polycrystalline diamond
   d. High-speed steel

9. Finish turning on a CNC lathe is commonly completed using an insert of what shape?

   a. 80 degrees Diamond
   b. 55 degrees Diamond
   c. 35 degrees Diamond
   d. Both b and c

10. When replacing an insert in a toolholder, it is important to do which of the following?

    a. Clean the insert pocket
    b. Examine the insert shim for damage
    c. Tighten the insert to proper torque requirements
    d. All of the above

Machining Mathematics

Objectives

1. Understand and apply the coordinate systems related to CNC milling and turning.
2. Use absolute and incremental values when defining part geometries.
3. Calculate spindle speed and feedrates for both milling and turning operations.
4. Use required algebraic formulas to calculate feeds and speeds.
5. Use trigonometry to calculate common machining problems.

In this section, we will present the direct impact that shop math has on successful machining and lay out the common formulas used in day-to-day standard operation practices. In the preceding section, "Cutting Tool Selection," you were introduced to cutting data provided by the cutting tool manufacturer and software tools used to calculate effective cutting conditions. Later in this text, you will be introduced to CAD/CAM where the same type of cutting tool data will be incorporated via onboard tool libraries with feeds and speeds and other calculators. While most calculations today are done by the methods just described, there is still a need to have knowledge of the formula necessary and be able to calculate manually using cutting data found in the *Machinery's Handbook*, especially in cases where the internet or CAD/CAM are not available.

The following are excerpts taken from *Programming of CNC Machines* (4th ed.), Ken Evans.

Applying the Rectangular Coordinate System

A thorough understanding of the machine tool coordinate system and the relationship it has to the part geometry coordinates (workpiece coordinates) is essential in the setup and programming processes. How this relates to CNC milling and turning will be presented first.

Coordinate Systems

Visualize a grid on a sheet of graph paper with each segment of the grid having a specific value. Now place two solid lines through the exact center of the grid and perpendicular to each other. By doing this, you have constructed a simple, two-dimensional coordinate system. Carry the thought a little further and add a third imaginary line. This line passes through the same center point as the first two lines but is vertical; that is, it rises above the first two lines but is vertical; that is, it rises above

and below the sheet on which the grid is placed. This additional line, which is called the Z-axis, represents the third axis in the three-dimensional coordinate system.

Two-Dimensional Coordinate System

A two-dimensional coordinate system, such as is used on 2-axis CNC lathes, uses the X and Z axes for measurement. The X-axis runs perpendicular to the workpiece and the Z-axis is parallel with the spindle center line. When working on the lathe, we are working with a workpiece that has only two dimensions, the diameter and the length. On engineering drawings or blueprints, the front view generally shows the features that define the finished shape of the part for turning. In order to see how to apply this type of coordinate system, study Figures 1-80, 1-81 and 1-82.

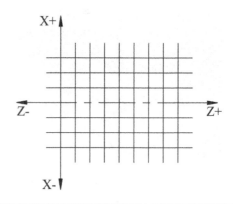

Figure 1-80. Two-dimensional coordinates system

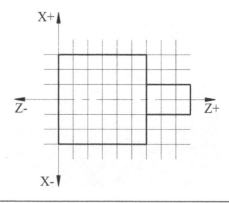

Figure 1-81. Drawing overlaid on 2D coordinate system

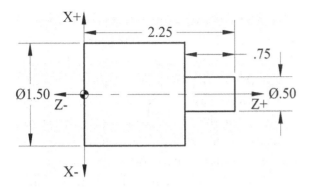

Figure 1-82. Two-dimensional turned part drawing

Think of the cylindrical work piece as if it were flat or as shown in the front view of the part blueprint. Next, visualize the coordinate system superimposed over the engineering drawing or blueprint of the workpiece, aligning the X-axis with the center line of the diameter shown. Then align the Z-axis with the end of the part, which will be used as an origin or zero point. In most cases, the finished part surface nearest the spindle face will represent this Z-axis datum and the center line will represent the X-axis. Where the two axes intersect is the origin or zero point. By laying out this grid, we can now apply the coordinate system and define where the points are located to enable programmed creation of the geometry from the blueprint. Another point to consider is that, on a lathe, the cutting takes place on only one side of the part or the radius, because the part rotates and is symmetrical around the center line. In order to apply the coordinate system in this case, all we need is the basic contour features of one-half of the part (on one side of the diameter); the other half is a mirror image. When given this program coordinate information, the lathe will automatically produce the mirror image.

Three-Dimensional Coordinate System

Although the mill uses a three-dimensional coordinate system, the same concept (using the front view of the engineering drawing or blueprint) can be used with rectangular workpieces. As with the lathe, the Z-axis is related to the spindle. However, in the case of the three-dimensional rectangular workpiece, the origin or zero point must be defined differently. In the example shown in Figure 1-83, the lower left-hand corner of the workpiece is chosen as the zero point for defining movements using the coordinate system. The thickness of the part is the third dimension or Z-axis. When selecting a zero point for the Z-axis of a particular part, it is common to use the top surface.

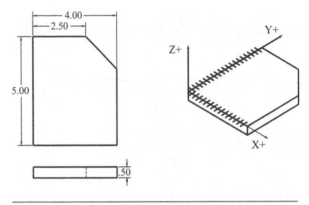

Figure 1-83. Three-dimensional milled part coordinate system overlay

The Polar Coordinate System

Draw a circle on a piece of graph paper so that the center of the circle is at the intersection of two lines and the edges of the circle are tangent to any line on the paper. This will help in visualizing the following statements. Let's consider the circle center as the origin or zero point of the coordinate system. This means that some of the points defined within this grid will be negative numbers. Now draw a horizontal line through the center and passing through each side of the circle. Then draw a vertical line through the center also passing through each side of the circle. Basically, we've made a pie with four pieces. Each of the four pieces or segments of the circle is known as a quadrant. The quadrants are numbered and progress counterclockwise. In quadrant 1, both the X- and Y-axis point values are positive. In quadrant 2, the X-axis point values are negative, and the Y-axis point values are positive. In quadrant 3, both the X- and Y-axis point values are negative. Finally, in quadrant 4, the X-axis point values are positive while the Y-axis values are negative. This quadrant system is applied in all cases regardless of the axis of rotation. The drawings in Figure 1-84 illustrate the values (negative or positive) of the coordinates, depending on the quadrant in which they appear.

not moved from its starting point, the angular measurement is known as 0.0 degrees. On the other hand, if the radius line has circled once around the zero point, the angular measurement is known as 360 degrees. Therefore, the movement of the radius determines the angular measurement. If the direction in which the radius rotates is counter-clockwise, angular values will be positive. A negative angular value (such as –90 degrees) indicates that the radius has rotated in a clockwise direction.

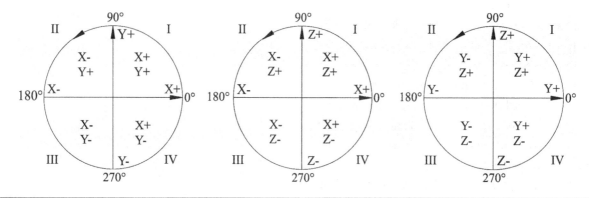

Figure 1-84. Polar coordinate system quadrants

Although the rectangular coordinate system can be used to define points on the circle, a method using angular values may also be specified. We still use the same origin or zero point for the X- and Y-axes. However, the two values that are being considered are an angular value for the position of a point on the circle and the length of the radius joining that point with the center of the circle. To understand the polar coordinate system, imagine that the radius is a line circling around the center origin or zero point. Thinking in terms of hand movements on a clock, the three-o'clock position has an angular value of 0.0 degrees and is counted as the *starting point* for the radius line. The twelve-o'clock position is referred to as the 90 degrees position, nine-o'clock is 180 degrees, and the six-o'clock position is 270 degrees. When the radius line lies on the X-axis in the three-o'clock position, we have at least two possible angular measurements. If the radius line has

Note: A 90-degree angle (clockwise rotation) places the radius at the same position on the grid as a +270 degrees (counterclockwise) rotation.

Sometimes the engineering drawing or blueprint will not specify a rectangular coordinate, but will give a polar system in the form of an angle for the location of a feature. With some basic trigonometric calculations, this information can be converted to the rectangular coordinate system. The same polar coordinates system applies regardless of the axis of rotation, as is shown once again in Figure 1-84. When rotation is around the X-axis, the rotational axis is designated as A; the Y-axis, the rotational axis is designated as B; and the Z-axis, the rotational axis is designated as C. These are considered additional axes and are known as the *fourth axis*.

All operations of CNC milling machines are based on three axes: X, Y and Z and rotational axes, A, B, and C, when available. On vertical milling machines, the spindle axis is perpendicular to the surface of the worktable (refer to Figure 1-15). On CNC lathes, the spindle axis is also the workpiece axis (refer to Figures 1-16 and 1-17).

Machine Coordinate System

When using CNC machines, any movement of tool location is controlled within the machine coordinate system. The origin of this coordinate system is called *machine zero*, a fixed point established by the manufacturer that is the basis for all coordinate system measurements. On a typical lathe, machine zero is usually located at the maximum travel distance from the spindle center line in the X-axis and the same maximum travel distance from the face of the spindle nose for the Z-axis. This machine zero point establishes the coordinate system origin for operation of the machine and is commonly called *machine home* (home position). Upon startup of the machine, all axes need to be moved to this position to establish the coordinate system origin (commonly called *homing the machine* or *zero return*). Today, some machines are equipped with absolute encoders so that homing is no longer necessary at machine startup. The operator's manual supplied with the machine should be consulted to identify where this location is and how to properly home the machine (covered in detail in Chapters 2 and 3). From machine zero, we can determine the values of the coordinates for fixture offsets that, in turn, determine the position of the points commanded in a CNC program.

Workpiece Zero

The fixture offset origin is called *workpiece zero*. The X, Y and Z points can be located anywhere within the machine work envelope, and it is the basis for programmed coordinate values used to produce the workpiece. This coordinate location is established within the part program by a special code (G54–G59) and the coordinates are referenced from their distance from the machine zero point. The code number in the program identifies the location of offset values to the machine control where the exact coordinate distance of the X, Y, and Z axes of workpiece zero is in relation to the machine zero. All dimensional data on the part will be established by accurately setting the workpiece zero. A way of looking at the workpiece zero is like another coordinate system within the machine coordinate system, established by the home position. In order to better understand this concept, this situation can be illustrated with a rectangular plate in which the workpiece zero origin is described at the corner (P1) in Figure 1-85. On milling machines, workpiece zero is frequently located on the corner of the workpiece or in alignment with the datum features of the workpiece.

The application of workpiece zero is advantageous to the programmer because the input values of X, Y and Z in the program can be taken directly from the drawing. If the program is used another time, the values of coordinates X and Y (assigned to functions G54–G59) will need to be inserted again prior to automatic operation. For all CNC machines, we follow certain principles based on the engineering design to define the method of selecting workpiece zero for the part program. At the beginning of the program, we input the value of the distance between machine zero and the selected workpiece zero by employing function G54–G59 for machining and turning centers. These measured values are input in offset registers in the control for G54–59.

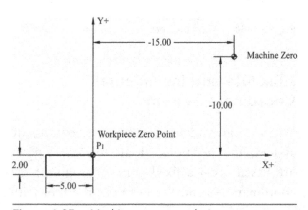

Figure 1-85. Machine zero to workpiece zero

Tool Offsets

Tool offsets are also considered to be zero points and are compensated for with tool gauge length and diameter offsets. The tool-setting point for the mill is the distance from the spindle face to the tool tip, and the distance from the tool tip to the spindle center line.

With turning centers, workpiece zero in the direction of the Z-axis is most often on the face surface of the workpiece, and the center line axis of the spindle in the direction of the X-axis.

The tool-setting point for a lathe has two dimensions: 1) the distance on part diameter from the tool tip to the center line of the tool turret, and 2) the distance from the tool turret face to the tool tip illustrated in Figure 1-86. To determine the coordinates of these points, the rectangle has been placed in such a manner that each side is parallel to one axis of the coordinate system.

coordinates from point P1–P10 are illustrated in Figure 1-87.

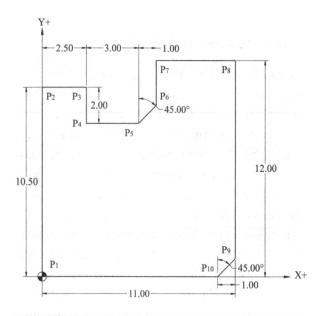

Figure 1-87. Absolute and incremental coordinate system points

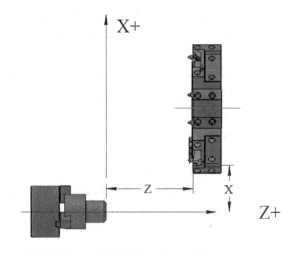

Figure 1-86. Machine zero to workpiece zero

Absolute and Incremental Coordinate Systems

When programming in an absolute coordinate system, the positions of all the coordinates are based upon a fixed point or origin of the coordinate system. For example, the tool path

	X	Y
P_1	0.0	0.0
P_2	0.0	10.5
P_3	2.5	10.5
P_4	2.5	8.5
P_5	5.5	8.5
P_6	6.5	9.5
P_7	6.5	12.0
P_8	11.0	12.0
P_9	11.0	1.0
P_{10}	10.0	0.0

Programming with an incremental coordinate system is based upon the determination of the tool path from its current position to its next consecutive position and in the direction of all the axes. Sign determines the direction of motion. Based on the drawing from the previous example, we can illustrate the tool path in an incremental coordinate system starting and ending at P1.

	X	Y
P_2	0.0	10.5
P_3	2.5	0.0
P_4	0.0	–2.0
P_5	3.0	0.0
P_6	1.0	1.0
P_7	0.0	2.5
P_8	4.5	0.0
P_9	0.0	–11.0
P_{10}	–1.0	–1.0
P_1	–10.0	0.0

Each cutting tool has a number of factors that affect proper function:

Cutting speed. Cutting speed is the rate at which the circumference of the tool moves past the workpiece in surface feet (sf/min) or meters per minute (m/min) to obtain satisfactory metal removal. The cutting speed factor is most closely related to the tool life. Many years of research have been dedicated to this aspect of metal-cutting operations. The workpiece and the cutting tool materials determine the recommended cutting speed. *Machinery's Handbook* is an excellent source for information pertaining to determining proper cutting speed. If incorrect cutting speeds, spindle speeds, or feedrates are used, the results will be poor tool life, poor surface finishes, and even the possibility of damage to the tool and/or part. Refer to the formulas in Chart 1-3 to calculate the cutting speed (V).

Spindle speed. When referring to a milling or a turning operation, the spindle speed of the cutting tool or chuck must be accurately calculated relating to the conditions present. This speed is measured in revolutions per minute (RPM) or r/min and is dependent upon the type and condition of material being machined. This factor, coupled with DOC, gives the information necessary to find the horsepower required to perform a given operation. In order to create a highly productive machining operation, all these factors should be given careful consideration. Refer to the formulas in Chart 1-3 to calculate RPM.

Feedrate. Feedrate is defined as the distance the tool travels along a given axis in a set amount of time, generally measured in inches per minute (in/min, formerly known as IPM) for milling or inches per revolution (in/rev, formerly known as IPR) for turning. This factor is dependent upon the selected tool type, the calculated spindle speed, and the DOC. Refer to the *Machinery's Handbook* and cutting tool manufacturer data for the chip load recommendations and refer to the formulas in Chart 1-3 to calculate feedrate for the metal-cutting operation.

Many machine controllers have a feature that allows automatic calculation of feeds and speeds that is based on appropriate operator input of the cutting conditions. This is also true when using CAM, where feeds and speeds data can be extracted from available tool material or part materials machining libraries.

Depth of cut. The depth of cut (DOC) is determined by the amount of material to be removed from the workpiece, cutting tool flute length or insert size, and the power available from the machine spindle. Always use the largest DOC possible to ensure the least effect on the tool life. Refer back to the prior section "Cutting Tool Selection" for factors used to determine the best DOC parameters. Cutting speed, spindle speed, feedrate and DOC are all important factors in the metal-cutting process. When properly calculated, the optimum metal-cutting conditions will result. Refer to the *Machinery's Handbook*, tool and insert ordering catalogs, and online applications from the tool and insert manufacturers for more information on recommended depths of cut for particular tooling.

End of excerpts from *Programming of CNC Machines* (4th ed).

Manual calculations are seldom used today with all of the resources available online and HELP menus within the controller software that are easily interfaced with CAM programs and used by inputting the well-defined required cutting parameters.

Notes on Speed and Feed Calculation Adjustments

If the correct cutting tool and work holding are selected and the proper speeds and feeds are programmed correctly, the result will be a quality product. When these criteria are met, excessive noise should not be generated during cutting. Remember, vibration is your enemy. Proper clamping of the workpiece is the first priority and excessive tool extension is another culprit that creates vibration. If these considerations are found to be acceptable then first try decreasing the spindle RPM in 10 percent increments until the harmonics are tamed down. Then adjust the

Common Machining Formulas	
Imperial	**Formula**
Cutting Speed (V)	$V = RPM \times Diameter \div 3.82$
Revolutions/Minute (RPM)	$RPM = V \times 3.82 \div D$
Feed Rate Inches/Minute (IPM)	$IPM = IPT \times RPM \times n$
Feed/Tooth (IPT)	$IPT = IPM \div n \times RPM$
Symbol Definitions: D = Diameter and n = Number of flutes	
Metric	**Formula**
Cutting Speed (Vc)	$Vc = 3.1416 \times D \times (n) \div 1000$
Revolutions/Minute (n)	$n = Vc \times 1000 \div 3.1416 \times D$
Feed Rate Millimeters/Minute (Vf)	$vf = (n) \times (z) \times (fz)$
Feed/Tooth (fz)	$fz = vf \div (n) \times (z)$
Symbol Definitions: D = Diameter and z = Number of flutes	

Chart 1-3. Machining Formulas

feedrate up or down while watching the spindle load meter on the controller. For the best results, try to keep the load in the 30–60 percent range. The lower the better, but the machine should easily handle these parameters. Do note, however, that it is not uncommon to run spindle load much closer to red-line and then back off 10–20 percent in high production environments. Surface finish, accuracy and tool life all factor into the consideration.

Tap-Drill Calculations

Most shops have wall charts available that include decimal equivalent and tap-drill sizes, or a machinist may keep a small plastic copy from the *Machinery's Handbook* in their toolbox. In the rare case they are not available, the following basic formulas will help you:

■ To obtain the tap-drill size (75 percent of full thread form) for any size of thread, subtract the thread pitch from the major diameter of the thread. For example, for a thread size of ¼-20 UNC 2B, take 1 divided by 20 to get the thread pitch which = .050 inch. The major diameter for this thread is .250. Now subtract the .050 inch from the .250 and you get .200 inch for the drill size.

Threading Calculations

Calculating the proper feedrate for tapping holes on a mill and threading on a lathe are critical to success. Follow these simple guidelines:

■ To calculate the feedrate in inches per minute (IPM) or (in/min) for milling, divide the RPM by the threads per inch (TPI).

■ To calculate the feedrate in inches per revolution (IPR) or (in/rev) for turning, divide the one (1) by the TPI.

Trigonometry Applications

The following charts contain formulas that are used to calculate right and oblique triangles in cases where the information is not available on the engineering drawing for programming and inspection purposes. For example: individual coordinates for bolt circle center points; center-to center dimensions; chord distances; milling cutter diameter offsets for cutting angular surfaces; tool nose radius compensation for cutting tapered surfaces and drill point angle compensations. Once again, these types of calculations are mostly handled today from within CAM programming, but it is helpful to know for one-off situations that may arise.

Drill Point Tip Calculations

When the intent is to drill all of the way through a part, the drill calculations must be done to add enough distance to the programmed value to accomplish this. One could use Charts 1-4 and 1-5 to find the formula to complete the calculation or use Chart 1-6.

Summary

Mathematical equations were presented here for manual use to calculate the required feeds and speeds to perform effective material cutting and to compensate for tool tip and cutter diameter offsets when needed. Remember to consult the *Machinery's Handbook* and manufacturer online data for extensive data for cutting speeds and feeds that are proven to produce quality results.

Known Sides and Angles	Unknown Sides and Angles			Area
a and b	$c = \sqrt{a^2 + b^2}$	$A = \arctan \dfrac{a}{b}$	$B = \arctan \dfrac{b}{a}$	$\dfrac{a \times b}{2}$
a and c	$b = \sqrt{c^2 - a^2}$	$A = \arcsin \dfrac{a}{c}$	$A = \arccos \dfrac{a}{c}$	$\dfrac{a \times \sqrt{c^2 - a^2}}{2}$
b and c	$a = \sqrt{c^2 - b^2}$	$A = \arccos \dfrac{b}{c}$	$B = \arcsin \dfrac{b}{c}$	$\dfrac{b \times \sqrt{c^2 - b^2}}{2}$
a and $\angle A$	$b = \dfrac{a}{\tan A}$	$c = \dfrac{a}{\sin A}$	$B = 90 - A$	$\dfrac{a^2}{2 \times \tan A}$
a and $\angle B$	$b = a \times \tan B$	$c = \dfrac{a}{\cos B}$	$A = 90 - B$	$\dfrac{a^2 \times \tan B}{2}$
b and $\angle A$	$a = b \times \tan A$	$c = \dfrac{b}{\cos A}$	$B = 90 - A$	$\dfrac{b^2 \times \tan A}{2}$
b and $\angle B$	$a = \dfrac{b}{\tan B}$	$c = \dfrac{b}{\sin B}$	$A = 90 - B$	$\dfrac{b^2}{2 \times \tan B}$
c and $\angle A$	$a = c \times \sin A$	$b = c \times \cos A$	$B = 90 - A$	$c^2 \text{x} \sin A$ x $\cos A$
c and $\angle B$	$a = c \times \cos B$	$b = c \times \sin B$	$A = 90 - B$	$c^2 \text{x} \sin B$ x $\cos B$

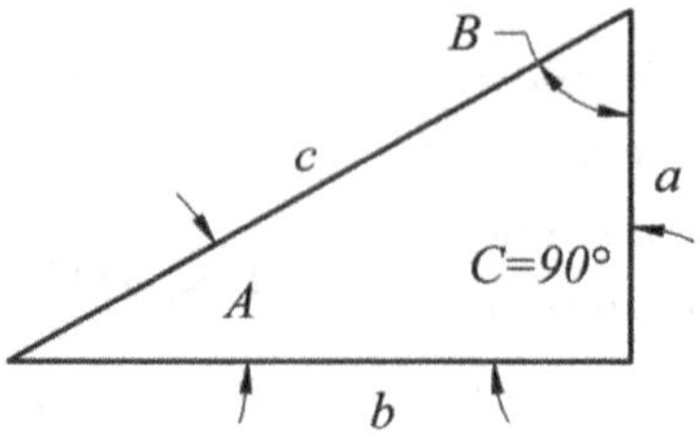

The triangle diagram has vertices with angle B at top, side c on the hypotenuse, side a on the right, $C = 90°$, angle A at bottom left, and side b along the bottom.

Chart 1-4.   Right Triangle Trigonometry

Known Sides and Angles	Unknown Sides and Angles			Area

Known Sides and Angles	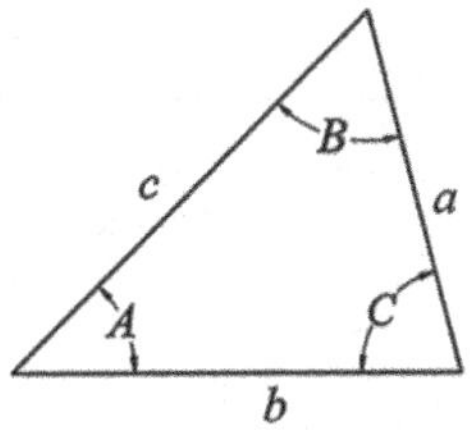 Unknown Sides and Angles			Area
All three sides a, b, c	$A = \arccos \dfrac{b^2 + c^2 - a^2}{2bc}$	$B = \arcsin \dfrac{b \times \sin A}{a}$	$C = 180° - A - B$	$\dfrac{a \times b \times \sin C}{2}$
Two sides and the angle between them $a, b, \angle C$	$c = \sqrt{a^2 + b^2 - (2ab \times \cos C)}$	$A = \arctan \dfrac{a \times \sin C}{b - (a \times \cos C)}$	$B = 180° - A - C$	$\dfrac{a \times b \times \sin C}{2}$
Two sides and the angle opposite one of the sides $a, b, \angle A$ ($\angle B$ less than 90)	$B = 180° - \arcsin \dfrac{b \times \sin A}{a}$	$C = 180° - A - B$	$c = \dfrac{a \times \sin C}{\sin A}$	$\dfrac{a \times b \times \sin C}{2}$
Two sides and the angle opposite one of the sides $a, b, \angle A$ ($\angle B$ greater than 90)	$B = 180 - \arcsin \dfrac{b \times \sin A}{a}$	$C = 180° - A - B$	$c = \dfrac{a \times \sin C}{\sin A}$	$\dfrac{a \times b \times \sin C}{2}$
One side and two angles $a, \angle A, \angle B$	$b = \dfrac{a \times \sin B}{\sin A}$	$C = 180° - A - B$	$c = \dfrac{a \times \sin C}{\sin A}$	$\dfrac{a \times b \times \sin C}{2}$

Chart 1-5. Oblique Triangles

Drill Point Tip Compensation Calculations		
Tool Point Angle	Calculation Factor x Tool Diameter	Point Depth Compensation Amount
135°	0.207	Point Depth Compensation Amount
118°	0.3	Point Depth Compensation Amount
90°	0.5	Point Depth Compensation Amount
82°	0.575	Point Depth Compensation Amount
60°	0.866	Point Depth Compensation Amount

Chart 1-6.   Drill Point Tip to Compensation Calculations

Machining Mathematics, Study Questions

For questions 1–4 refer to the *Machinery's Handbook* for required cutting data for calculations.

1. An internal threading operation is required on a CNC lathe to make a 1-8 UNC thread in a plain carbon steel (1018) part. A HSS drill .875 inch in diameter is used to prepare the hole. What would the cutting speed (V) range and RPM range be for this operation?

    a. V = 80/100 and RPM = 349/437
    b. V= 35/100 and RPM = 153/437
    c. V = 55/125 and RPM = 240/546
    d. V= 200/350 and RPM = 873/1528

2. In this example, the material is 303 series stainless steel. A .5625 inch diameter hole is to be drilled (HSS tool) through a plate that is 1.25 inches thick. Identify the cutting speed (V) range and calculate the RPM range best suited for this operation.

    a. V = 30/80 and RPM = 204/543
    b. V = 250/450 and RPM = 1698/3056
    c. V = 35/55 and RPM = 236/374
    d. V = 20/40 and RPM = 136/272

3. A 4340 alloy steel plate 4.0 inches square requires a 2.0 inch diameter hole to be machined through the center. A predrilling operation uses a 1.25-inch diameter HSS drill and a finishing operation uses a .875 inch diameter four-fluted HSS end mill to circle mill out the remainder of material. What is the cutting speed (V) and RPM ranges for the drill?

    a. V = 75/140 and RPM = 229/428
    b. V = 80/100 and RPM = 244/306
    c. V = 250/450 and RPM = 764/1375
    d. V = 30/80 and RPM = 92/244

4. A five-tooth 3.0 inch diameter uncoated carbide face mill is used to machine a 4340-alloy steel bar that is 2.0 inches wide and 6.0 inches long. There are two depth passes of .080 inch each required to bring the part to size. Use 350 for the cutting speed (V) and calculate the RPM for this cut.

    a. RPM = 350
    b. RPM = 646
    c. RPM = 446
    d. RPM = 997

5. Negative rotation direction for the polar coordinate system is:

    a. Clockwise
    b. Counterclockwise
    c. Incremental
    d. Either a or b

6. In which quadrant of the two-dimensional coordinate system are all of the values positive?

    a. 1
    b. 2
    c. 3
    d. 4

7. In order to program the following part (Figure 1-88), it will be necessary to identify the absolute rectangular coordinate location for the center point for each hole. List the X and Y coordinate values for the starting hole at the one o'clock position. The angular value for this hole is 70 degrees.

    a. X.513, Y1.410
    b. X1.359, Y.634
    c. 1.500, Y.645
    d. X-.513, Y1.410

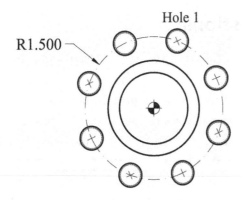

Figure 1-88. Absolute coordinate points for hole 1

8. In order to inspect the part in the following drawing to specification, a center-to-center dimension is required. Use the data given in Figure 1-89 to calculate what this dimension would be.

   a. .75 inch
   b. .8385 inch
   c. .875 inch
   d. .8835 inch

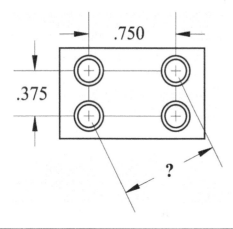

Figure 1-89. Calculate center-to-center distance

9. Calculate the amount of tool travel necessary to allow for the drill point plus .090 to drill through a .875- inch thick plate using the drill shown in Figure 1-90 and with a drill point angle of 135 degrees.

   a. .090 inch
   b. .875 inch
   c. .0938 inch
   d. 1.058 inch

Figure 1-90. Drill point compensation

Understanding Engineering Drawings

Objectives

1. Identify view types used in engineering drawings.
2. Distinguish between line types used to identify features and dimensions.
3. Understand the process of development for the drawing using CAD software.
4. List three different types of dimensioning systems.
5. Identify datum features on the drawing.
6. Locate feature control frames on the drawing.

The information given on the engineering drawing or blueprint will include the material, overall shape and the dimensions for part features (Figure 1-91). The geometry determines the type of machine (mill or lathe) to be used to produce the part. By studying the engineering drawing or blueprint, material and operations (drilling, milling, boring, etc.) can be identified. The cutting tools and work-holding method can also be determined. Occasionally, the geometry will require multiple machines to manufacture the part, and, thus, additional operations will be necessary. An engineering drawing or blueprint may be thought of as a map that defines the destination. This destination is the end product. Planning is required to develop a path to the destination (end product).

Engineering Drawing or Blueprint Relationship to CNC

The American Society of Mechanical Engineers established engineering drawing and related documentation practices that are outlined in ASME Y14.100, Engineering Drawing Practices. It presents a method for communicating part dimensional values in a uniform way, on the engineering drawing or blueprint. The drawing information will be translated to the coordinate system in order for dimensional values and part features to be manufactured.

Typically, on the engineering drawing or blueprint, datum features are identified as primary (A), secondary (B) and tertiary (C). The

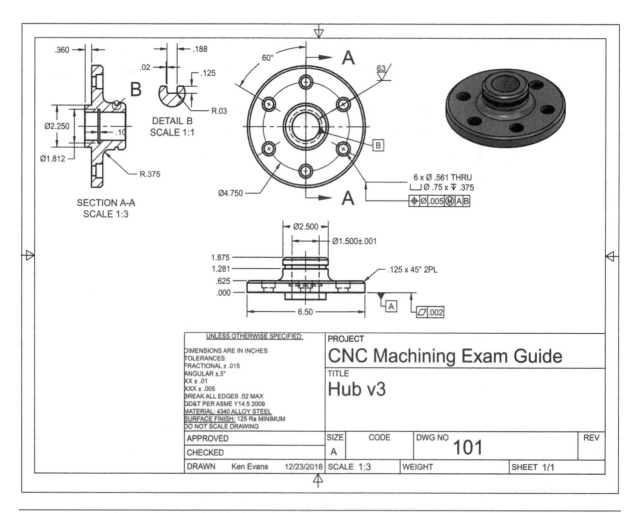

Figure 1-91. Engineering drawing example

concept of the datums is described in detail in the next section. Dimensions for the workpiece are derived from these datum features. On the drawing, the point where these three datum features meet is called the origin or zero point for the part. This type of dimensioning is called *baseline dimensioning*. When possible, this same origin point should be used for workpiece zero. This allows the use of actual engineering drawing or blueprint dimensions within the part program and often results in fewer calculations. It is common that engineering drawings are developed using an absolute dimensioning system based on datum dimensions derived from the same fixed point (origin or zero point). Occasionally, some features may be dimensioned from the location of another feature. An example of this is a row of holes exactly one-half inch apart. This type of dimensioning is called *chain dimensioning* and uses relative or incremental values. For the best results, when using manual or CNC equipment, a thorough knowledge of engineering drawing or blueprint reading is imperative.

Process Planning for CNC

Certain steps must be followed in order to produce a machined part that meets specifications given in an engineering drawing or blueprint. These steps need to be organized in a logical sequence to produce the finished part in the most efficient manner. Before machining begins, it is essential to go through the procedure called *process planning*. The following are the steps in the process:

1. Study the engineering drawing or blueprint to identify raw material.
2. Select the proper raw material or rough stock as described in the engineering drawing or blueprint.
3. Study the engineering drawing or blueprint and determine the best sequence of individual operations needed to machine the required geometry.

4. Transfer the information onto planning charts: Operation Sheet, CNC Setup Sheet and Quality Control Check Sheet.
5. While the part is still mounted on the machine, use in-process inspection to check dimensional values as they are completed.
6. Make necessary corrections and deburr.
7. Perform a 100 percent dimensional inspection when the part is finished and log the results of the first article inspection on the Quality Control Check Sheet (see Figure 1-285).
8. Take corrective action if any problems are identified.
9. Begin production.

From CAD to Engineering Drawing
Creation of the Drawing

Once the model has been created by the engineer in the design program, the drawing can be made. Step-by-step instructions are given here for the model depicted in Figure 1-92 using Autodesk Fusion 360. Other CAD programs operate in a similar fashion. Descriptions will be given for each aspect of the process to help you form an understanding of how to read and interpret engineering drawings. Open the model in Fusion 360 and click the down arrow on the Model button from the toolbar.

Figure 1-92. Completed design model

Step 1: Select Drawing and From Design from the choices (Figure 1-93).

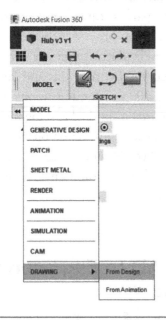

Figure 1-93. Step 1: Creating a model from design

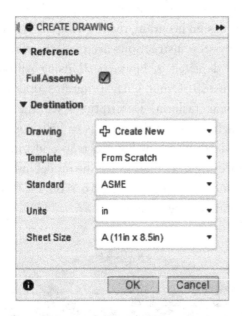

Figure 1-94. Create drawing dialog

Step 2: CREATE VIEW dialog enables selection of the standard, units and sheet size for the drawing.

In the CREATE DRAWING dialog, the standard used in the USA is ASME and the units are set to inches. ISO and millimeters

(mm) are available for designs where metric units are used. In this case, the print size is set to A, which is 11 inches × 8.5 inches. Use the drop-down to access any of the other options available. For the ASME standard they are A, B, C, and E and for the ISO standard they are A4, A3, A2, A1 and A0 in portrait or landscape orientations. The size is dependent on how much room is needed to communicate enough detail to eliminate ambiguity and clutter. When setting, match those in Figure 1-94 and click OK. The basic drawing border is set and the default title block is inserted, and the DRAWING VIEW dialog is displayed (Figure 1-95). Observe the settings in the figure and note that the orientation for the base view defaults to Front. The style options are: Visible Edges, Visible and Hidden (default setting), Shaded and Shaded with Hidden Edges. Next, the scale for the drawing is given in a ratio. In our example, we set a custom value of 1:3. This is totally dependent on how much space is needed to fully define the model and all of its details. The last part of the dialog lists options for Edge Visibility. We will leave the settings as they are. The settings can be changed later if needed by double-clicking on the view.

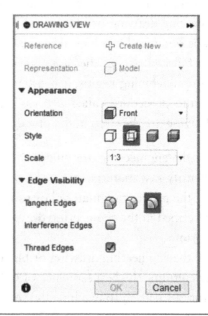

Figure 1-95. Drawing view dialog

Step 3: DRAWING VIEW dialog. Once the dialog setting is adjusted to specific needs, move the mouse into the drawing area and you will see the prompt "Place Base View." Move the cursor and view to a spot that is a few inches above the border of the title block and left-click, then press the ENTER key to accept the placement. (The final location can be moved later if adjustment is needed.) The Appearance setting controls the displayed part. In this case, the style of Visible and Hidden Edges is selected to create line types (Object and Hidden) and are automatically established for each view.

Step 4: Set up each of the orthographic and isometric projected views. To do this, press the icon labeled "Projected View" (second panel from the left on the toolbar). Follow the cursor prompt, "Select the parent view" and left-click the base view you just inserted. Move the mouse vertically above the base view to place the Top View. Follow the cursor prompt "Place the projected view" or press ENTER (left-click will also accept the placement).

Next, create a 3-View drawing. Move the mouse horizontally to the right of the base view to place the Right-Side View. Accept the placement as previously described. (This step is omitted for the drawing of our example because the Front and Right-Side views would be identical.)

Move the mouse diagonally to the right to place the Isometric View. Accept the placement. Press the escape (ESC) key to end placing views. This arrangement of views is called *third-angle projection* and it is most commonly used in the United States. Double-click on the Isometric View and the DRAWING VIEW dialog will reappear. Select Shaded for the style and click Close. If you examine the drawing closely, you will notice the Object (solid) and Hidden lines (dashed) are displayed in each view and that all views are projected exactly in line with the base view. Any of the existing views can be moved, if needed, by selecting the view (the view will be outlined and a center point square will be visible) and then left-click on the center point square and reposition where necessary. You will notice the associated views move correspondingly. **Save your work often.** Before adding dimensions, we will add some additional views so that some specific areas of the part can better be defined.

On the toolbar, the third icon from the left is used for the creation of a Section View. It allows the creation of a full, half, offset or aligned section view by using a section line to cut through a drawing view in order to reveal what is inside and obscured from view. This makes it easier to dimension. In our example, the internal groove width and diameter will be better visualized by using this method.

To create a Section View, left-click the icon on the toolbar. Then, follow the cursor prompt to "Select the parent view." In this case, it will be the Top View. You will see the letter A with an arrow pointing toward it appear at the cursor location. Align the cursor with the center point on the Top View and then slide the cursor directly below it just outside the part edge (a guide line will be available to assist in alignment) and then left-click to accept the location. Now, slide the cursor above the top edge of the part an equal distance as the bottom location and press the ENTER key to accept the location. The new view will now be attached to the cursor; move it to the left to position it far enough away to allow room for dimensions to be added and click OK to accept. The label for the view will be placed below with its scale (in this case, it matches our drawing scale). A second section view on the same drawing would sequence the letter identification to B-B, and so on.

Note: The direction the cutting plane arrows point match the viewing direction of the section and the crosshatch lines indicate the view is a Section View. A cutting plane that crosses through the entire part is considered a *full section.*

On the toolbar, the fourth icon from the left is used for the creation of a Detail View. It allows the creation of an enlarged (scaled) portion view of a specific small area of the drawing that would be difficult to clearly see and dimension. In similar fashion to section views, when more than one is needed on the drawing, the letter identification is sequenced to the next letter.

To create a detail view, left-click the toolbar icon and follow the "Select Parent View" prompt. In this case, we will select the section view as the parent. In the DRAWING VIEW dialog, change the scale to 1:1. Then use the cursor to specify the center point of the circular area to magnify. In this case, the midpoint of the horizontal edge of the groove is selected. Use the mouse to move outward from the center point to create a circle that encompasses the area to specify the detail boundary. Left-click to accept the size and then drag the mouse to a position to place the newly created detail view.

On a cylindrical part, such as this, it is obvious where the center line is, but we will add one to both the Front view and Section view for absolute clarity. To create a center line, left-click on the Center Line icon on the toolbar (seventh from the left). Follow the prompt and select two straight edges and a center line will be created between them. The length of the center line will be dependent on the objects selected. In the case of the Section view, pick the two horizontal edges of the inner diameter. In the case of the Front view, select the same two vertical hidden lines that represent the inner diameter. Press Escape (ESC) when done. In both of these cases, the line needs to be extended. To do this, select one of the center lines just created and drag the arrowhead end to an acceptable length (extended beyond the edge to the part by at least $\frac{1}{16}$ inch). Repeat for the second center line.

A *center mark* is often used to identify the center of cylindrical objects, holes, fillets or rounded edges.

To create a Center Mark, left-click the icon on the toolbar (eighth from the left). Select the largest diameter in the Top View. The size of the center mark is dependent on the diameter chosen.

A center mark can be added to a pattern of holes such as those in the Top View with a point circle diameter, commonly called *bolt circle diameter*. To create a Center Mark Pattern, left-click the icon on the toolbar (ninth from the left), the CENTER MARK PATTERN dialog will appear. Use the cursor to select the desired features, the pattern of 6 holes. Press OK when done.

Now that the necessary views have been added to our drawing, we can begin to define more about the model with dimensions. You can create linear, aligned, angular, radius and diameter dimensions. How you go about this is by activating the DIMENSIONS icon on the tool bar. Follow the prompt to select objects, points, edges, existing dimensions, or select two points. Then place the dimension by dragging it into position and left-click to accept the dimension. The dimension and extension lines are created automatically. The dimension precision level is established in the settings but can be changed by double-clicking on the actual dimension and changing it in the DIMENSION dialog box. When dimensioning diameters and radii, a leader line will be automatically connected and the diameter symbol or letter R will be placed before the dimensional value. (All dimensioning symbols are covered in detail in the next section, "Basic Geometric Dimensioning and Tolerancing.")

By using the drop-down for the DIMENSIONS icon on the toolbar, you can access:

- Ordinate, Baseline and Chain dimensioning styles.
- Ordinate dimensioning allows definition of view origins where all dimensions in sequence are referenced from. In the example (Figure 1-91), the Front view utilizes this method for all of the vertical dimensions.
- Baseline dimensioning allows selection of an existing linear dimension, then you specify a point to create a new dimension. Continue specifying points to create multiple dimensions all originating from the same base point.

- Chain dimensioning allows selection of an existing dimension, then you specify a point to create a new dimension. Continue specifying points to create multiple side-by-side dimensions.

Oftentimes, ordinary text needs to be added to the drawing for NOTES and such that cannot be communicated otherwise. For this example, the tolerances information in the upper-left portion of the title block will be completed. To create text, press the TEXT (A) icon on the toolbar then specify the corners for the text box, then enter the text. When finished, click outside the text editor. Symbols can be added to any text entered in this way. To edit existing text, double-click to open the editor dialog. The data entered is called the Tolerance Block (Figure 1-96).

To create a *leader* line where needed, specify the arrowhead location, specify the location for the text, then enter the text. When finished, click outside the text editor. The chamfers (.125 × 45 degrees × 2PL) in the Front view of our example drawing are defined with a leader.

Some features of a part may require special surface finish requirements that must be communicated on the drawing. To create the symbol and give it the required value, press the Surface Texture icon. Then, select an edge, or an existing dimension, or an existing feature control frame symbol, then place the surface texture. Complete the necessary information in the

SURFACE TEXTURE dialog and press OK. The 1.500 diameter in the Front view of our example drawing includes the surface texture symbol with a value of 63 micro inches.

Datums (described in detail in section "Basic Geometric Dimensioning and Tolerancing") are physical features of parts to make repeatable measurements from. They provide clarity required for proper feature orientation.

To create a datum symbol, press the DATUM IDENTIFIER icon and follow the prompt. Then, select an edge, an existing dimension, or an existing feature control frame symbol and place the datum identifier. Follow the cursor prompts for placement. In the sample drawing (Figure 1-91), there are primary (A) and secondary (B) datums identified. Most prismatic designs require primary, secondary and tertiary datums.

In the section titled Basic Geometric Dimensioning and Tolerancing, the components that make up feature control frames are covered in detail. In the sample drawing, a feature control frame is added to the Front view by pressing the Feature Control Frame icon. Select an edge or an existing dimension and place the frame. In this case, the bottom surface of the part is selected (DATUM A) and the mouse is moved to the right, left-click to accept the end position of the extension line, then the mouse is moved downward enough to attach an arrowhead to the extension line. Press the ENTER key to accept the placement.

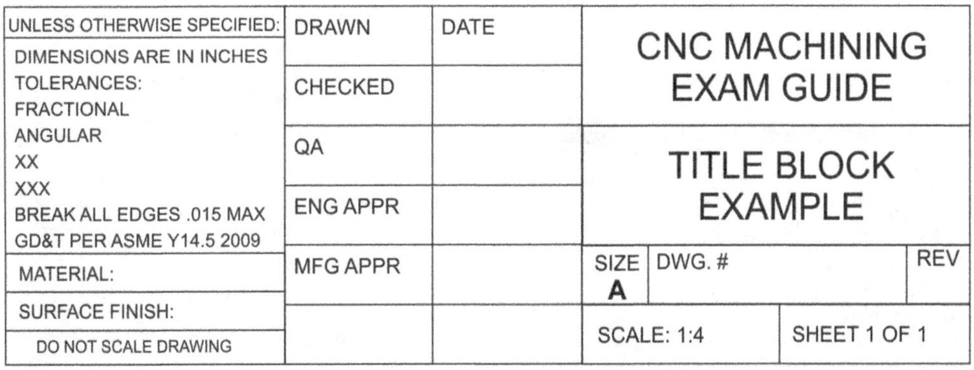

UNLESS OTHERWISE SPECIFIED:	DRAWN	DATE	CNC MACHINING EXAM GUIDE		
DIMENSIONS ARE IN INCHES					
TOLERANCES: FRACTIONAL	CHECKED				
ANGULAR XX	QA		TITLE BLOCK EXAMPLE		
XXX BREAK ALL EDGES .015 MAX GD&T PER ASME Y14.5 2009	ENG APPR				
MATERIAL:	MFG APPR		SIZE A	DWG. #	REV
SURFACE FINISH:					
DO NOT SCALE DRAWING			SCALE: 1:4	SHEET 1 OF 1	

Figure 1-96. Title and tolerance block

The FEATURE CONTROL FRAME dialog is displayed and the Geometric Symbol field is changed to Flatness and First Tolerance field is set to .002 inch. Using the same steps, another feature control frame is added to the dimensions for the bolt circle of holes. To start, left-click on the dimension line and move downward, left-click, move to the right to align with the end of the dimension leader, left-click and press ENTER to accept the location. Change the geometric symbol to Position and then use the drop-down in the First Tolerance field to set it to a diameter symbol. Input .005 inch in the Second Tolerance field. In the First Datum field, use the drop-down to set the modifier for Maximum material condition (M). Input the letter "A" into the Second Datum field and then the letter "B" into the Third Datum field. Press Close to end entry.

As designs are manufactured, there are findings that result in necessary changes to the drawing. These revisions are driven by engineering change orders and must be tracked carefully to avoid using outdated information. This is indicated in the lower-left corner of the title block. Generally, the first revision is the letter "A" and then progresses in sequence from there.

Output styles can be set to the formats: .pdf (default), .dwg and .CSV (if there are any tabular data in the drawing). These files can be printed or viewed electronically by operators. (The current movement is away from paper copies.)

Model-Based Definition

Model-based definition (MBD) is the inclusion of product and manufacturing information within the 3D model in order to communicate required data to aid the entire manufacturing process, including QC. The information will include: dimensions, tolerances, geometric tolerances, surface finish requirements and materials to be used. MBD uses the 3D model as opposed to printed engineering drawings or blueprints and is commonly displayed at a terminal or PC at or near the manufacturing center.

Summary

In this section, we introduced some basic blueprint reading techniques by building an actual engineering drawing using the Fusion 360 CAD/CAM software. There are many additional techniques that are important in communicating design intent that are covered in the next section, Basic Geometric Dimensioning and Tolerancing, and in complete detail in *Hammer's Blueprint Reading Basics* (4th ed.), by Charles A. Gillis.

Understanding Engineering Drawings, Study Questions

1. The most widely used projection system is Third Angle Projection. Which are the three common views depicted on drawings in this system?

    a. Front, top and right side
    b. Front, bottom and right side
    c. Front, top and left side
    d. Top, front and right side

2. What line type is used to describe the part boundaries?

    a. Hidden
    b. Center
    c. Object
    d. Leader

3. Where are engineering drawing revision data displayed on the drawing?

    a. In the Notes
    b. In the Title Block, REV frame
    c. In the Tolerance Block
    d. In the Bill of Materials

4. When a dimension does not include a tolerance, how do you know what the tolerance is?

    a. Notes
    b. By the number of decimal places given in the dimension (i.e., xx)
    c. From Tolerance Block of the Title Block
    d. b and c

5. What type of view is used to see a cross-section of the part?

    a. Section view
    b. Detail view
    c. Auxiliary view
    d. Interrupted view

6. What type of view is used to enlarge a small area of the print for detail dimensioning?

    a. Detail view
    b. Section view
    c. Auxiliary view
    d. a and b

7. What determines the viewing direction when a cutting plane line is used?

    a. Front view
    b. From the view the cutting plane is in
    c. The arrowhead direction
    d. b and c

8. Extension lines correspond with?

    a. Center lines
    b. Dimension lines
    c. Phantom lines
    d. Hidden lines

9. What is the type of dimensioning where all values originate at a common origin?

    a. Ordinate dimensioning
    b. Baseline dimensioning
    c. Polar dimensioning
    d. Chain dimensioning

10. Surface finish callouts on engineering drawings are values given that represent

    a. Roughness average
    b. Flaws
    c. Roughness height
    d. a and c

Basic Geometric Dimensioning and Tolerancing

Objectives

1. Identify dimensioning symbols and interpret their meaning.
2. Define the meaning and understand the application of geometric characteristic symbols.
3. Understand the proper use of datums.
4. Interpret feature control frames.
5. Define the meaning of orientation, location, and runout tolerances.

Excerpts from *Interpretation of Geometric Dimensioning and Tolerancing, Based on ASME Y14.5-2009*, with permission from the authors, Daniel E. Puncochar, Ken Evans.

Today's manufacturing complexities have increased, and, as a result, so has the need for engineering drawings of parts and their assemblies. These engineering drawings require tolerances, that is allowances, for variation. This tolerance is specified as a plus/minus tolerance. This plus/minus tolerancing of the *coordinate dimensioning* system works quite well for many applications. Today, however, the need for interchangeability of parts in assemblies that are manufactured from around the world requires a tolerancing method that ensures that parts being brought together at an assembly plant fit together properly. Industries such as aerospace, automotive, energy and oil, medical, agriculture and tool and die are direct beneficiaries of the geometric dimensioning and tolerancing (GD&T) method of tolerancing. GD&T incorporates the idea of positional tolerancing, which provides a means for locating round features within a round tolerance zone rather than the traditional square tolerance zone. The concept was adopted by the military and became part of the military standards and later the Unified American Standard Association standard, ASA Y14.5. This standard was released in 1956 and was accepted by the military. Later, ASA became the American National Standards Institute (ANSI). ANSI later published a complete system of symbology for geometric form and positional tolerances, "Dimensioning and Tolerancing." In 1983, ANSI Y14.5m-1982 was released. This standard clarified some of the old practices and moved a little closer to the practices of the International Organization for Standardization (ISO), a primarily European standard. In 1995, ASME Y14.5-1994, from the American Standard of Mechanical Engineers, was released. This new standard further clarified requirements within the standard and again moved more in line with the ISO standard. The most current standard, ASME Y14.5-2009, is a revision of the 1994 standard and was adopted in 2009. Its purpose is to further standardize design and functional requirements in order to aid in manufacture on a global scale. My opinion is that ultimately, coordinate dimensioning should be replaced with GD&T for everything except for features of size. The new ASME Y14.5-2009 standard is further in line with the needs of the international community. Today, geometric GD&T is used by the majority of manufacturing companies around the world.

What Is a Standard?

A standard is a model or rule with which other similar things being manufactured are to be made or compared; GD&T symbols, principles, and rules are the model that is provided internationally. This system was created to improve communications, control and productivity in manufacturing throughout the world. Standards are critical to all of us, and they have become increasingly important as our technologies continue to develop.

Why GD&T

GD&T adds clarity and contributes its many advantages to our coordinate system of dimensioning. The old system of coordinate dimensioning was lacking in many respects. A part of the designer's intent was always left to interpretation by the craftsman (i.e., dimension origin, form profile and orientation). Probably the most significant difference between the two systems is the location of round features; the coordinate system had a square or rectangular (linear) tolerance zone, which allowed some good parts to be rejected. In our world of high technology, high cost, and transfer of parts around the world, we cannot tolerate the misinterpretation that is possible with the coordinate dimensioning system and its square tolerance zone.

GD&T Is Not a Replacement

The coordinate dimensioning system is not being replaced entirely with GD&T. GD&T is specified to enhance the coordinate dimensioning system as required per design. When the advantages of GD&T can be utilized, they are simultaneously specified. GD&T, a system of symbols, provides a means of completely specifying uniformity and describing the designer's intent. These symbols eliminate most of the drawing misinterpretation by not having notes in drawing margins and by having complete descriptions of features and design requirements.

Complete Specification

A complete specification of design requirements is made possible with symbols that communicate clearly the design intent. These symbols also allow the designer to specify maximum tolerances for parts that must be assembled with other parts. These maximum tolerances also ensure the interchangeability of parts. The use of symbols for complete specification is becoming more important with the growing interrelated ownerships of companies around the world. GD&T is an international common symbolic language controlled by standards. Today, the majority of US manufacturing companies are applying GD&T to their drawings.

GD&T Advantages

There are many reasons for specifying geometric tolerancing wherever design integrity must be controlled and communicated completely to others. Two key principles for applying GD&T are the function and the relationship of parts in an assembly. Probably the most advantageous part of GD&T is the method of specifying feature location. In the past, features were located with the coordinate dimensioning system.

The coordinate dimensioning system is a method of tolerancing that uses a plus/minus tolerance. Plus and minus tolerances are specified for lengths, widths, diameters, shapes and locations. An illustration of how drawings may be dimensioned and toleranced with the coordinate system is shown in Figure 1-97. This method of tolerancing permits the length and diameters to vary by a plus/minus value. It also allows the maker of the part to put the center hole wherever he or she desires. By looking at the drawing, we can only assume that the hole is centered. This is an example of a drawing being left open to misinterpretation by anyone reading it. With the proper use of GD&T, ambiguity is minimized.

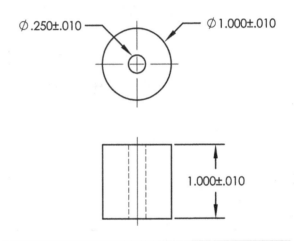

Figure 1-97. Dimensions and tolerances with the coordinate system

A similar situation exists where hole or pin locations are specified with the coordinate dimensioning system. An example of how holes are specified is shown in Figure 1-98.

The tolerance as specified establishes a square tolerance zone based on the plus/minus .005-inch tolerance in the X and Y directions. There is no consideration for the actual mating part size. The tolerance zone is .005 inch on each side of center location creating a square tolerance zone regardless of the actual mating part size. An illustration of how the tolerance zone appears is shown in Figure 1-99.

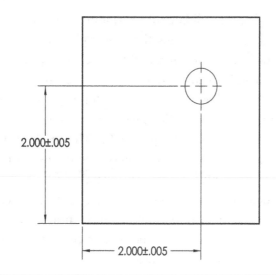

Figure 1-98. Specifying holes with the coordinate dimensioning system

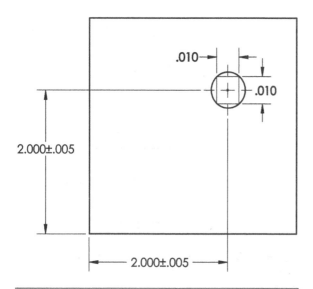

Figure 1-99. Ten-thousandths square tolerance zone

The axis of the hole or pin must be positioned within that square zone in order for the feature to be located properly. The feature may lean or slant an uncontrolled amount as long as the axis stays within the square zone. The designer only assumes that the feature will be produced nearly perpendicular to the material it is put into. If the center axis of the hole were in an extreme corner of this zone, the feature location is still acceptable; that radial measurement is .007 inch, as shown in Figure 1-100.

In the coordinate dimensioning system, the only place there is a .007-inch measurement is

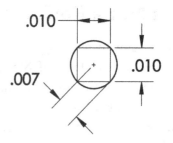

Figure 1-100. Radial measurement of the square tolerance zone

from the center to any corner. That .007-inch should be usable all around the desired true position, as illustrated in the example in Figure 1-101. GD&T provides a method of specifying a tolerance zone that takes the shape of the feature into consideration if so desired by the designer. GD&T also allows consideration for the feature's actual local size for calculating total tolerance. This concept is presented later in Tolerances of Location and Position Theory.

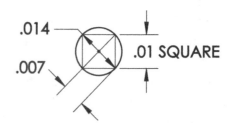

Figure 1-101. Cylindrical tolerance zone vs. square

Another advantage of GD&T is international uniformity in describing designer intent. The symbolic method of specifying design intent eliminates most misinterpretation of drawing notes. In the past, drawings usually contained a list of notes that were intended to explain certain requirements. These notes were all subject to misinterpretation. With the available symbols, designers can more readily specify exact design requirements. The proper application of geometric tolerances ensures the interchangeability of parts. GD&T is the common language used throughout industry internationally.

Basic Geometric Dimensioning and Tolerancing, Study Questions

1. Drawings are the primary _____ tool between designers and manufacturing.

2. GD&T is a system made up of _____ primarily.

3. The GD&T standard is one of the standards maintained by the _____.

4. GD&T adds to the coordinate dimensioning system when specific _____ is required.

5. GD&T does not _____ the coordinate dimensioning and tolerancing system.

6. GD&T can best be described as a _____ dimensioning and tolerancing system.

7. GD&T is used to control the _____ of a part feature and its relationship to other features.

8. GD&T is also used to control feature _____ and _____.

9. The key principles of GD&T are _____ and _____.

10. Two advantages of the GD&T system are maximum _____ and ensured _____ of mating parts.

11. The total amount that a part size may vary is a size _____.

12. A common method used to specify a tolerance for the nominal size of a feature is _____ values.

13. GD&T should be used for all dimensioning except for features of _____.

Part 1, GD&T Symbols and Abbreviations

Geometric dimensioning and tolerancing (GD&T) is a language of symbols. This section will introduce dimensioning, geometric characteristic and modifying symbols, and related terms as well as provide an application for each of them. Throughout the remainder of this section you will see these symbols applied to various drawings in combination with other symbols. The symbols presented here are the ones used to specify geometric characteristics and dimensional requirements on engineering drawings, which are in accordance with ASME Y14.5-2009. Figure 1-102 is an example of the symbols used in designs today.

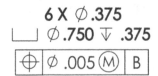

Figure 1-102. Sample of symbols in use

One of the advantages of GD&T is that it is an international language of symbols that generally eliminates the need for drawing notes. However, in limited situations it may be necessary for the designer to add a short note to aid in conveying the design requirements. In Figure 1-105 is an example of when a note may be specified. Figure 1-103 is an example of a note, BETWEEN, where the symbology as shown has replaced the need for the note.

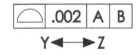

Figure 1-103. Sample note

Dimensioning Symbols

The symbols in this chart are described next and are used throughout the remainder of this text (Chart 1-7).

DESCRIPTION	SYMBOL
DIAMETER	⌀
RADIUS	R
SPHERICAL DIAMETER	S⌀
SPHERICAL RADIUS	SR
CONTROLLED RADIUS	CR
TIMES OR PLACES	X
COUNTERBORE	⌴
SPOTFACE	⌴SF⌴
COUNTERSINK	⌵
DEPTH	⤓
CONICAL TAPER	▷
SLOPE	◺
SQUARE SHAPE	□
REFERENCE DIMENSION	(6.00)
DIMENSION ORIGIN	⊶⊕
ARC LENGTH	⌒
ALL AROUND	⌀
ALL OVER	⌀
CHAIN LINE	— – – —

Chart 1-7. Dimensioning Symbols

Diameter. The term is very commonly used in many aspects of our lives. Most of us are familiar with the abbreviations D and DIA. Now with GD&T, we also have a symbol for diameter; it is a circle with a slash through it, as shown Chart 1-7. The diameter symbol is used to describe cylindrical features and tolerance zones. This symbol always precedes the feature size or tolerance specification. Figure 1-104 is an example of how this symbol appears in application. *Note:* The absence of this symbol before either the feature size or tolerance indicates a non-cylindrical feature or tolerance.

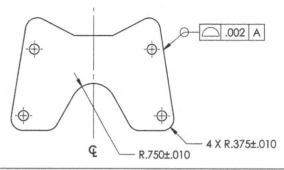

Figure 1-105. Application of the radius symbol

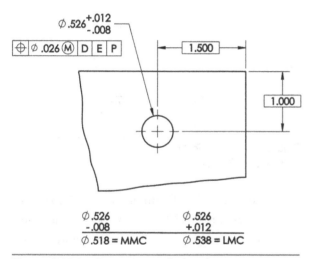

Figure 1-104. Diameter symbol in use

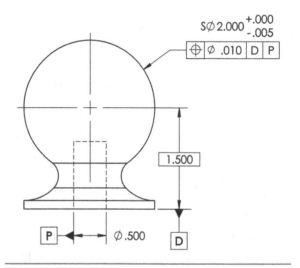

Figure 1-106. Spherical diameter symbol

specified for round features. It specifies the diameter of these features. The symbol is specified before the round feature size.

Radius. Radius, like diameter, is another term that is used on engineering drawings. Radius is not abbreviated; the letter "R" is used as the symbol. Radius is applied to rounded features or to designs that require the removal of sharp edges or fillets between two adjacent surfaces. The letter "R" precedes the radial dimension (Figure 1-105).

Spherical Diameter. This term was formerly abbreviated as SD; the new symbol is shown above in Chart 1-7 and in Figure 1-106. Spherical diameter is

Spherical Radius. This term is abbreviated SR; there is no symbol. Spherical radius is applied to round features. The abbreviation is specified before the radial value of the feature. See the application in Figure 1-107.

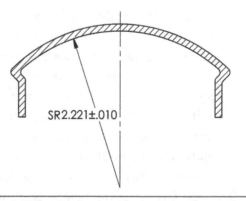

Figure 1-107. Application of spherical radius symbol

Controlled Radius Tolerance. The abbreviation for a controlled radius is CR. There is no symbol. Radius is specified when a fair curve without reversals is desired. The zone is created by two arcs tangent to the adjacent surface (Figure 1-108). The actual feature surface must lie within the tolerance zone.

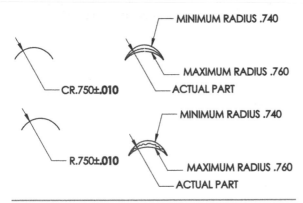

Figure 1-108. Application of controlled radius symbol

Times or Places. This term is not really new to drawings; they just have a little different application now. There is no abbreviation; the symbol is the capital letter "X" immediately following the number of places (e.g., 6 X). The designer may specify the number of places or times for holes in a pattern or other repeating feature, for example, as shown in Figure 1-102.

Counterbore/Spotface. The terms *counterbore* and *spotface* were formerly abbreviated as CBORE and SF on engineering drawings. The symbol used now is shown in Chart 1-7. The symbol is preferred with a GD&T application. Counterbores or spotfaces may be specified for features that require a recessed or flat mounting surface for fasteners. If no depth or remaining material value is specified the spotface depth is to be machined to a minimum cleanup. A typical callout for a counterbore or spotface using the symbol is shown in Figure 1-109.

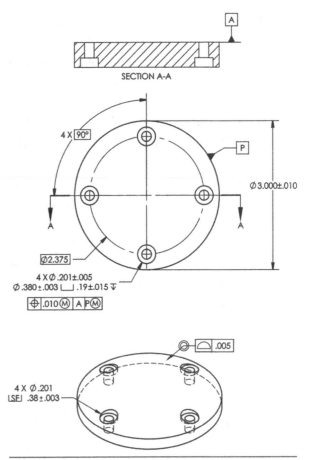

Figure 1-109. Typical callouts for counterbore (upper) and spotface (lower)

Countersink. Countersink was formerly abbreviated as CSINK and is now indicated by the symbol as shown in Figure 1-110 and Chart 1-7. Countersinks are specified for features that require a conical recessed surface for flathead screws. The symbol precedes the diametrical value, tolerance and angular requirement. Countersinks are typically specified as shown in Figure 1-110.

Depth/Deep. The words DEPTH or DEEP were formerly written out on drawings, but have been replaced by the symbol in Chart 1-7. No abbreviation is used. Depth or deep is specified for features that do not pass completely through a part. The symbol may be specified as in Figure 1-102 for counterbores, or it may be specified for blind holes measured to the beginning of the drill point taper.

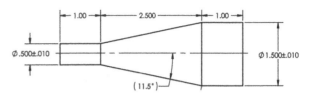

Figure 1-113. Conical taper using diameter and length specification

Slope. There is no abbreviation, and the symbol is shown in Chart 1-7. Slope is primarily specified to control flat parts with tapers. Slope is not specified as degrees, but as a ratio of height differences from one end of the flat taper to the other end. An example of how slope is specified is shown in Figure 1-114.

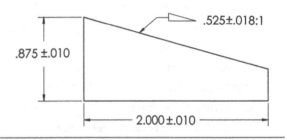

Figure 1-114. Specifying slope

Square. Square features may be identified with a symbol as shown above and in Chart 1-7. There is no abbreviation. This symbol may be specified on a drawing to indicate the feature is square.

Reference Dimension. The term was formerly abbreviated with the dimension REF and is now represented with the value in parentheses, that is, (.875). It can still be used in NOTES on drawings as REF or spelled out REFERENCE. Reference dimensions are specified for a relationship between features in a flat pattern application. It is not used to define parts. Reference dimensions may be the sum of several dimensions, the size or thickness of material, and specify travel of moving parts, etc. See Figure 1-115. Use of this method should be minimized.

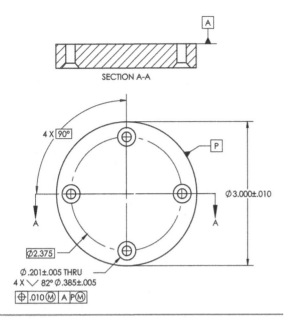

Figure 1-110. Typical use of the countersink symbol

Conical Taper. There is no abbreviation for Conical Taper; the symbol is shown in Chart 1-7. It is used to describe cylindrical tapers not specified in degrees but as a ratio. There are three methods of specifying a conical taper. First, it may be specified with basic dimensions for the diameters and the taper (Figure 1-111); there may be a feature size and profile tolerance combined with a profile of the surface (Figure 1-112); and the diameters along with the length may be toleranced (Figure 1-113).

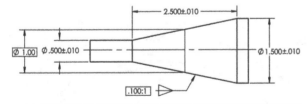

Figure 1-111. Conical taper with basic dimensions and ratio

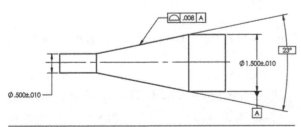

Figure 1-112. Conical taper using profile of surface symbol

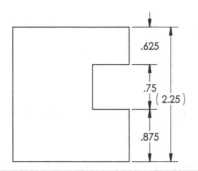

Figure 1-115. Specifying reference dimension

Dimension Origin. The term dimension origin is not abbreviated, and its symbol is shown in Chart 1-7. This symbol is used to identify the surface or feature where the dimension originates. Some designs are complex, thus making it difficult to determine where dimensions are to begin. In these situations, the designer specifies where the dimension is to originate. Usually the part will be different if made by starting dimensions from a different surface of feature than intended. The example in Figure 1-116 illustrates such a part.

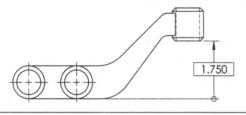

Figure 1-116. Application of the dimension origin symbol

Another example of using the dimension origin symbols is in Figure 1-117. Here it specifies where the angular measurement begins when an angle is specified with linear and angular dimensions. This tolerance zone will widen from the apex of the angle as the distance increases.

This symbol may also be utilized to specify a basic angle for a surface. In Figure 1-118, the symbol indicates where the dimension originates to establish the basic 45-degree angle. This tolerance zone will remain an equal width within parallel boundaries for the length of the feature surface.

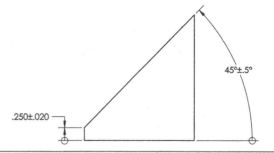

Figure 1-117. Dimension origin symbol used for angles

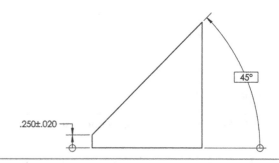

Figure 1-118. Specifying a basic angle dimension with origin symbol

Arc Length. Arc length is a term used to describe the length of a curved surface. The symbol is shown in Chart 1-7, and there is no abbreviation. This symbol is placed over the dimension. The arc length symbol is specified when it is required to measure along the actual part surface. When this symbol is specified, a linear measurement across the arc is not permitted. See the example in Figure 1-119.

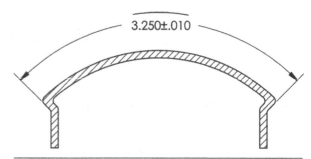

Figure 1-119. Specifying arc length symbol

All Around. All-around is a term that is not abbreviated. The symbol is shown in Chart 1-7. This symbol is specified along with profile specifications given in the feature control frame. It means that the profile tolerance applies all around the controlled feature and replaces the words ALL AROUND in NOTE form. This symbol should only be applied when a uniform profile tolerance is required all around the feature; also, the tolerance is the same around the feature. Figure 1-105 shows an example of how the uniform tolerance applies.

All Over. The term All Over has no abbreviation and the tolerance associated with it applies to the entire three-dimensional surface profile of the part. The symbol is shown in Chart 1-7 and in Figure 1-109.

Chain Line. Chain line is a term used to describe or identify a specific area, surface, or portion of a part for special treatment.

There is no abbreviation, but the term has a symbol which is shown in Chart 1-7, a heavy phantom line parallel to the line or surface. The chain line symbol is applied when the designer requires only a limited portion of surfaces or areas to be treated differently from the rest of the part.

Geometric Characteristic and Modifying Symbols

Geometric characteristic symbols correspond with various tolerance types according to feature application; they are the first item listed in feature control frames, followed by the tolerance, a modifying symbol and associated datums. The geometric characteristics symbols are outlined in Chart 1-8.

CHARACTERISTIC	SYMBOL	TYPE OF TOLERANCE	APPLICATION
STRAIGHTNESS		FORM	INDIVIDUAL FEATURES
FLATNESS			
CIRCULARITY			
CYLINDRICITY			
PROFILE OF A LINE		PROFILE	INDIVIDUAL FEATURES OR RELATED FEATURES
PROFILE OF A SURFACE			
ANGULARITY		ORIENTATION	RELATED FEATURES
PERPENDICULARITY			
PARALLELISM			
POSITION		LOCATION	
CONCENTRICITY			
SYMMETRY			
CIRCULAR RUNOUT		RUNOUT	
TOTAL RUNOUT			

Chart 1-8. Geometric Characteristic Symbols

Modifying Symbols

The modifying symbols in Chart 1-9 are described next and used throughout the remainder of the text.

DESCRIPTION	SYMBOL
MAXIMUM MATERIAL CONDITION	Ⓜ
LEAST MATERIAL CONDITION	Ⓛ
PROJECTED TOLERANCE ZONE	Ⓟ
FREE STATE	Ⓕ
TANGENT PLANE	Ⓣ
UNEQUALLY DISPOSED PROFILE	Ⓤ
INDEPENDANCY	Ⓘ
STATISTICAL TOLERANCE	⟨ST⟩
CONTINUOUS FEATURE	⟨CF⟩
BASIC DIMENSION	6.00
BETWEEN	X◄──►Y
TRANSLATION	▷

Chart 1-9.   Modifying Symbols

The next two symbols and abbreviations introduced are those for the modifiers. The modifiers are maximum material condition (MMC) and least material condition (LMC). These may modify a specified tolerance, as their name indicates. Regardless of feature size (RFS, no symbol is used) is implied for all geometric tolerances, unless otherwise specified.

Maximum Material Condition and Maximum Material Boundary.   The symbol for *maximum material condition* is a capital "M" in a circle as shown in Chart 1-9. Maximum material condition may be

abbreviated MMC. The term MMC is used to describe the maximum condition of the actual mating size. For example, a hole is a feature of size that is permitted to vary in size within the limits of a plus/minus size tolerance. For holes or any internal feature, MMC is the smallest actual mating size for that feature. In other words, the maximum material remains for the piece the hole was put in. An example of a hole size specification is shown in Figure 1-99. When the hole is produced at its smallest actual mating size diameter of .518 inch, the material is at MMC. Maximum material condition is also applicable to external features of size, such as pins, tabs and splines. With these features, MMC is equal to the largest actual mating size permitted by the size specification. An example of a cylindrical pin size that permits an MMC diameter of .512 inch is shown in Figure 1-120. MMC can be thought of in terms of the smallest hole and largest pin, or it can be thought of in terms of weight: smallest holes—more weight; largest pins—more weight in part.

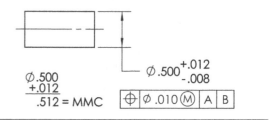

Figure 1-120.   Pin size at MMC

The application of this modifier would include designs where clearance is required for assembly. Maximum tolerances are achieved as the feature departs from MMC. A minimum tolerance is stated and then a modifier symbol is added following the tolerance when applicable to permit an increase in tolerance equal to the departure from MMC.

Least Material Condition and Least Material Boundary. The next modifier is *least material condition* (LMC). The symbol is a capital "L" in a circle as shown in Chart 1-9. As with MMC, the symbol is used on drawings. The term is used to describe the opposite condition as maximum material condition. This modifier is specified for features of actual mating size to describe the least material condition. For example, by using the same hole callout as in Figure 1-121, the LMC is equal to a diameter of .538 inch. Here the least amount of material remains in the material after the hole is machined. The application of this modifier would include designs where a part is located, or a minimum material thickness must be maintained, such as holes near material edges or bosses with holes bored in them where the wall thickness is critical. This modifier provides the same advantages as MMC, but in the opposite direction.

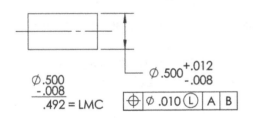

Figure 1-121. Pin size at LMC

Least material condition also applies to external features of size. For external features subject to size variation LMC means these features contain the least material within the specified size limits. In Figure 1-121, an example of a pin at least material condition is shown.

Projected Tolerance Zone. Projected tolerance zone is not abbreviated; the symbol for it is a capital "P" in a circle as shown in Chart 1-9. The projected tolerance zone describes a tolerance, usually for a fixed fastener application, to prevent interference between mating parts. The projected tolerance zone specifies a tolerance of location through the projected height. As shown in Figure 1-122, the axis of the hole must be controlled within a specified tolerance above the part the hole is in. The tolerance zone is projected to a height equal to the thickness of the mating part. The purpose of this control is to prevent the hole for the fixed fastener from interfering with the mating part.

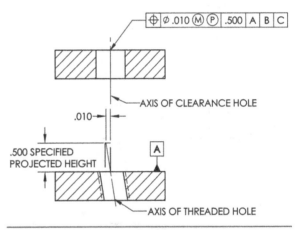

Figure 1-122. Projected tolerance zone

Free State. When an individual form or location tolerance is applied to a feature in the free state, the symbol, a capital "F" in a circle as shown in Chart 1-9, is placed in the feature control frame following the specified tolerance and any modifier (Figure 1-123). There is no abbreviation. This symbol may be specified for parts such as plastic, nylon, rubber, etc., that are not rigid enough to hold their form when clamped and therefore, must be inspected in the same state.

Figure 1-123. Using free state modifier

Tangent Plane. When a specified tolerance applies to the tangent plane of a feature, the symbol, a capital "T" in a circle as shown in Chart 1-9, is placed in the feature control frame following the specified tolerance for the feature (Figure 1-124). There is no abbreviation. The plane contacts the high points of the surface and shall lie within two parallel planes that are the stated tolerance apart.

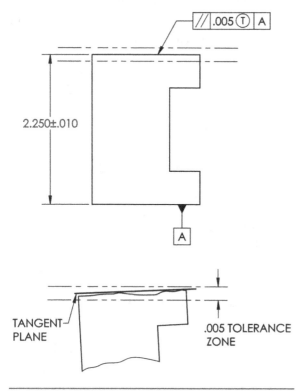

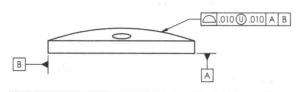

Figure 1-124.   Using tangent plane modifier

Unequally Disposed Profile. The symbol refers to a unilateral and unequally disposed profile tolerance. The symbol is a capital "U" in a circle as shown in Chart 1-9. Figure 1-125 shows an example of the unequally disposed profile in use.

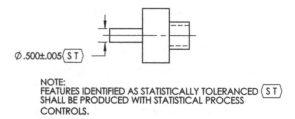

Figure 1-125.   Unequally disposed profile tolerance

Independency. The symbol refers to a feature of size at MMC or LMC that does not require perfect form. The modifier is placed directly next to the tolerance (Figure 1-126). No feature control frame is needed.

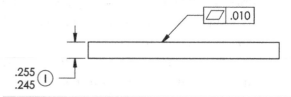

Figure 1-126.   Independency tolerance modifier in use

Statistical Tolerance. The symbol for *statistical tolerance* is shown in Chart 1-9, and is introduced to indicate that a tolerance is based on statistical tolerancing. There is no abbreviation. When a tolerance is a statistical geometric tolerance, the symbol is placed in the feature control frame following the stated tolerance and any applicable modifier (Figure 1-127).

Figure 1-127.   Using the statistical tolerance modifier

Statistical tolerancing may be applied to increase individual feature tolerance. Tolerances for individual features of an assembly are determined mathematically by dividing the tolerances among the features. When this assignment is mathematically restrictive when adding them, a statistical tolerance would be specified. Statistical tolerances on dimensions are designated as shown in Figure 1-128.

Note: A drawing note must be added to the drawing.

Figure 1-128.   Designating statistical tolerances

If it is necessary to specify both the statistical limits and the arithmetic stacking limits (Figure 1-129) when the feature may be produced with SPC, a note must be added to the drawing.

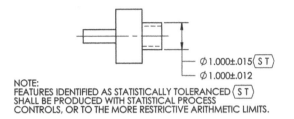

NOTE:
FEATURES IDENTIFIED AS STATISTICALLY TOLERANCED (ST) SHALL BE PRODUCED WITH STATISTICAL PROCESS CONTROLS, OR TO THE MORE RESTRICTIVE ARITHMETIC LIMITS.

Figure 1-129. Specifying both statistical and arithmetic stacking limits

Statistical tolerancing may also be specified for other features such as holes. Figure 1-130 illustrates how statistical tolerancing may be applied to a hole specification.

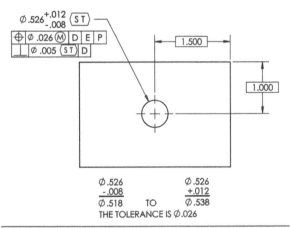

Figure 1-130. Applying statistical tolerancing to a hole location

Continuous Feature. The continuous feature symbol, shown in Figure 1-131 and in Chart 1-9, refers to features that are interrupted that should be considered as a single feature.

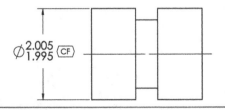

Figure 1-131. Continuous feature symbol

Basic

Basic is a term used to describe a theoretically exact size, shape or profile of surfaces either regular or irregular, or location of a feature or datum. Basic dimensions do not have a tolerance. Tolerances for basic dimensions are specified in feature control frames, or tooling tolerances apply under other conditions. The symbol for basic is a rectangle around the dimension.

Most frequently, basic is used to specify the exact desired position of features. Then, each feature is given a tolerance that allows some variation depending on the design requirement. Figure 1-132 shows an example of how basic may be specified. Basic dimensions may also be indicated in a note as follows

Note: Untoleranced dimensions locating true positions are basic.

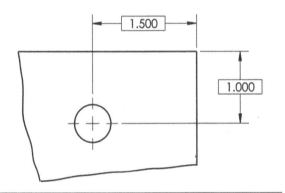

Figure 1-132. Applying the basic dimensioning modifier

Between

There are designs where the tolerance for the feature applies to only a portion of the feature. In these instances, the designer has the *between* symbol available as shown in Figure 1-132 and Chart 1-9.

Translation

The *translation* symbol, when included in the feature control frame, refers to a movable datum feature simulator that is free to translate.

Summary

This completes your introduction to the majority of the symbols and abbreviations associated with geometric dimensioning and tolerancing. There are other symbols that will be introduced in the next sections, "Datums," "Form/Orientation Controls," and "Tolerances of Location." The symbols and abbreviations introduced in this section will be applied to other drawings and examples throughout this text.

Part 1, GD&T Symbols and Abbreviations, Study Questions

Fill in the abbreviations and symbols that match the phrases. The abbreviations and symbols may be used more than once and some not at all.

Evaluation of Dimensioning and Modifying Symbols

A. X

B. Ⓛ

C. SR

D. S∅

E. ∅

F. 6.00

G. R

H. ⤓

I. ⌴

J. NO SYMBOL or ABBREVIATION

K. ◁

L. ⟶⌀

M. IMPLIED MATERIAL CONDITION

N. ▷

O. ⌀

P. Ⓟ

Q. Ⓕ

R. CF

S. (6.00)

T. ST

U. Ⓜ

V. 105

W. ∨

X. SF

Y. ⌀

Z. X⟷Y

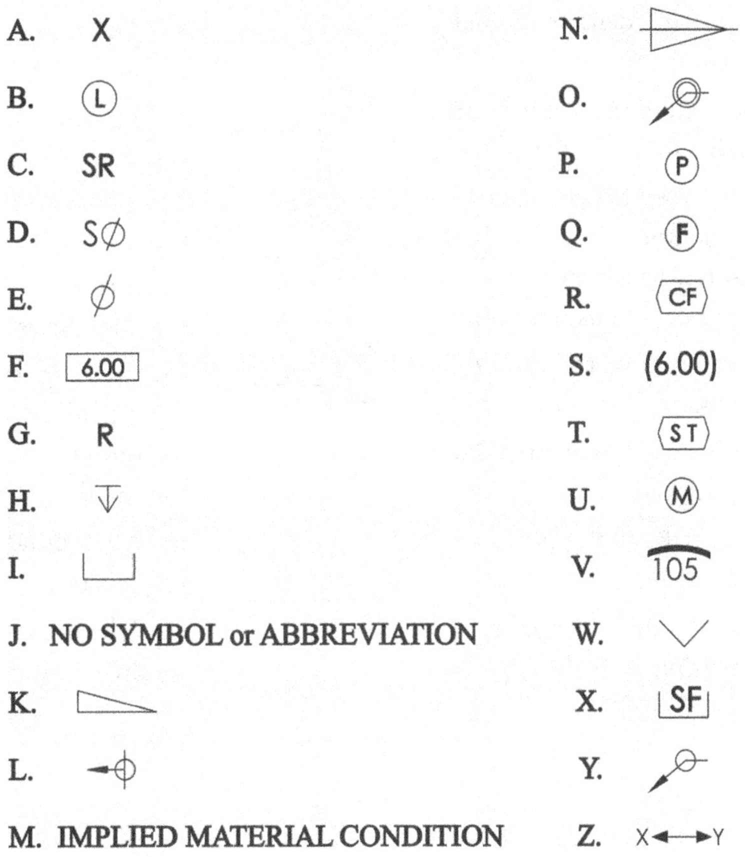

Chart 1-10. Dimensioning and Modifying Symbols

_____ 1. The symbol used to indicate Free State condition.

_____ 2. The condition of an internal feature when it measures the largest size within design limits or weighs the least.

_____ 3. The symbol specified to indicate diametrical or circular features.

_____ 4. Statistical tolerance symbol.

_____ 5. Dimensions that are for reference only.

_____ 6. Dimensions that are theoretically exact and do not have tolerances.

_____ 7. The symbol to indicate diametrical tolerance zones.

_____ 8. The symbol specified to indicate the number of times or places (instances).

_____ 9. A curved surface that is to be measured along the curve is specified with which symbol?

_____ 10. A spotface is specified with which symbol?

_____ 11. Parts or features requiring a rounded edge or corner are identified with which symbol?

_____ 12. Features that are spherical shaped are identified radially with which symbol?

_____ 13. The symbol used to indicate the depth of a feature.

_____ 14. The symbol that indicates the condition of an external feature when it measures the largest or weighs the most.

_____ 15. A symbol used to indicate the amount of taper on a flat part.

_____ 16. Symbol used to identify conical tapers.

_____ 17. Symbol used to indicate a countersink.

_____ 18. Indicates the origin of a dimension.

_____ 19. Symbol used to control the perpendicularity of a fastener to a given height.

_____ 20. A counterbore is specified with which symbol?

_____ 21. RFS, Regardless of Feature Size

_____ 22. The total movement of a dial indicator needle.

_____ 23. Features that are spherical shaped are identified with which symbol?

_____ 24. The symbol that indicates a three-dimensional control over an entire surface is?

_____ 25. An interrupted surface that is intended to be represented as one continuous feature is indicated by what symbol?

_____ 26. All Around symbol.

_____ 27. The symbol used to identify tolerance applies BETWEEN two points on a profile.

Part 2, GD&T Datums

Datums provide the framework from which part features are manufactured, and they are critical for all engineering drawings that are created for today's production. Parts produced today must be 100 percent interchangeable with their counterparts. The designer can ensure complete fit, function, and interchangeability by clearly specifying the design intent. Datums provide the clarity required for proper feature orientation, which originates from datum planes. Datums are specified on drawings so that repetitive measurements can be made consistently from design through the production and inspection processes. Datums are nothing more, nor less, than physical features of parts to make repeatable measurements from. These features are considered theoretically exact.

Datum Rational

Datum identification is required with ASME 14.5-2009. Previously, implied datums were permitted. Too often with implied datums, the same surfaces, edges, or features were not used throughout the design, manufacturing and inspection processes. This can be illustrated with a simple part. A drawing specification for a simple flat part with three holes in it is shown in Figure 1-133.

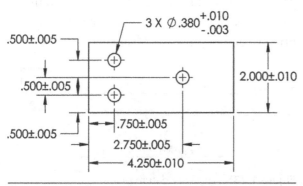

Figure 1-133. Simple part with no datums used

The potential exists that the part may not be oriented the same during inspection as it was through manufacturing, causing it to be rejected. To illustrate, consider that the part was manufactured using the left-hand end as the locating surface parallel with the Y-axis and the lower edge as the locating surface parallel with the X-axis. The holes were then machined in the part. After this, the part was set up for inspection using the same left end but the opposite longest surface to locate the part. When inspected in a different orientation than when machined, the potential of part rejection is increased due to dimension referencing causing the part to be out of design specification. Therefore, datums must be specified so the drawing is interpreted the same by all who read it.

What Is a Datum?

A datum is a theoretically exact line, surface, point, area or axis that is used as an origin for dimensions. These regions are considered perfect for orientation purposes only. During machining processes, the part is located against a theoretically perfect or exact datum surface where features of the part have been identified as datum features; the part is oriented and immobilized relative to the datum reference frame in their selected order of precedence. This orientation makes the geometric relationships that exist between features measurable and establishes what is called the *datum feature simulator*. The features used to establish datums may be:

- A plane
- A maximum material condition boundary (MMB concept)
- A least material condition boundary (LMB concept)

- A virtual condition boundary
- An actual mating envelope
- A mathematically defined contour

It is from these actual surfaces that measurements are made to check feature relationships. On the drawing, these datum features are identified with the datum feature symbol.

When the symbol for MMC or LMC is included after the datum in the feature control frame it refers to either maximum material boundary (MMB) or least material boundary (LMB). Determining the MMB or LMB is done by calculating the smallest (in case of external) or (largest in case of internal) value that will contain the feature, with consideration given for datum precedence for the datum feature simulator.

Datum Feature

The datum feature symbol is a square box that contains a capital block letter with a leader connecting it to the feature with a triangle. The triangle may be filled or not filled. Figure 1-134 shows a drawing illustrating the proper symbol and attachment. Any of the letters of the alphabet may be used except for I, O, and Q, which may be confused with numbers. The letters may be used in any order because alphabetical order is meaningless in this system. The important mental distinction that must be made is that a datum is theoretically perfect, whereas the datum feature itself is imperfect.

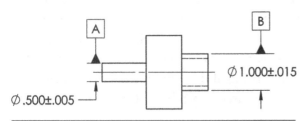

Figure 1-134. Datum feature symbols in use

The datum symbol may not be applied to center lines, center planes or axes.

Three-Plane Concept—Flat

Theoretical datum planes or surfaces are established from a perfect three-plane reference frame. This frame is assumed to be perfect with each plane oriented exactly 90 degrees to each other (orthogonal), referred to as the *datum reference frame*. This reference frame, with mutually perpendicular planes, provides the origin and orientation for all measurements. When physical contact is made between each datum feature and its counterpart in associated manufacturing or inspection equipment, measurements do not consider any variations in the datum features. These planes are identified as the primary, secondary, and tertiary datum planes. This is considered the order of precedence.

Each of these planes exists along the X, Y or Z linear planes. Further, each of these planes have a rotational attribute about the linear axis called U, V and W, respectively. This accounts for a total of *6 degrees of freedom*, all of which must be locked in place to properly orient the part for manufacturing and inspection purposes.

Primary

The primary datum is the one that is, functionally, usually the most critical feature or surface on the part. It is part to part interface, typically the largest surface when area is involved. The primary datum feature must contact the theoretically exact datum plane in a minimum of three points not in a line. The required contact is to prevent the part from rocking during manufacturing or inspection processes. This three-point contact is not difficult to achieve. Figure 1-135 shows an example of the primary datum plane establishment.

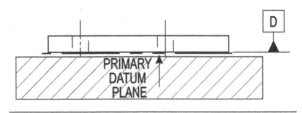

Figure 1-135. Primary datum plane

Remember that the fixture or surface plate that the part is clamped to or positioned on in order to be inspected is nearly perfect, but it is not perfect. In every case, no matter how irregular the part surface is, the highest points of the part contact the device. This is referred to as the *simulated datum*.

Secondary

The secondary datum plane must be at a 90-degree angle (perpendicular) to the primary datum plane. The secondary datum feature is usually selected as the second most functionally important feature. There is only a two-point minimum contact required for this plane. These two points establish the part in the second direction to prevent it from rocking about the primary datum plane. This plane may be a two-point stop, fence, or angle plate on processing or inspection equipment. The illustration in Figure 1-136 shows the secondary datum plane.

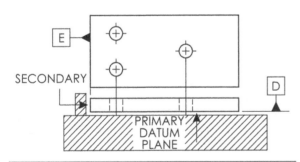

Figure 1-136. Secondary datum plane

Tertiary

The tertiary datum plane must be at exactly a 90-degree angle to the primary and secondary datum planes. The tertiary or third datum plane is also perpendicular to the other two planes. The part must contact this plane at least at one point. This contact is required for dimension origin and to prevent any back-and-forth movement along the third plane. The tertiary plane could be a locating or stop pin in a processing or inspection process.

All measurements, setups and inspections are to be made from these three mutually perpendicular planes. Figure 1-137 is an illustration of the theoretical datum reference frame.

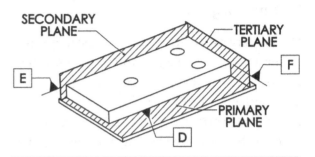

Figure 1-137. Theoretical datum reference frame

The fixture that could be manufactured to orient this part might look like the one illustrated in Figure 1-138. The part must contact three points for a primary datum. The secondary datum plane requires a two-point contact, which in this illustration are the sides of the pins. The tertiary datum plane must be in contact with the part in one place only.

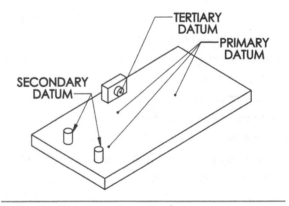

Figure 1-138. Part orientation fixture

There are times when a datum plane is separated by an obstruction or pocket and the design intention is that both sides make up the entire plane. In this case, a chain-line is added to the drawing in the appropriate view to indicate that both surfaces are required to establish the plane.

There is another instance where two separate surfaces with their own datum symbols may be equally important, when related to a feature, and are thus combined to make up the datum. In this case, the feature control frame will be given in Figure 1-162.

Datum Targets

Datum targets are also used to establish the datum reference frame and to establish orientation on irregular parts. They may also be used for large surfaces where it would be impractical to use an entire surface as the datum because of size, or it may be subject to warping, bowing, or other distortions. Usually datum targets are specified on castings, forgings, weldments or on any other application where it may be difficult to establish a datum. Datum targets may be points, lines, or areas of a part that provide an orientation–dimension origin. A datum target symbol is used to establish contact with datum simulators and to identify the datum planes.

Datum Target Symbol

The datum target symbol is a circle divided in half by a horizontal line. When the symbol is used to identify a circular target contact area, the top half will contain the diameter. Where it is impractical to specify the circular target area inside the upper half of the datum target symbol, the area may be specified outside the symbol and attached to the upper half with a leader line as shown in Figure 1-139.

Figure 1-139.   Datum target symbol

The contact area diameter is BASIC. When the symbol is used to identify any other datum target, the upper half remains open (see Figure 1-140). The lower half of the symbol always contains the datum target identification number. The identification consists of a letter (the datum reference

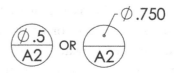

Figure 1-140.   Specifying datum target area

letter) and target number. Sufficient targets are specified to satisfy the three-plane datum concept and required points of contact.

Datum target symbols are attached to the datum target with a leader line. A solid leader line indicates the datum target is on the near side of the part; a dashed or hidden line type leader line indicates that the datum target is on the back or far side of the part as illustrated in Figure 1-141.

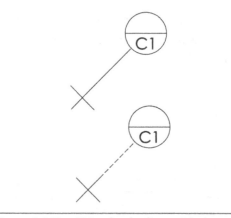

Figure 1-141.   Using leader lines with datum targets

Datum targets are dimensioned with BASIC dimensions or toleranced dimensions (see Figure 1-142). When BASIC dimensions are specified, tooling tolerances apply. BASIC dimensions do not have tolerance, but to aid in locating datum targets properly, some tolerance is required. Therefore, tooling tolerance (usually one-tenth of specified feature tolerance) is permitted. The feature tolerance used to calculate the tooling tolerance is the specified tolerance for the feature oriented from the datums. Usually, BASIC dimensions are specified because of subsequent machining and inspection processes.

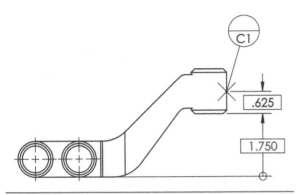

Figure 1-142. Datum target located with BASIC dimensions

Datum Target Point

A datum target point is identified with an "X," shown in Figures 1-141, 1-142 and 1-143. The X, or datum target point, is usually located dimensionally with BASIC dimensions. The target point, regularly identified, is a front view of the part. Where there is no front view, the point location is dimensioned on two adjacent views. Datum target points are normally simulated or picked up with the point of a cone or spherical radius pin. Tooling pins are spherical or pointed pins that are used to contact the part for holding and inspection and are located by target points given on the print. In cases where tooling area is given, flat ended pins

of the same size as indicated in the Target Area symbol are used.

The datum target line is identified with a phantom line in the direct view. The phantom line is shown in the front view and X in an adjacent view. These symbols are located with BASIC dimensions, as shown in Figure 1-143. Datum target lines are simulated with the edge contact of pins.

Datum Target Area

The datum target area is identified with a phantom line circle, with hatch lines inside the circle. The size is specified in the upper half of the datum target symbol. If it is a circular area, the size is preceded with the diameter symbol. Phantom lines and dimensions are specified to define the size and shape of target areas. Dimensioning of diameters is not required; the datum target symbol specifies the target area diameters as shown in Figure 1-144. Datum target areas are simulated with a flat nose pin. In the case of diameters, the pin is the diameter specified in the datum target symbol, using gauge tolerances for the pin diameter. The target area is indicated with hatch lines.

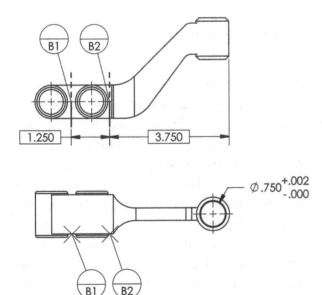

Figure 1-143. Datum target line represented by a phantom line

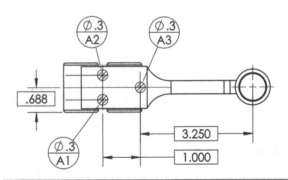

Figure 1-144. Use of datum target area symbols

Datum Translation

This modifier, shown in Chart 1-9, is used in the feature control frame to indicate that the BASIC location of the datum feature simulator is free to translate, within the tolerance, to accomplish full contact with the feature. It may be necessary, for clarity, to include direction of movement for the simulator.

Three-Plane Concept—Circular

Circular parts, like non-cylindrical parts, also require a three-plane concept for repeatable orientation. The primary datum plane is frequently one flat surface of the part. Then two planes (X and Y), intersecting at right angles, establish the axis. This axis is then used as the theoretically exact datum axis. The two intersecting planes provide dimension origins in the X and Y directions for related part features (see Figure 1-145).

There are only two datum features referenced for a part like the one in Figure 1-145. The primary datum plane is one datum feature and the others are the intersection or axis of the X and Y planes. All dimensioning, orientation and measurements originate from the axis and planes.

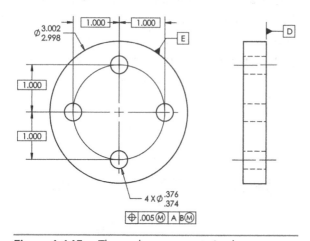

Figure 1-145.   Three plane concept-circular

The drawing in Figure 1-146 is an example of how the three-plane concept applies to a circular part. Even though the overall diameter of the part is shown as datum A, only the theoretical axis is used for orientation of related features. Also, BASIC dimensions are used to locate features from the theoretical planes.

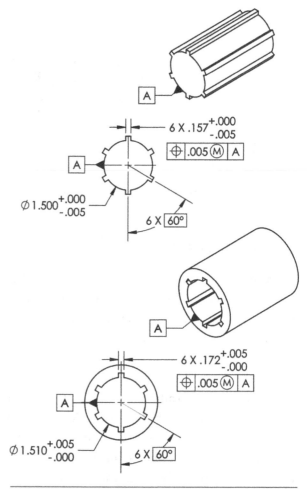

Figure 1-146.   Three-plane concept applied on a cylindrical part

Partial Datums

Occasionally, designs require a datum on a particular surface, but not necessarily the entire surface. Examples of such situations would be large parts, weldments, castings and forgings, and plastics. Some designs incorporating these parts will have partial datums specified. A partial datum is specified to reduce special treatment to an entire surface, such as machining or controlling straightness or flatness. A partial datum is specified with a chain line offset from the datum area, as illustrated in Figure 1-147. The chain line (see Chart 1-7) is drawn parallel to the surface and dimensioned in length specified in note form or by datum target.

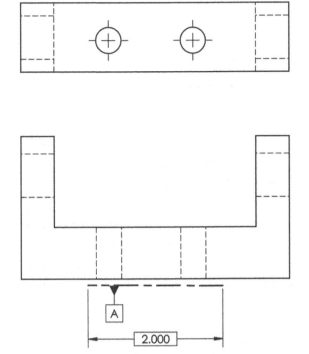

Figure 1-147. Partial datum–mathematically defined

Datums of Size

A *datum of size* is any feature subject to actual mating size variation based on size dimensions. A feature of size is a hole, slot, tab, pin, etc. Because variations are allowed by the size tolerance, it becomes necessary to determine whether MMC or LMC applies in each case. A datum feature of size is not a single point, line, or plane. Features that are datums and subject to actual mating size variation must be verified with a simulated datum (see Rule Four at the end of this section, Basic Geometric Dimensioning and Tolerancing). Figure 1-148 is a drawing of an external feature of size. The diameter of this part is subject to actual mating size variation. When features of this type are specified as datums, the material condition must be specified with the datum identification letter in the feature control frame (Rule Two). The effect of material condition and datum features of actual mating size variation is explained in detail in Part 5, GD&T Tolerances of Location.

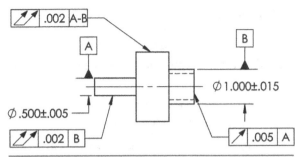

Figure 1-148. Datums that are an external feature of size

External Cylindrical

External features of size are verified with an adjustable chuck, collet, ring gauges, etc. They are used to simulate a true geometric counterpart of the feature and to establish the datum axis or center plane. The geometric counterpart (or actual mating envelope), is the smallest circumscribed cylinder that contacts the datum feature surface that determines the simulated axis. The axis of these irregular features must be established so that measurements and feature relationships may originate from them. The axis of the simulated datum (true geometric counterpart) becomes the datum axis for all related dimensions. This axis can be determined with a height gauge, coordinate measuring machine or any other similar instrument. External features are simulated, as illustrated in Figure 1-149.

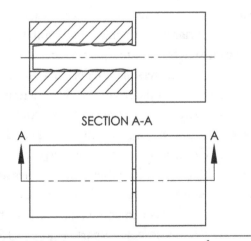

Figure 1-149. True geometric counterpart for external cylindrical features of size

Internal Cylindrical

Internal features of size are verified in a similar manner. If the feature of size is a hole, the true geometric counterpart or actual mating envelope is determined by the largest inscribed cylinder that will fit the hole. The cylinder must be an expandable pin, mandrel, gauge, etc. These are utilized to simulate a true geometric counterpart of the feature. The geometric counterpart axis becomes the datum axis for dimension origins and location of related features. This true geometric counterpart axis may be determined with a height gauge, coordinate measuring machine or similar instrument. Figure 1-150 illustrates how the true geometric counterpart axis is determined.

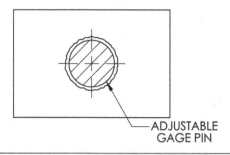

Figure 1-150. True geometric counterpart for internal cylindrical features of size

Non-Cylindrical

Internal features may also be specified as datum features. These features are subject to actual mating size variation and must be simulated with a true geometric counterpart to determine the datum plane, center line, median plane, etc. The true geometric counterpart plane is determined with two parallel planes separated to contact the corresponding surfaces of the specified datum feature. The true geometric counterpart may be a gauge block, an adjustable gauge or measuring instrument used to establish the datum plane, center line, etc.; measurements then originate from the true geometric counterpart. Figure 1-151 is an illustration of a datum feature simulated with a gauge block.

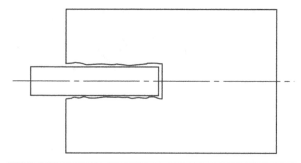

Figure 1-151. True geometric counterpart for non-cylindrical features size

Datum features of size must also have modifiers specified for them when associated with the positioning of features (Rule Two). An example of modifiers being specified is illustrated in Figure 1-152.

Figure 1-152. Specifying modifiers

When a datum reference letter (B in this example) is followed with a modifier, additional consideration must be made for that datum feature. According to the datum/virtual condition rule (see Rule Four), datum feature B must be used at its virtual condition even though it is modified to MMC. Datum reference letters identifying features of size are implied RFS if not modified with MMC or LMC and must be treated like any other datum feature of size. The datum is a true geometric counterpart established by an adjustable gauge to contact the datum feature as produced.

Summary

Datums are assumed to be theoretically exact in order to ensure repeatability from design to inspection. Datums are dimension origins used to establish measurements and feature-to-feature relationships. Datum features, on the other hand, are actual part features that include all variations and irregularities. It is the irregular features that contact the true geometric counterpart. In some instances where a feature is not well defined, the part might have to be adjusted in order to achieve the best fit. Datum features may be a point, line, surface, axis, center line, median plane, etc.

Datums are specified to convey the design intent clearly to all that read the drawing. Before datums were specified, assumptions were made about the intent of the design. Today, datums are specified for all parts within a design, based on the three-plane (Datum Reference Frame) concept for both circular and noncircular parts. The three-plane concept provides a solid repeatable orientation. Datum features are identified with a datum feature symbol or datum target symbol. Letters of the alphabet are used to identify the datum features. Datum features may also be only part of a surface, axis, center plane, etc. If so, the designer will indicate the partial feature with a chain line and give required dimensioning for location and length or area of the partial datum. Datums are located either with BASIC or toleranced dimensions. Features of actual mating size and patterns of features may also be specified as datum features. These features are permitted actual mating size variation, therefore, requiring adjustable gauging to determine the datum. The gauging provides a true geometric counterpart for dimension origins and feature relationship dimensioning.

Part 2, GD&T Datums, Study Questions

1. The ASME Y14.5-2009 Standard requires that datums are _____.

2. Datums are theoretically exact and are used for _____ origins and part _____.

3. List those features of parts that can be used as datums:

   _____ _____

   _____ _____

   _____ _____

4. Is datum reference letter alphabetical order in feature control frames important? _____

5. How many points of contact minimum are required for a primary datum plane? _____

6. The planes of a datum reference frame are assumed to be at _____ degrees basic to each other.

7. Datum target areas are identified with a phantom line _____ with crosshatch lines inside.

8. The upper half of the datum target symbol contains the area _____ when the symbol is attached to a datum target area.

9. Datum targets may be _____ with adjustable gauges, pins, collets, etc.

10. Datums of size are features associated with a dimension and _____.

11. Datums of size are _____ with adjustable gauges, pins, collets, etc.

12. Datums of size also need additional consideration when the _____ are specified with them.

13. Datums are specified on drawings to _____ a clear intent of the design.

14. The minimum point contact required in the three-plane concept is to eliminate part _____.

15. When a datum target area is specified, a _____ nose pin the size of the specified area is required.

Part 3, GD&T Feature Control Frames

This section deals specifically with feature control frames, which are rectangular boxes with multiple compartments. These compartments contain the symbols, tolerances, and datum reference letters discussed earlier. Where applicable, the tolerance is preceded by the diameter symbol and followed by a material condition symbol. The datum reference letters may also be followed by a material condition symbol. The symbols, when combined in a specific sequence in the feature control frame, provide a specific control instruction for the feature or group of features to which it is attached. The contents of feature control frames must always be specified in a standard arrangement. Each feature control frame relates specific tolerancing information for manufacturing and inspection. Feature control frames may be single, combined, or composite.

Feature Control Frame Definition and Symbols

The feature control frame provides a specified control for single or multiple features. The rectangular frame is constructed, as required by the designer, to control one specific feature or group of features. A feature control frame must contain at least a geometric characteristic symbol and a tolerance value. Feature control frames are read from left to right and line-by-line in the case of composite control frames. A feature control frame may contain the following symbols and tolerance:

- Geometric characteristics
- Diameter symbol
- Tolerance
- Tolerance modifier

- Datum reference letter(s)
- Datum modifier

Figure 1-152 shows an example of a feature control frame that was specified to control the position of a feature or group of features. The first symbol in a feature control frame is the geometric characteristic symbol.

Figure 1-152. Feature control frame specified to control the position of a feature or group of features

Attachment

Feature control frames may be attached to features in various ways. The method of attachment picked by the designer determines the effect of the control specified for that feature or group of features. Feature control frames may be attached to a surface, axis, or center line. With each method of attachment, the feature control is limited to only that portion of the part or feature to which the frame is attached. For example, if a feature control frame is attached to a surface extension line, then only that surface is controlled.

Surface

Feature control frames that control surfaces either control the entire surface or just the surface indicated. For example, round features are controlled all around. The control applies all around because it is usually too difficult to orient the part for control on one side. Generally, a designer will want the same control to apply all around the part. In Figure 1-153, an example of

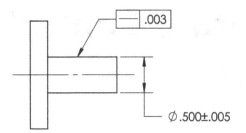

Figure 1-153. Feature control frame applied to the surface of a circular part

how a feature control frame is attached to the surface of a circular part is shown.

Axis or Center Line

Feature control frames associated with round or width-type features are attached to extension lines of that feature. The interpretation means the axis or center line of the feature is controlled with no regard for the feature surface. The designer is specifying the required control so that fasteners will pass through parts or so that parts will mate with each other. The specification of such callouts means the axis or center line orientation—not the sides of the feature—is the critical concern. Examples of how an axis or center line is controlled are shown in Figure 1-154.

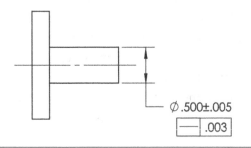

Figure 1-154. How an axis or center line is controlled

The feature control frame is usually attached to the controlled feature with one of four methods. These methods are as follows, with an example in Figure 1-155:

■ The feature control frame is placed below a dimension pertaining to a feature. The leader is from the dimension.

■ A leader from the feature control frame runs to the controlled feature.

■ A side or end of the feature control frame is attached to an extension line from the feature. The feature surface must be a plane.

■ A side or end of the feature control frame is attached to a feature-of-size dimension extension line.

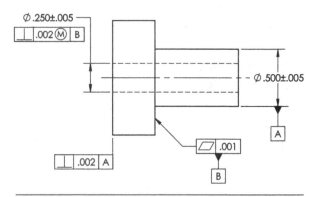

Figure 1-155. Attaching feature control frames to the controlled feature

Geometric Characteristic Symbol

The feature control frame consists of many compartments that contain information specified by the designer. The first compartment of the frame always contains the geometric characteristic symbol for Form, Profile, Orientation, Location, or Runout. Figure 1-156 provides an example of a feature control frame for Form and Orientation.

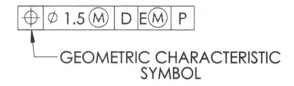

Figure 1-156. Feature control frame

Tolerance

The next compartment always contains a tolerance (Figure 1-157). The tolerance is either a diameter or a width. If the tolerance is cylindrical, the diameter symbol will precede the specified tolerance. The tolerance is always a total tolerance and not a plus/minus tolerance, as with the coordinate method of dimensioning.

Figure 1-157. Compartment containing a tolerance

Datum Reference Letters

When required, the datum reference letter or letters follow the stated tolerance (see Figure 1-158). These letters are not always required, and the number of letters may vary from one to three, depending on the datum reference frame required. The alphabetical order is insignificant; the order from left to right is what establishes the order of precedence for the datum reference frame. The first letter identifies the primary datum plane, the second letter identifies the secondary datum plane, and the third letter identifies the tertiary datum plane.

Figure 1-158. Datum reference letters

Modifiers

Modifiers are required in certain situations, or they may be specified in other cases depending on design requirements. The modifier symbols appear in feature control frames, as shown in Figure 1-159.

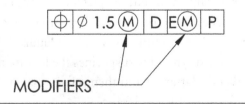

Figure 1-159. Modifier symbols

How to Read Feature Control Frames

Left to Right

Feature control frames are read from left to right. If the feature control frame is a composite or combined symbol, you must read the first line and perform that requirement. Proceed with the next line or lines performing each requirement in succession. Feature control frames do not specify an order of processing, but state the final requirement for the design. Figure 1-160 illustrates how to read two feature control frames. Note that a feature control frame is required for each feature or group of features to be controlled.

Note: There are conditions that require modifiers to be specified, whereas in other cases they are implied. See Rule Two, at the end of this section, Basic Geometric Dimensioning and Tolerancing, for more information.

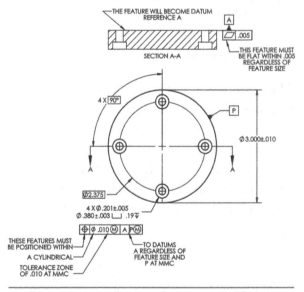

Figure 1-160. Reading two feature control frames

Datum Reference Letter Precedence

The Three-Plane Concept

Datum reference letters are arranged in the feature control frame for a specific orientation of the part. The datum reference letters specify the three-plane, Datum Reference Frame concept. The first datum reference letter in the feature control frame determines the primary datum, the second letter identifies the secondary datum, and the third letter identifies the tertiary datum. Feature control does not always require three datum reference letters, as shown in Figure 1-160. The designer will specify the number of letters (datum references) required for proper part orientation and feature control. The datum reference letters are also specified in the feature control frame from left to right in their order of precedence (see Figure 1-161).

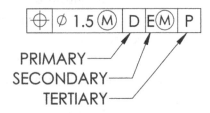

Figure 1-161. Datum reference letters in order of precedence

Two Datum Features

There are some designs that require the identification of two features as datum features to establish a single datum plane, axis, etc. In such cases, each feature is identified with a datum reference letter (Figure 1-162). These two letters then share the same compartment in the feature control frame and are separated by a dash.

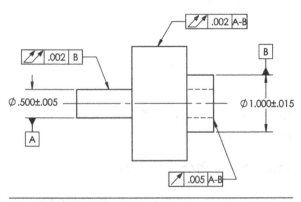

Figure 1-162. Two datum features

As illustrated in Figure 1-162, the center portion of this part is controlled in relation to datum A-B. (The .500 and 1.000 diameters establish datum A-B through the part.) These two features establish a single datum axis through the part.

Feature Control Frame Types

Feature control frames may be combined in many different ways. Regardless of the combinations used, there are only three types of feature control frames. Some frames provide control of a single characteristic; others are many combinations of combined feature control frames. A third type includes those that are interpreted one line at a time. Once the first line of a combined or composite feature control frame is used, then the next line comes into effect. The following sections give some examples of the various combinations that may be created to control the design intent properly.

A single-feature control frame provides only one control for a feature or pattern of features. The control may be specified to control a surface or to control a feature. Figure 1-163 illustrates two examples of single-feature control frames.

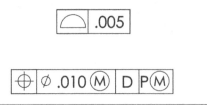

Figure 1-163. Single-feature control frames

Combined-Feature Control Frames

Combined-feature control frames are just what their name indicates. They are two or more frames joined together, or they may be a feature control frame and datum reference symbol combined. These symbols are also interpreted one line at a time. For example, as shown in Figure 1-164, a combined symbol first specifies the feature to be parallel to datum D and then refines that surface to a flatness of less tolerance.

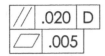

Figure 1-164. Combined-feature control frame

Another application is a profile tolerance. Figure 1-165 illustrates a combined-feature control frame that specifies a surface control and then refines that control to any individual line along the profile with a lesser tolerance.

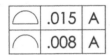

Figure 1-165. Combined-feature control frame with refinement

Another type of combined-feature control frame is where a feature is given a specified tolerance and then identified as a datum feature. This may be common practice where one feature or group of features must be in a specific relationship to another feature. Figure 1-166 illustrates two different combined-feature control frames.

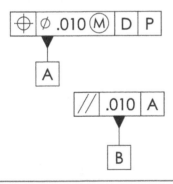

Figure 1-166. Combined-feature control frames with relationship

Composite-Feature Control Frames

Composite-feature control frames are specified to provide a maximum tolerance to orient or position a feature and then refine that feature to a tighter tolerance. A composite-feature control frame will contain only one geometric

characteristic symbol; a combined-feature control frame may contain two different symbols. With composite controls, the first line of the feature control frame is considered to be the largest orientation or location tolerance allowable for the feature(s). Then, once the feature is within this tolerance, it is refined to a closer tolerance to control it to ensure assembly with a mating part. An example of a composite-feature control frame is shown in Figure 1-167. This composite control is most frequently specified with location tolerances.

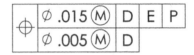

Figure 1-167. Composite-feature control frame

General Rules

Most standards have limited rules that must be observed at certain times. GD&T has rules that must be observed by designers and those who interpret drawings. These rules, in brief, are as follows:

Rule One: When only a tolerance of size is specified, that tolerance controls both size and form.

Rule Two: Regardless of Feature Size (RFS) is implied for all geometric tolerances and the former symbol is no longer used. Maximum Material Condition (MMC) or Least Material Condition (LMC) must be specified on the drawing where it is applicable.

Note: Circular runout, total runout, concentricity and symmetry are applicable only on an RFS basis and cannot be modified to MMC or LMC.

Rule Three: For screw threads, splines and gears, the tolerance and datum reference originate from the pitch cylinder axis.

Rule Four: A virtual condition exists for features of size and datum features of size.

For complete details on general rules, refer to Chapter 5, in *Interpretation of Geometric Dimensioning and Tolerancing, Based on ASME Y14.5-2009*, Puncochar & Evans 2011.

Summary

The feature control frame is specified for each feature or group of features. These feature control frames provide one instruction concerning the form, profile, location, or runout of features. This means that each feature control frame relates specific tolerancing information for manufacturing and inspection. The feature control frame contains the information for proper part orientation in relation to the specified datums. The datum reference letters in the feature control frame denote the datum precedence in relation to the three-plane, Datum Reference Frame concept. Feature control frames may be attached to features in one of four ways. The selected method of attachment by the designer controls how features are controlled. The feature control frame may be constructed as single, combined, or composite. Regardless of the type of feature control frame, they are read from left to right and one line at a time. When one line is read and applied, that line is finished. The information is only used once for a feature.

Part 3, GD&T Feature Control Frames, Study Questions

1. Feature control frames are rectangular boxes that contain specific information to _____ a feature or group of features.

2. Feature control frames may be single, _____, or composite.

3. Feature control frames may be attached to a _____ axis or center line.

4. The first symbol in a feature control frame is a _____.

5. The feature control frame must always contain a specified _____.

6. Feature control frames are always read from _____ to _____.

7. When datum reference letters are specified in a feature control frame, the first letter always identifies the _____ datum.

8. When reading a composite-feature control frame, you always read one _____ of it at a time.

9. Feature control frames consist of a various number of _____ that contain symbols.

10. Is the alphabetical order of the datum reference letters important in the feature control frame? _____

11. Letters that cannot be used to identify datums in a feature control frame are _____ and _____.

Part 4, Form, Orientation, Profile and Runout Tolerances

The portion of the GD&T standard that provides methods of controlling part features is the form and orientation controls. These controls are specified for features critical to function and interchangeability where tolerances of size and location do not provide adequate control. When form and orientation tolerances are specified, the tolerance for some features may increase with the use of modifiers. Form tolerances are most frequently applied to single features or portions of a feature. These controls are specified without a datum reference because the features are not controlled in relation to another feature. Orientation tolerances control features in relation to one another; therefore, a datum reference is required. All dimensions and tolerances apply at free state unless otherwise specified.

Tolerances of Form

Refer to Chart 1-8 as you read this section.

Straightness Definition. Straightness is the condition where one, line element of a surface or axis follows is in a straight line within the tolerance stated.

Straightness Tolerance. Each line element or axis must lie within the limits of size for the feature. The tolerance zone may be a width or a diameter. The feature must be within the stated size limits at each cross-sectional measurement.

Straightness Surface Control. When a surface is to be controlled, the feature control frame is attached to the surface with a leader or extension line. In the case of cylindrical features, the entire surface must be checked.

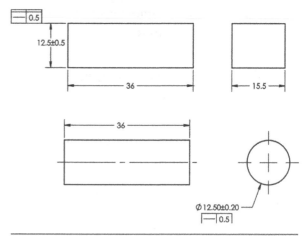

Figure 1-168. Applying straightness control to flat and cylindrical features

All elements of the surface must first be within the specified size tolerance, and then within the limits of the straightness tolerance zone, which is also within the size limits (tolerance). Figure 1-168 shows examples of how the straightness control may be applied to flat and cylindrical features.

Straightness Tolerance Interpretation. The tolerance zone is a space between two parallel straight lines that may make contact with the surface of the feature. The tolerance zone for both flat and cylindrical features is applied along the entire surface. This surface may be measured or verified with a dial indicator or any other digital readout. The feature surface may take any shape such as barreling, waving, concaveness (waisting), etc., as long as it meets the size requirements, and then falls within the specified form tolerance at full indicator movement (FIM). Figure 1-169 illustrates how the features may appear.

Note: The term full indicator movement (FIM) is a new term for older terminology. This new term replaces total indicator movement (TIM) and total indicator reading (TIR). There is no symbol for full indicator movement; it is abbreviated FIM. This abbreviation does not appear on drawings. FIM is understood or implied for certain form controls, which control cylindrical features.

An example of how FIM is used on a drawing is shown in Figure 1-148 and means that the stated tolerances of .002 and .005 are FIM. For example, FIM is the complete movement of a needle on a dial indicator. To measure the variation of the .500 inch and the 1.000-inch diameters, a dial indicator would be rested on each surface and then that surface rotated 360 degrees. During this complete rotation, the dial indicator reading (FIM) must not exceed the stated tolerance.

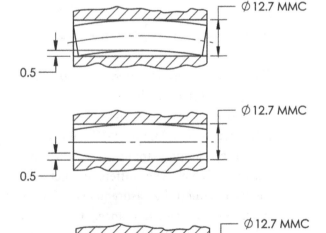

Figure 1-169. Surface shapes acceptable within size and form tolerances

Straightness control specified for a flat part or surface would have a tolerance zone like those shown in Figure 1-168. The

major difference is that for round features, the tolerance applies all around, whereas for a flat feature, the tolerance applies only to the surface indicated as in Figure 1-168. The tolerance will apply only to the top surface of the part.

Straightness Axis Control. To control an axis, the feature control frame is specified below the diameter feature size (Figure 1-170); furthermore, the diameter symbol precedes the geometric tolerance in the feature control frame. The control is specified in a view where the axis is shown as a straight line. This type of attachment means that the axis must be within a specified cylindrical tolerance zone. The use of this control does not require perfect form at MMC. In most cases, the geometric tolerance is less than that of the size tolerance.

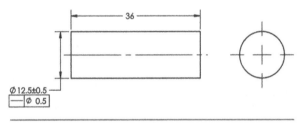

Figure 1-170. Axis control

Flatness Definition. Flatness is the condition of a surface where all elements are in one plane.

Flatness Tolerance. Flatness tolerance provides a zone of a specified thickness defined by two parallel planes in which the surface or center plane must lie. The feature control frame may be attached to the feature with an extension line of the controlled surface, or attached with a leader pointed to the controlled surface.

Flatness Application. Figure 1-171 illustrates a proper specification of the flatness form control.

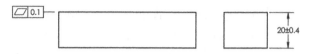

Figure 1-171. Flatness form control

Flatness Tolerance Interpretation. Flatness tolerance is the specified distance between two parallel planes of which the upper limit plane must contact the actual feature surface (Figure 1-172). The other plane then should be the stated tolerance from the first and below all surface area irregularities. The actual surface may be verified with a dial indicator. The indicator should be zeroed for the highest or lowest point on the surface. Then the surface must be checked sufficiently in all directions to ensure that it is within the specified tolerance. The readings obtained are FIM and must not exceed the stated FIM tolerance in the feature.

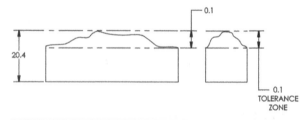

Figure 1-172. Flatness tolerance

Circularity Definition. Circularity is roundness. It is a condition of a cylindrical surface other than a sphere; at any cross-sectional measurement during one complete revolution of the feature, all points of the surface are perpendicular at an equal distance from a common axis. Circularity of a sphere is a condition where all points of the surface intersected by any plane passing through a common center are equal distance from that center.

Circularity Tolerance. Circularity tolerance provides a circular zone in which all points of a cross-section or slice of the

surface must lie. The tolerance zone is two concentric circles that are the stated tolerance apart. The feature control frame is usually specified in the end view. The tolerance zone must be within the size limits of the feature.

Circularity Application. The circularity tolerance is applied to compare the circular elements or slices of cylindrical features. Figure 1-173 illustrates the proper application of a circularity tolerance. This tolerance may be specified for any cylindrical feature such as cones, spheres, or cylinders that require only line control around the feature. Circularity may also be specified for internal features that are circular in cross-section.

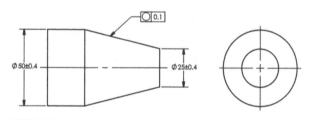

Figure 1-173. Applying a circularity tolerance

Circularity Tolerance Interpretation. The circularity tolerance is the space between two concentric circles that is the stated tolerance apart. Figure 1-174 illustrates how the tolerance applies to an external feature. (*Note:* The difference between the two diametral measurements is 0.2, or twice the specified tolerance.) Circularity tolerance is a radial tolerance. The larger circle must make contact with the actual surface of the controlled external feature. Then the smaller circle would have to be the stated tolerance away from the larger one or the same as the feature's smallest permissible diameter.

The opposite is true for internal circular features. This tolerance zone is applicable to each cross-sectional element of the feature. The tolerance zone must be perpendicular to the controlled feature axis. All elements of the controlled feature must be within the specified size limits.

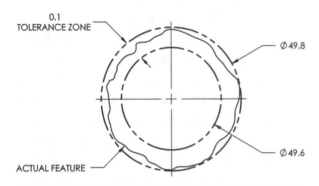

Figure 1-174. Applying circular tolerance to an external feature

Controlled features may be verified with several instruments. The primary concern, however, is how the feature is measured. For example, if a V-block is used, the measurement may include unwanted variables that may not be noticed, such as lobing, out-of-straightness and the composite effects of a diametrical reading. If possible, the measurements should be made in relation to the axis because the specified tolerance is on the radius. In this way, all readings on the indicator will be radial, as the tolerance is intended. Regardless of the method of measurement, sufficient measurements must be made to ensure feature acceptance. The specified tolerance is implied FIM for each circular element of the controlled feature.

Cylindricity Definition. Cylindricity is the condition of an entire feature surface during one revolution in which all surface points are an equal distance from a common axis.

Cylindricity Tolerance. Cylindricity tolerance provides a zone bounded by two concentric cylinders in which the controlled

surface must lie. Cylindricity tolerance is a radial tolerance. The feature control frame is attached to the feature with a leader.

Cylindricity Application. Cylindricity tolerance is specified in addition to the specified feature size and tolerance. It remains the same size for all possible actual mating sizes and is not an addition to the feature size or tolerance. The cylindricity tolerance is a composite control. It controls feature circularity, straightness, and any taper of the entire cylindrical feature. Figure 1-175 illustrates how cylindricity is applied.

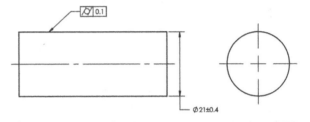

Figure 1-175. Applying cylindricity

Cylindricity Tolerance Interpretation. Cylindricity tolerance is the space identified by the geometric tolerance between two concentric cylinders the stated tolerance apart. The largest tolerance cylinder must make contact with the actual surface of the maximum diameter per tolerance of the controlled external feature. Then, the smaller tolerance cylinder is the stated tolerance away from the minimum diameter per tolerance of the large cylinder. Neither of the cylinders may exceed the feature size limits. The tolerance zone is an equal distance from the controlled feature axis (radial measurement) for the length of the feature. The controlled feature may be verified with several different measuring devices. This control, however, is composite, and the method of measurement is important. If functional gauging is not used, then an inspection method must be used that will detect excessive taper, out-of-straightness and out-of-roundness in relation to the feature axis. The measuring device must not indicate more than the stated tolerance. Figure 1-176 shows an example of an external feature and the specified tolerance zone.

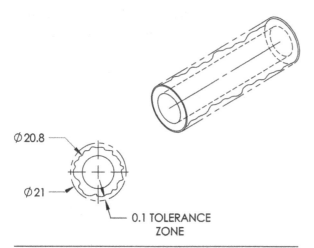

Figure 1-176. An external feature and the specified tolerance zone

Tolerances of Orientation

Perpendicularity Definition. Perpendicularity is the condition of an entire surface, plane, or axis at a right angle to a datum plane or axis.

Perpendicularity Tolerance. Perpendicularity tolerance provides a zone defined by two parallel planes, two parallel lines, or a cylinder parallel to a datum. The controlled feature surface, plane, or axis must lie within the specified tolerance zone. Perpendicularity is an orientation control; therefore, a datum reference is required, and the orientation tolerance is implied RFS if not modified to LMC or MMC. Remember, only features of size can be modified—not surface or line features. The feature control frame is specified in drawing a view where the relationship of features is shown.

Perpendicularity Application. Perpendicularity tolerance is specified for designs that require one feature to be perpendicular to another; therefore, a datum reference is required. Perpendicularity controls all surface error including flatness and angularity. This control is specified in addition to actual mating size requirements and may be specified in four different methods. A perpendicularity tolerance may be specified for 1) a surface perpendicular to a datum plane, 2) an axis perpendicular to an axis, 3) an axis perpendicular to a datum plane, or 4) line elements of a surface perpendicular to a datum axis. Only the first of these applications will be explained here. For complete details, see *Interpretation of Geometric Dimensioning and Tolerancing, Based on ASME Y14.5-2009*, Puncochar & Evans, 2011.

Feature Surface Perpendicular to A Datum Plane. The feature surfaces perpendicular to a datum plane is the most frequent application for perpendicularity. Figure 1-177 provides an illustration. The feature control frame, which is specified in a drawing view where the relationship between features appears, may be attached to the feature with an extension line or leader.

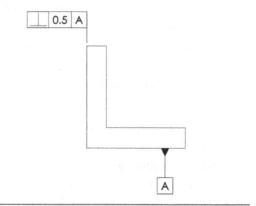

Figure 1-177. Feature surface perpendicular to a datum plane

Perpendicularity Tolerance. This perpendicularity tolerance provides a zone defined by two parallel planes that are the distance of the specified tolerance apart. The tolerance zone must be within the limits of the feature size. All elements of the controlled feature must lie within the perpendicularity tolerance zone.

Perpendicularity Tolerance Interpretation. The controlled feature must first meet the feature size requirements and then the geometric orientation tolerance. The datum reference feature must be placed in contact with the datum plane. Then the tolerance zone must be established at exactly 90 degrees to the datum. All elements of the controlled surface must lie within the parallel planes of the tolerance zone, as in Figure 1-178. *Note*: Datum error is not additive to the feature being controlled.

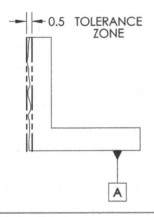

Figure 1-178. Parallel planes tolerance zone

The actual feature surface may be verified in a number of ways. A simple check can be made with a dial indicator or coordinate measuring machine. The datum feature must be placed in contact with the datum plane. Then, with a measuring instrument, make contact with the controlled surface. Zero the measuring device on a high point and continue to measure the entire surface. The FIM may not exceed the tolerance specified in the feature control frame. Enough of the

surface must be measured to ensure that design requirements are met.

Angularity Definition. Angularity is the condition of an axis or plane other than 90 degrees to another datum plane or axis.

Angularity Tolerance. Angularity tolerance provides a zone defined by two parallel planes that are a stated tolerance apart and at the specified basic angle to the datum reference. The controlled feature surface, plane, or axis must lie within this zone. Angularity is another of the orientation tolerances; therefore, a datum reference is always required. The tolerance may be modified if a feature of size is being controlled. If modifiers are not specified, Rule Two governs the tolerance. The feature control frame is specified in a drawing view where the angular relationship is shown. This relationship must be specified with a basic angle. The datum reference feature irregularities do not affect the controlled feature surface or axis.

Angularity Application. A control is specified to control features that are required to be at an angle other than 90 degrees in relation to another feature plane or axis. Angularity tolerance controls surface, plane or axis errors within the limits of the tolerance zone. Angularity also controls flatness and straightness. The angularity tolerance is in addition to the feature size tolerance. The tolerance provides a zone that controls feature origin in relation to another datum feature. Figure 1-179 illustrates a surface application.

When angularity is specified for an internal feature such as a slot or hole, the tolerance applies only in the view and relative to the datums indicated. The feature is not controlled in any other direction. The tolerance may be modified depending

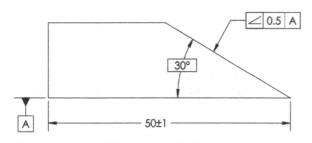

Figure 1-179.   Angularity tolerance

on the final requirement of the feature. If angularity is specified for a feature of size, it is usually a refinement of a location control. Figure 1-180 illustrates an axis control.

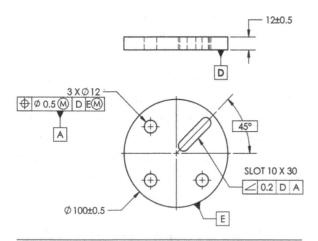

Figure 1-180.   Angularity tolerance applied to a slot

Angularity Tolerance Interpretation. The controlled feature must meet all other tolerance, and then the orientation control for angularity. The datum reference feature irregularities are not considered when measuring angularity. The tolerance is established at the specified basic angle to the datum. The outer plane of the tolerance zone contacts the highest point(s) on the controlled surface. The inner plane is at the specified tolerance from the first. The controlled surface may take any form through this tolerance zone. The beginning of the angle must be within the length requirement of the part, and may be curved, bowed, twisted or at a different angle.

Figure 1-181 illustrates how a tolerance zone for a surface is applied. The surface may be verified with a dial indicator or coordinate measuring machine.

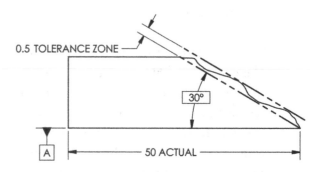

Figure 1-181.   Applying a tolerance zone for a surface

The control of an axis or plane is slightly different than the control of a surface. The major difference is the tolerance zone. The tolerance only applies in the view that it is specified in and relative to the indicated datums. The actual tolerance zone, then, is a slice the thickness of the specified tolerance that passes through the controlled feature. When viewing the feature as shown in the drawing view with the orientation control, the end of the tolerance would be seen. This zone extends the total length of the part. Figure 1-182 illustrates the tolerance zone for a slot or hole.

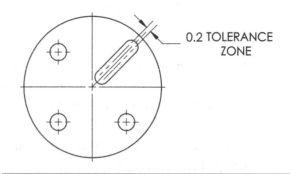

Figure 1-182.   The tolerance zone for a slot or hole

Angularity may be verified with several methods. If a modifier were specified, then a gauge would be the most effective way to verify correct feature angularity. If the feature is toleranced at RFS, an expandable gauge pin, dial indicator, or coordinate measuring machine might be used to determine correct angularity.

Parallelism Definition. Parallelism is the condition of a surface, center plane, or axis that is an equal distance at all points from a datum plane or axis.

Parallelism Tolerance. Parallelism tolerance provides a zone defined by two parallel planes, lines, or a cylinder parallel to a datum plane or axis. The surface elements or axis of the controlled feature must lie within this zone. The feature control frame is specified in a drawing view where the parallel relationship is shown. This feature control frame must contain a datum reference letter because parallelism is an orientation tolerance.

Parallelism Application. Parallelism tolerance is specified for designs where a plane, surface, or axis must be controlled for parallelism in addition to the feature size tolerance. A datum reference is required with this orientation control. Parallelism also controls flatness of a surface and straightness of an axis within the limits of the tolerance. When parallelism is specified for a surface, the controlled surface must be within the specified size limits. Parallelism orientation may be specified to control three different conditions: 1) a surface parallel to another surface, 2) a cylinder parallel to a surface, and 3) a cylinder parallel to a cylinder. Only the first of these applications will be explained here. For complete details, see *Interpretation of Geometric Dimensioning and Tolerancing, Based on ASME Y14.5-2009*, Puncochar & Evans 2011.

Parallelism Surface Parallel to Another Surface. This application is the most frequent of the parallelism orientation control (Figure 1-183). The feature control frame is attached to the controlled feature with an extension line in a drawing view where the relationship is shown.

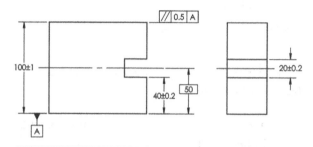

Figure 1-183. Surface parallel to another surface

Parallelism Tolerance. The tolerance for the controlled feature is the space between two parallel planes that are a specified tolerance apart and parallel to the datum plane. This zone must be within the limits of feature size. All elements of the controlled surface must be contained within the tolerance zone.

Parallelism Tolerance Interpretation. The controlled feature must first meet the actual mating size limit requirements, and then the geometric orientation tolerance. The datum reference feature must contact the simulated datum plane. Any irregularities in the datum reference surface are not additive to the controlled feature. The outer tolerance zone plane must be established from the outermost surface elements and must be parallel to the datum plane. The inner plane is then the specified orientation tolerance apart from the outer plane. All controlled surface elements must be within the limits of actual mating size and the parallelism tolerance zone. Figure 1-184 illustrates how the tolerance zone would apply to the actual mating size of a feature.

The surface may be verified with the basic instruments of inspection, such as

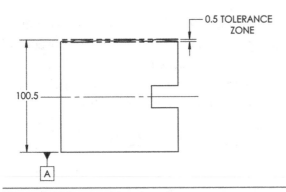

Figure 1-184. Applying the tolerance zone to a feature's actual mating size

a coordinate measuring machine or dial indicator. The datum surface must have the minimum three-point contact with the datum plane. The controlled surface must be measured in many directions and across the entire surface until it is ensured of acceptance.

Tangent Plane. When it is desirable to control a feature surface by contacting points of the surface, the tangent plane symbol is specified in the feature control frame following the stated tolerance. Figure 1-185 illustrates this specification.

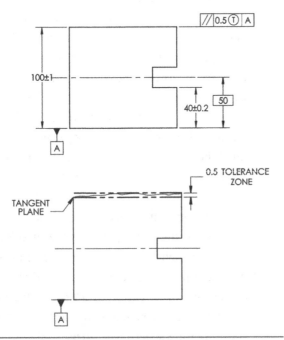

Figure 1-185. Applying the tangent plane symbol and tangent plane tolerance zone

Tangent Plane Tolerance Interpretation. The controlled surface of the feature must lie within two parallel planes 0.5 inch apart. A plane contacting the high points of the surface must lie within this established zone. Figure 1-186 illustrates this specification.

Tolerances of Profile

Profile tolerancing is a method of specifying control of deviation from the desired profile along the surface of a feature.

Profile Tolerance. Profile tolerances may be specified either as a surface or line profile. The tolerance provides a uniform zone along a desired true profile (a bilateral zone) or a zone defined by a phantom line either inside or outside the basic true profile of the part (a unilateral zone). The surface of the controlled feature must lie within this zone.

The feature control frame is usually attached with a leader to the surface to be controlled. The feature control frame is specified in a drawing view where the profile is shown. The feature control frame must contain a datum reference and may also have a note added beneath stating: BETWEEN two points (X and Y) or include the symbol for ALL OVER.

Profile Application. Profile tolerance is specified for designs where the surface is to be controlled within a given basic shape. Most frequently, profile tolerances are specified for irregular features that are difficult to control with other form or orientation tolerances. However, they may also be specified to control the all-around shape of stampings, burned parts, etc. The basic profile of a part is described with basic dimensions, radii, arcs, angles, etc., from the datum. Then the specified profile tolerance controls the amount of deviation in relation to the datum reference(s). Profile tolerances may be specified to control either a surface or line element of a feature. These two applications are discussed next.

Surface Profile. Surface profile is a method of specifying a three-dimensional control along the entire surface to be controlled. This control is usually applied to parts having constant cross-section, surfaces of revolution, weldments, forgings, etc., where an ALL OVER requirement may be desired. Surface control may apply ALL AROUND. Surface profile requirements may also be specified for coplanar surfaces (Figure 1-187). Surface profile controls actual mating size as well as shape; therefore, actual mating size and geometric form are verified simultaneously. Figures 1-186 and 1-187 illustrate the proper application of surface profile tolerancing.

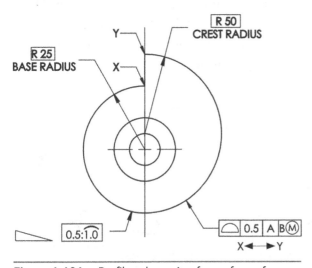

Figure 1-186. Profile tolerancing for surface of revolution

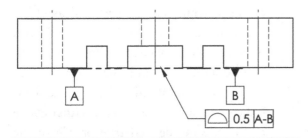

Figure 1-187. Profile tolerancing for coplanar surfaces

Surface Profile Tolerance. Surface profile tolerance may be specified as an equal or unequal tolerance on either side of the desired basic profile. If the design requires an unequal application of a bilateral tolerance, a phantom line is included along the basic profile, and the amount of tolerance between the basic profile and the phantom line is specified. Then the balance of the tolerance is applied to the other side of the basic profile. Phantom lines are also specified when the tolerance is unilateral; all tolerance is applied to one side or the other of the basic profile.

Surface Profile Tolerance Interpretation. Profile tolerance for a surface control is implied to be bilateral. If the design requires the tolerance to be unequal or unilateral, phantom lines are added to the drawing view indicating how the tolerance is to be applied. The controlled feature's size and shape are toleranced with the profile tolerance. Surface profile tolerance is a three-dimensional zone. The tolerance applies perpendicular to the true profile at all points along the controlled surface. If the tolerance is bilateral, the actual surface of the feature may vary both inside and outside of the true basic profile. If the tolerance is unilateral, then the actual feature may vary only to the inside or the outside of the basic true profile as indicated with a phantom line. Figures 1-188, 1-189, 1-190, and 1-191 illustrate the various tolerance zones permitted by surface profile tolerancing. The tolerance zone is established in relationship to the basic true profile of the feature, not the actual feature surface.

The actual part surface may be verified with several techniques. The actual part may be compared to a master part, an overlay may be used, optical comparison

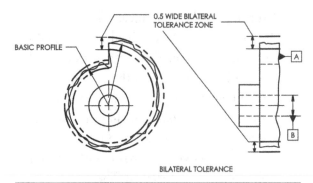

Figure 1-188.　Bilateral profile tolerance

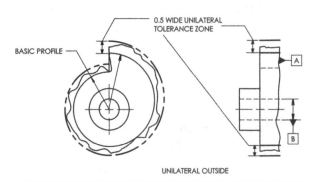

Figure 1-189.　Unilateral outside profile tolerance

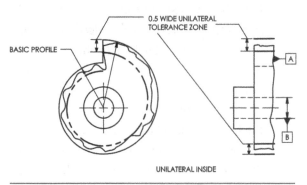

Figure 1-190.　Unilateral inside profile tolerance

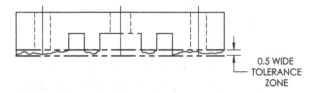

Figure 1-191.　Bilateral profile tolerance applied to coplanar surfaces

can be made, or a dial indicator or coordinate measuring machine can be used. Verification will depend largely on the accuracy required.

Line Profile. Line profile tolerancing is a method of specifying a two-dimensional control for a single line element along the true profile of a surface. This control is usually specified for the shape of cross-sections or cutting planes of parts. The control is most frequently specified for manufactured parts for trucks, automobiles, and marine uses (impellers, body parts, propellers, etc.). This control should be used where blending is required. Line profile, like surface profile, may be specified between points or all-around. Line profile tolerancing is usually a refinement of some other geometric control, form and size control. The application of line profile is illustrated in Figure 1-192. In this feature control frame, the combined geometric tolerances are applied. First, the entire surface (profile of a surface) is required to be within the 0.5 mm tolerance. Then, each cross-sectional line, (profile of a line) is further refined to a lesser tolerance requirement.

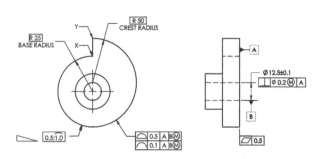

Figure 1-192.　Line profile tolerancing

Line Profile Tolerance. The tolerance for line profile may be specified as either bilateral or unilateral. The bilateral tolerance is implied by the absence of phantom lines to indicate the tolerance zone. The unilateral tolerance zone is indicated with a phantom line either inside or outside the desired true profile of the controlled part.

Line Profile Tolerance Interpretation. The actual surface must lie within the boundaries of the specified zone. The tolerance applies perpendicular to the line profile at all points along the controlled surface. Line profile control may require datum references. When specified, the feature must be properly oriented when applying the specified tolerance. The profile tolerance controls both actual mating size and form. The tolerance zone is established in relationship to the true profile of the controlled feature, not the actual surface. All points along the controlled surface must lie within the specified tolerance, as shown in Figure 1-193. Line profile may be verified with the same method used to verify surface profile.

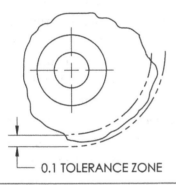

Figure 1-193. True profile relationship to tolerance zone

Tolerances of Runout

Runout is a composite form and location control of permissible error in the desired part surface during a complete revolution of the part around a datum axis.

Runout Tolerance. Runout tolerances may be specified as either total or circular. The specified tolerance is the deviation permitted in relation to the controlled features axis. It provides a zone between two concentric cylinders for total runout control and between two concentric circles for circular runout control. The surface or all points on a cross-sectional line must lie within the specified tolerance zone. The tolerance is specified in a drawing view where the controlled feature(s) are shown. The feature control frame is attached to the controlled feature with a leader or associated with the feature size callout. A datum reference is required.

Runout Application. Runout tolerance is specified for designs where rotation is involved, such as shafts, pulleys and bearing surfaces. Runout may also be specified for coaxial features. This control, however, is restrictive for manufacturing because the tolerance is always RFS. Location controls may be a better choice for coaxiality, depending on design requirements. Runout tolerances control the amount of radial deviation for a line or surface of parts that are circular in profile.

Total Runout. *Total runout* is specified to control feature surfaces that are manufactured with an axis. This control is more stringent than *circular runout*. It provides a composite control of all surface elements in relation to a datum axis. The controlled surface may be at right angles to or around the datum axis. This control is specified when the composite effect of all surface elements together is critical to the final assembly. Figure 1-194 illustrates an application of total runout control.

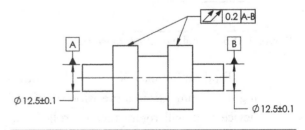

Figure 1-194. Total runout control

Total Runout Tolerance. Total runout tolerance is specified in the feature control frame and is implied to be radial—RFS and FIM. The feature control frame is attached directly to the controlled feature. The actual feature must not exceed the boundary of perfect form at MMC. The controlled surface must lie within two concentric cylinders that are the stated tolerance apart and an equal distance from the datum axis. The entire actual feature must lie within this tolerance zone simultaneously.

Total Runout Tolerance Interpretation. Total runout tolerance establishes a cylindrical tolerance zone that is the width of the specified tolerance in the feature control frame. This cylinder of tolerance is an equal distance from the datum axis all around. The tolerance is applied simultaneously to all circular and longitudinal elements in one setup during a complete revolution of the controlled feature. The tolerance is composite and cumulative. The feature is controlled for taper, coaxiality, circularity, cylindricity, straightness, angularity, flatness, perpendicularity and profile. Figure 1-195 illustrates the actual tolerance zone for a controlled feature.

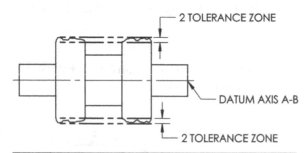

Figure 1-195. Total runout tolerance zone

Total runout may be verified by mounting the datum axis in a precision rotational device that will rotate the controlled feature(s) around the datum. The feature may be mounted on a functional diameter or

mounted on centers. When the method of measurement is selected, an indicator is set up in contact with the controlled surface and zeroed. The indicator is not re-zeroed during the measuring operation. The controlled feature is measured parallel to the datum for circular control and perpendicular to the datum for surfaces perpendicular to the datum. The part must be rotated 360 degrees during the measuring operation, and the indicator reading must not exceed the tolerance stated in the feature control frame. Sufficient measurements must be made to satisfy the drawing requirement.

Circular Runout. *Circular runout* is specified to control only surface elements of features that are circular in cross-section or surfaces perpendicular to a datum axis. Circular runout is only a line-by-line control of a surface. Each line is completely independent of the other. The tolerance is implied FIM in relation to a feature datum axis. This control is specified when the part function is not critical to rotational speeds. Circular runout is applied in Figure 1-196.

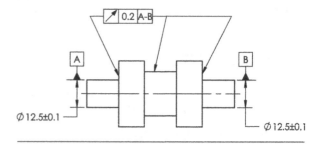

Figure 1-196. Applying circular runout

Circular Runout Tolerance. The tolerance for circular runout is specified in the feature control frame and is always implied RFS. The tolerance is radial and normally implies FIM. The feature control frame is attached to the controlled feature with a leader or with the feature size callout. The actual feature surface must not exceed the boundary of perfect form at MMC. Each cross-sectional slice of the surface must lie within the two concentric circles established by the tolerance zone.

Circular Runout Tolerance Interpretation. Circular runout tolerance is a width between two concentric circles the size of the stated tolerance. Each circular element independent of the next must lie within the RFS tolerance zone. This tolerance zone is an equal distance from the datum axis all around. The tolerance controls the cumulative variations in circularity and coaxiality of controlled features around a datum axis. For features perpendicular to a datum axis, the tolerance controls circular elements on the plane. Figure 1-197 illustrates how each circular element is controlled by the specified tolerance.

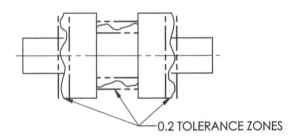

Figure 1-197. Circular runout for features perpendicular to a datum axis

Circular runout may be verified similar to total runout. The difference is that the indicator does not have to be moved along or over the controlled surface. The indicator may be re-zeroed for each measurement. The feature must be rotated a complete 360 degrees for each measurement. To ensure acceptability, several independent measurements along the controlled surface should be made.

Free State Variation. *Free state variation* is a term used to describe distortion of a part after removal of forces applied during manufacturing. Variation or flexibility is due to thin cross-section, weight, and stress relief. In some cases, parts are required to meet drawing requirements while in free state, whereas others are to be restrained. Restraining forces are those exerted in assembly or functioning of the part. If free state variation must be controlled, it may be accomplished by specifying tolerances subject to free state variation or by specifying tolerances on features to be restrained.

Specifying Geometric Tolerances on Features Subject to Free State Variation. Where an individual form or location tolerance is specified for a feature at free state, the maximum allowable tolerance is specified in the feature control frame. This tolerance is followed by any applicable modifiers and then the free state symbol. Figure 1-198 illustrates such a requirement.

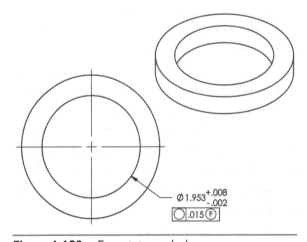

Figure 1-198. Free state symbol

When the free state symbol is given with the geometric tolerance, it indicates that the part must be inspected in a restrained condition while located on the datums. Because these datums may be subject to free state variation, the maximum restraining force must be specified. This force or method of restraint is specified on the drawing in the form of a NOTE. The restraining note may specify a particular type of clamp, bolt and torque value, as well as partial assembly to achieve actual fit, etc.

Summary

The GD&T form and orientation controls control the required shapes of features. In this summary, each of the controls will be grouped to aid in remembering what and how they control. The groups will address those controls that control lines and those that control surfaces or areas. All of the controls will be compared to the two basic form controls—straightness and flatness. Straightness only controls one line element at a time—or, the straightness of an axis. Flatness controls all elements or points of a surface simultaneously in all directions.

The remaining controls can all be related to straightness and flatness. Consider circularity and cylindricity: circularity is a line control around features circular in form. It is like straightness around a feature. Cylindricity controls the complete surface area of circular features. Cylindricity is like flatness rolled around a feature.

Perpendicularity may be compared to both straightness and flatness depending on the features being controlled. Perpendicularity may be specified to control four different types of feature-to-datum relationships. When the control is specified for a plane or surface, the control can be related to flatness. When the Perpendicularity control is specified for line element, axis, or median plane, it can be related to straightness.

Angularity is similar to perpendicularity because it controls all angles other than 90 degrees. Some of the angular relationships will be surfaces, and others will be cylinders or non-cylindrical features. Therefore, angularity can be related to flatness when controlling surfaces. When angularity is specified for cylinders or non-cylindrical features, a line or median plane is being controlled like straightness.

Parallelism can also be related to either straightness or flatness depending on the type of feature being controlled. When surfaces are controlled, it is similar to flatness where all elements must be in one plane of given thickness. When features with a median plane or axis are controlled with parallelism, the control is similar again to straightness where a line must be straight within a given tolerance.

Profile tolerances are specified to control either surfaces or lines. These may be features that are irregular where flatness or straightness do not apply. Surface profile tolerance provides a thickness zone around or over the controlled surface like flatness control would provide. Profile of a line is like straightness where only a line element is compared to the true basic profile. Each line element is considered individually.

Runout tolerances also provide for two types of control, where total runout requires measurement of an entire circular surface in one setup similar to flatness. Circular runout is only a cross-section or slice of a circular feature control similar to straightness. Only one line element at a time is measured and compared to the specified tolerance.

These comparisons provide a means to identify the control provided by each of the geometric form and orientation controls. There are some differences between controls that should be summarized. The primary difference between a runout control and cylindricity is that runout—either total or circular—controls surface-to-axis, whereas cylindricity controls axis-to-axis. Another difference between controls is that of perpendicularity and runout. Perpendicularity is primarily specified to control non-cylindrical features where runout is specified to control cylindrical features.

In the end, the design requirements will always dictate the control to be specified. Parts are not commonly designed to be used independently of others. Instead, they are designed for function and relationship in a final assembly.

Part 4, Form, Orientation, Profile and Runout Tolerances, Study Questions

1. Flatness is a form control that controls surface elements in all _____ within a specified tolerance.

2. Circularity control applies to feature surfaces during one complete revolution as measured _____ of a surface or axis is in a straight line.

3. Straightness is the condition where one line _____ of a surface or axis is in a straight line.

4. Form and orientation tolerances permit features to vary within the _____ of the tolerance zone.

5. Cylindricity controls the _____ surface of the features.

6. Per unit control is specified to prevent the continuation of feature _____ or abruptness of the controlled feature.

7. For tolerances of perpendicularity, the zone established by the specified _____ must be within the limits of feature size.

8. The tolerance boundary for a cylindrical feature axis is diametrical when the _____ symbol is specified.

9. Feature control frame _____ determines whether a tolerance is applied to the median plane, center line, or axis of a controlled feature.

10. Form and orientation tolerances are _____ to be RFS.

11. When a form or orientation tolerance is specified for a feature in relation to a datum feature, the datum feature is _____ to be theoretically exact.

12. With the application of GD&T, there are two tolerances allowed: _____ and _____.

13. Angularity is the condition of a surface or _____ at an angle other than 90 degrees from a datum.

14. Parallelism is the condition of a surface or axis an equal _____ at all points from a datum plane or axis.

15. For non-cylindrical features, angularity tolerance is a _____ and not an angular tolerance zone.

16. Total runout is always implied _____.

17. A _____ tolerance is applied on either side of the basic profile.

18. A _____ profile is specified to control a line element of a surface.

19. Profile tolerance is a method of specifying control of deviation from the desired basic _____ along the surface of a feature.

20. Runout is a composite form and location control of permissible error in the desired part surface during a complete _____ of the part around a datum axis.

21. Orientation tolerances need not be referenced to _____.

Part 5, GD&T Tolerances of Location

This section introduces the *tolerances of location*. These tolerances or geometric controls are concentricity, symmetry, and position. The location controls are specified to control the relationships between features or between features and a datum feature. The relationships are toleranced at the axis, center line, or center plane. Positional tolerance then provides the permissible variation in the specified location of the feature or group of features in relation to another feature or datum. A tolerance of location is applied to at least two features, of which one must be a feature of size (meaning one is a datum). Because one of the features must be a feature of size, the modifier principles apply. General Rule Two requires the designer to specify modifiers for all features, tolerances, and datums of size. The advantages of the modifiers can be used to their greatest extent with tolerances of location involving part interchangeability and functionality of mating parts. GD&T's advantages are best realized when position and modifiers are specified.

Concentricity Definition

Concentricity is the condition where the median points of all diametrically opposed elements of a figure or revolution (or correspondingly located elements of two or more radially disposed features) are congruent with the axis (or center point) of a datum feature.

Concentricity Tolerance

Concentricity tolerance is a diametral zone in which the axis of the controlled feature must lie.

This zone must coincide with the axis (center point) of the datum features. Concentricity is a very restrictive geometric control. A specified tolerance controls the amount of eccentricity error, parallelism of axis, out-of-straightness of axis, out-of-circularity, out-of-cylindricity, and any other possible errors in the feature axis. This tolerance controls all possible errors at the feature axis. Therefore, it is difficult to verify and may be excessively expensive to produce. The actual feature axes must lie within the specified tolerance zone. It may be more effective to use runout or position tolerances.

Concentricity Application

Concentricity is considered when critical axis-to-axis control is required for dynamically balanced features. This control is selectively specified, because features may be controlled with runout or position. Runout only controls a surface-to-axis. Position offers all the advantages of GD&T. Concentricity is an axis-to-axis control. Concentricity is normally specified for high-speed rotating parts, rotating mass, axis-to-axis precision, or any other feature critical to function. Figure 1-199 illustrates the proper application of concentricity.

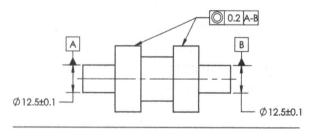

Figure 1-199.   Applying concentricity

Concentricity Tolerance Interpretation

Concentricity tolerance zone is diametral around and parallel to the datum axis. Verification is difficult. Concentricity is the determination of the controlled features axes in relationship to the datum axis. A radial differential measurement is the most accurate method of determining the actual controlled feature axis. These measurements must be taken opposite of each other. The tolerance zone for the part in Figure 1-199 is illustrated in Figure 1-200.

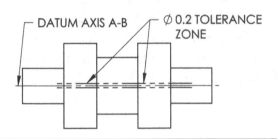

Figure 1-200. Concentricity tolerance zone

For concentricity, the feature surfaces must be measured diametrically opposed to each other to determine the midpoints relative to the axis of the datum. For coaxial features located with a positional tolerance, the location of the axis of the actual mating envelope of the feature is controlled relative to the axis of the datum feature.

Symmetry Definition

Symmetry is the condition where a feature or part has the same profile on either side of the center plane (median plane) of a datum feature.

Symmetry Tolerance

The tolerance is applied equally on either side of the controlled feature center line. The implied modifier restricts the tolerance to the specified amount only.

Symmetry Application

Symmetry is specified for features to be located symmetrically with respect to the median plane of a datum feature. It may also be specified for a feature in a common plane with a datum plane. This symbol is specified extensively throughout industry. Figure 1-201 illustrates an application of symmetry.

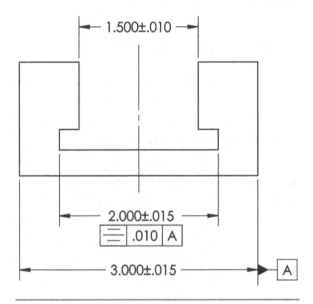

Figure 1-201. Applying symmetry

Symmetry Tolerance Interpretation

The specified tolerance establishes a tolerance zone that is the specified width, with half of the tolerance on either side of the datum feature center line. This width zone allows the feature to vary from side-to-side or angularly within the tolerance zone. The tolerance zone size is not permitted to vary with feature size. Figure 1-202 illustrates how the interior of the part could vary in relation to the exterior, which is the datum feature. The controlled feature is permitted a maximum of .005 inch shift to the side in either direction.

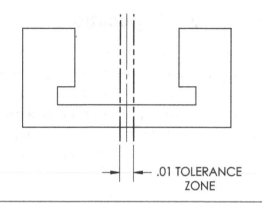

.01 TOLERANCE
ZONE

Figure 1-202.   Symmetry tolerance zone

Position Introduction

Position is one of the most effective and used controls in GD&T. This control provides the designer with the ability to specify clearly all design requirements and intentions. Through clearer specification, higher production yields are possible, interchangeability is ensured and definable quality requirements are attained.

The coordinate dimensioning system is not *replaced* with GD&T; it is *enhanced* by it. The two methods of dimensioning and tolerancing may be specified simultaneously on engineering drawings. The coordinate method doesn't provide the same tolerance advantages; system comparisons are illustrated in Figures 1-203 to 1-209.

Position is always specified for features of size, and datums must be specified. Position may be specified to control feature locations, relationships, coaxiality, concentricity, and symmetry of cylindrical and non-cylindrical features of size. Because position is specified for size features, the modifier principles must be considered. The datum reference frame becomes an important concept when position is specified in relationship to datums. Position is always specified in conjunction with Basic dimensions from specified datums or between interrelated features. Basic dimensions establish true position.

Positional tolerance may be explained either in terms of the internal surface of a hole, slot, etc., or in terms of the axis, center plane or center line. This section explains position in terms of feature centers. The specified positional tolerance defines a zone within which the center of the feature of size is permitted to vary from true position (theoretically exact) in relation to another feature or datum. True position is established by Basic dimensions from specified datum features and between interrelated features. True position is an axis, center line, or center plane of a feature as defined by Basic dimensions. The tolerance specified in the feature control frame is either a diameter or a width located equally around true position.

The tolerance zone may use geometric modifiers. Application of the modifiers allows the specified tolerance to increase an amount equal to the features size departure from Maximum Material Condition (MMC) or Least Material Condition (LMC).

Position Theory

Design requirements for assemblies with interfacing mating parts usually relate one feature to another. These features—holes and pins, or holes for floating fasteners—relate to each other in 360 degrees of location to each other. The coordinate dimensioning method of tolerancing does not permit 360 degrees flexibility. The coordinate dimensioning system makes no provisions for feature size variations in relation to feature location. GD&T takes into consideration both size and location when determining feature or part acceptability.

Position Theory Application

The following is an explanation of the reasoning behind the use of position tolerancing versus the coordinate dimensioning system. For discussion and illustration purposes, consider the requirements in Figure 1-203 for a .530 inch plus .008 inch, minus .002 inch diameter hole in a part. The hole is to be located within plus or minus .010 to the center of the part. This translates to a .020 square tolerance zone around true position, which the actual axis of the hole must lie within.

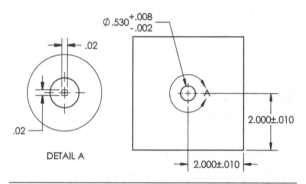

Figure 1-203. Square tolerance zone

Axis Location

The axis of the hole must lie within the square tolerance zone of .020 to be an acceptable part. To further expand upon this, assume that five parts were inspected, and the center point axis of each hole is illustrated in relation to the .020 tolerance zone in Figure 1-204. The illustration shows that two parts are acceptable and three are out-of-tolerance. Note in Figure 1-204, only the tolerance zone is depicted.

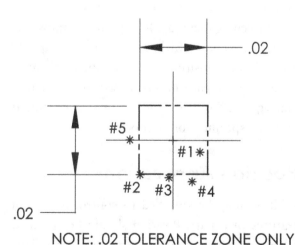

NOTE: .02 TOLERANCE ZONE ONLY

Figure 1-204. Parts rejected due to square tolerance zone

Diagonal Measurement

The concept behind positional tolerancing is to use a circular tolerance zone rather than a square one. The circular zone allows for 360 degrees of axis or feature movement. To achieve the circle, the diagonal measurement from center to a corner becomes the radius for a circular zone. In this illustration, the total diagonal measurement is .028 inch. Therefore, if part number two is acceptable, then part three should also be acceptable because it is closer to true position. By specifying a circular tolerance zone, four of the five parts are acceptable. Figure 1-205 illustrates the two tolerance zones. The shaded area indicates the amount of tolerance gained with the circular tolerance zone.

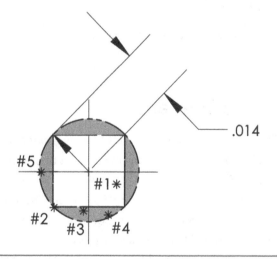

Figure 1-205. Tolerance gained with a circular tolerance zone

Circular Tolerance

In Figure 1-206, the illustrated approach to tolerancing appears to be more logical, but another consideration must be made. The designer intended for the holes to vary in the X and Y directions, not diagonally. Therefore, the circular tolerance zone must not exceed the original intention of .010 inch. The tolerance zone must be an inscribed circle .020 inch in diameter. Figure 1-206 illustrates the proper comparison between the coordinate and geometric tolerance zones. This illustration appears to have reduced the allowable variation in feature location; in fact, it has for a feature of minimum size. When the modifier principles are considered, the tolerance is not lost, but gained.

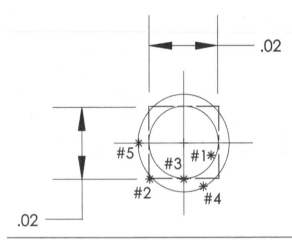

Figure 1-206. Square and circular tolerance zone comparisons

Bonus Tolerance

The major advantage of GD&T is a greater tolerance based on actual mating size. In this example, assume parts four and five were produced at the largest permissible size, .538 inch. The preceding .060 inch tolerance zone has increased to .070 inch. This practice is possible with the application of the modifier principles. The ASME Y14.5M-1994 standard states, in part, that "the tolerance is limited to the specified value when the feature is at MMC. Where the actual mating size departs from MMC, an increase in tolerance is allowed equal to the amount of departure." Figure 1-207 illustrates how the bonus tolerance zone accepts parts four and five when they are made larger than the specified MMC size. All of the parts are acceptable if the axes remain in this .070-inch tolerance zone for the thickness of the part. This tolerance zone can be considered to be a tube (see Figure 1-209). The axis of the controlled feature must lie in the tube in any shape or manner as long as it is

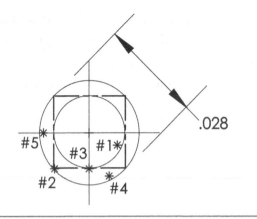

Figure 1-207. Gained tolerance allows more acceptable parts

in there. For non-cylindrical features, the zone would be like a box or rectangle.

Position Definition

Position is the condition where a feature or group of features is located (positioned) in relation to another feature or datum feature.

Position Tolerance

Location tolerance zones are either cylindrical or non-cylindrical; this determination is made in the feature control frame. The tolerance zone is cylindrical if the diameter symbol precedes the specified tolerance. The absence of the symbol indicates a width zone. The controlled feature(s) axis or center line must lie within the allowable tolerance zone, which is distributed equally around true position. The specified tolerance zone may increase in size based on the actual mating size. Tolerance zone increase is permitted with the specification of modifiers.

Position Application

Position may be specified to control nearly all features of a part. Position should be specified whenever the design requirements permit. This control provides an opportunity to utilize many of the advantages of GD&T. Figure 1-208 illustrates a simplified application of position.

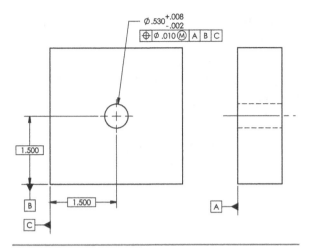

Figure 1-208. A simplified application of position

Position Tolerance Interpretation

The tolerance in this example is .030 in diameter when the feature (hole) is at MMC, .528 inch. The tolerance zone is cylindrical because the designer specified the diameter symbol preceding the tolerance. Figure 1-209 illustrates how the tolerance zone is determined. The tolerance zone in this position example is permitted an increase in size by .010 inch. This is available based on the feature plus/minus size tolerance. The tolerance zone diameter is determined by actual mating size. Before the feature is produced, only the minimum and maximum tolerance zone sizes can be determined. If this example contained a group of features, each of them may have a different size tolerance zone based on the actual mating size.

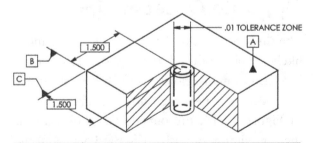

Figure 1-209. Cylindrical/tubular tolerance zone

Positional tolerance specified at MMC may be explained in terms of the surface of a hole or the axis of a hole.

1. *In terms of the surface of a hole.* While maintaining the specified size limits of the hole, no element of the surface is permitted to be inside the theoretical boundary located at true position.

2. *In terms of the axis of a hole.* Where a hole is at MMC (minimum diameter), its axis must lie within a cylindrical tolerance zone whose axis is located at true position. This zone diameter is equal to the positional tolerance specified for the hole.

Note: In certain cases of extreme form deviation (within limits of size) or orientation deviation of the hole, the tolerance in terms of the axis may not be exactly equivalent to the tolerance in terms of the surface. In such cases, the surface interpretation shall take precedence. The length of the tolerance zone is equal to the feature length, unless otherwise specified.

Position tolerance in this book is explained in terms of the axis. In any case, the MMC-LMC surface interpretation take precedence over the axis interpretation.

Position of Multiple Cylindrical Features

The advantages of GD&T and position can best be explained when two or more features are positioned. When a group of features are to be controlled with position, a tolerance for location is specified and appropriately modified. The feature is then related to something, which may be an edge, surface, or other feature(s); these are the datum reference features. Specification in this manner ensures a clearer intent of the design. The datum reference frame is utilized for the relationships as required for part function. The rules are applied as required for the specified controls and controlled features. The geometric control—position—brings all of these concepts together better than any of the other controls. An illustration of how multiple features are specified and controlled with position is shown in Figure 1-210.

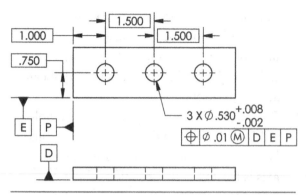

Figure 1-210. Multiple features are specified and controlled with position

Feature Location

The effects of feature size, MMC, and virtual condition can best be illustrated when the explanation begins with the features perfectly located at MMC. Figure 1-211 illustrates these holes with the gauge inserted. Note that the gauge pins are perfectly centered in the holes when they are in this condition. There is clearance between the pins and holes because the gauge is at virtual condition. For this multiple-feature example, the gauge pins are .498 in diameter. When all features are located perfectly as they are here, there will be an equal distance between each gauge pin and the hole. This gauge is simulating the fit or assembly of a mating part.

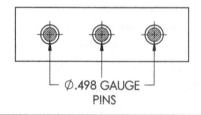

Figure 1-211. Holes with gauge inserted

Opposite Offset at MMC

The gauge will allow all acceptably located features regardless of their shape, location and size (must be verified separately). Each feature may be off true position in a different direction and

be of different sizes. Figure 1-212 illustrates a possible hole arrangement. These features are at their maximum tolerance location when at MMC. The axis of each feature is .010 inch from true position.

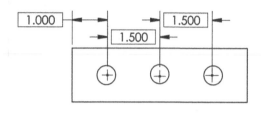

NOTE: ALL HOLES ARE .528
DIAMETER AND OFFSET .010

Figure 1-212. Hole center axis offset at MMC

The gauge will accept this part when the features are offset in various directions. Figure 1-213 illustrates the part and gauge. Note there is line contact between the gauge and part.

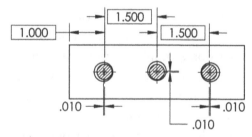

NOTE: HOLES ARE ALL .528 DIAMETER
GAUGE PINS ARE .518 DIAMETER

Figure 1-213. Gauge allows acceptance for hole centers offset at MMC

Opposite Offset at LMC

The three features (holes) in the figure are more likely to be produced at some size larger than MMC. Normally, holes and other features are produced at or near their nominal size or larger. In Figure 1-214, the outside two holes are at their maximum opposite offset at LMC, and the center one is at the maximum vertical offset at MMC. When the holes are produced at LMC, as they are in this figure, the advantage is maximum tolerances (or bonus tolerance). Here an internal feature is produced at its largest specified size. In this condition, the additional feature size increase beyond MMC may be added to the specified location tolerance. Therefore, for the two outside holes, the location tolerance is .020 inch, a bonus of .010 over MMC.

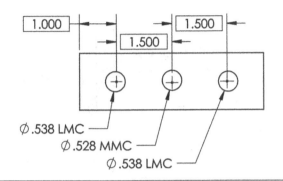

Figure 1-214. Holes produced at LMC vs. MMC

The gauge will accept the part in Figure 1-213. All features are within specified size and location tolerances. Figure 1-215 illustrates the part and gauge. Note that there is a tangency line contact between the gauge and features. The examples illustrated in Figure 1-212 and 1-214, or any other conceivable combination, illustrate part acceptance when manufactured to the specifications. The LMC method is used to protect wall thickness on parts like thin sheet-metal parts or thick castings. Again, the feature axis must lie within the tube cylindrical zone of tolerance for the thickness of the part.

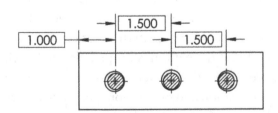

NOTE: GAUGE PINS ARE .518 DIAMETER

Figure 1-215. Gauge allows acceptance for hole centers offset at LMC and MMC

Patterns Positioned from A Datum of Size

Positional tolerances may be specified to control a pattern of features in relationship to another feature. This type of specification is used when pattern-to-feature relationship is more critical than pattern-to-edge. The single feature in this type of application becomes the locating feature for the pattern. Most frequently, the locating feature is a feature of size. Features of size must have a modifier specified for them in the feature control frame. These locating features are also controlled by Rule Four, the datum/virtual condition rule. The specification of modifiers in these designs provides the full advantages of GD&T.

Regardless of Feature Size

A locating feature (datum) for a related pattern is shown in Figure 1-216. In this example, the locating datum feature for the pattern is modified to *regardless of feature size* (RFS). This is restrictive, but some designs do require the restriction based on the function of the final assembly.

To begin, the center hole must be located as specified by the basic dimensions and feature control frame. The hole is to be produced at 1.500 inches, plus .008 inch or minus .002 inch. The hole is allowed a .010 inch diametral tolerance zone for location. In this example, the positional tolerance zone for the hole is modified to MMC. This means that the hole has a .010 inch tolerance when it is produced at 1.498. If the hole is produced at 1.508 inches, the tolerance zone increases to .020 inch. In this case, the hole was measured to be 1.502 inch, so the positional tolerance is .014 inch.

Next, the pattern of eight holes must be produced in relationship to the center hole, which has become datum feature X. Datum feature X is implied RFS because no modifier is in the feature

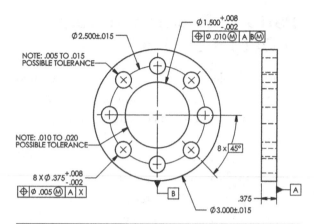

Figure 1-216.   Pattern located regardless of feature size

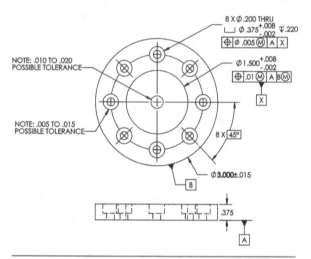

Figure 1-217.   Possible tolerance zones for pattern

control frame for the eight holes. These eight holes are to be .375 inch, plus .008 inch or minus .002 inch. They each have a positional diametral tolerance of .005 inch at MMC. Figure 1-217 illustrates the possible tolerance zones for all of the features.

In this example, as presented with datum feature X at RFS, the pattern shift is restricted to zero. The shift is zero because regardless of the feature location or size, it becomes the dimension origin for the true position of the pattern of eight holes.

Pattern Tolerance

When a pattern of features is located from an RFS datum, the pattern is not permitted any shift from the datum feature. Here the datum feature

axis becomes the origin for the pattern-locating dimensions. Figure 1-218 illustrates the dimensioning for the pattern from datum feature X. All of these dimensions must be basic.

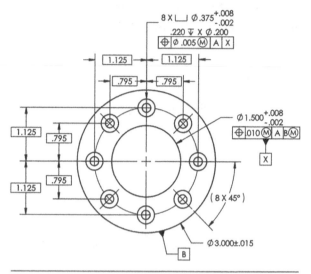

Figure 1-218.   Basic dimensioning from datum X

These basic dimensions establish the true position locations for the pattern. When the basic true position of the pattern in relationship to datum feature X is established, the individual tolerance for each feature in the pattern allows variation from true position. Each feature has an individual tolerance based on the actual mating size. This tolerance may vary from .005 inch to .015 inch because of the MMC modifier following the .005 inch positional tolerance. Each feature in the pattern is permitted to shift or vary. Verification of this part would require the establishment of the datum axis. This axis must meet the drawing specifications. After verifying an acceptable datum axis, proper pattern location must be measured. If the pattern is properly located, each feature within that pattern must be verified for proper location and orientation (perpendicularity to datum surface A). Verification may be accomplished, depending on required accuracy, with a coordinate measuring machine, paper gauging, or one hard gauge that has an adjustable pin to fit the datum feature. The pin must be adjustable because of RFS.

Maximum Material Condition

When any of the common datums in multiple patterns of features are specified at MMC, there is an option whether the patterns are to be considered as a single pattern or as having separate requirements. If no note is added under the feature control frames, the patterns are treated as one. When the patterns are to be treated individually, the notation SEP REQT (separate requirement) is placed beneath each feature control frame. The MMC modifier specified for a datum feature of size permits the pattern to vary depending on datum size. When the datum feature is at MMC, the pattern variation is restricted. However, as the datum feature increases in size, the amount of pattern orientation shift/rotation also increases. Tolerancing patterns in this manner are more common than RFS. Figure 1-219 specified positional tolerances at MMC for the same part discussed with RFS.

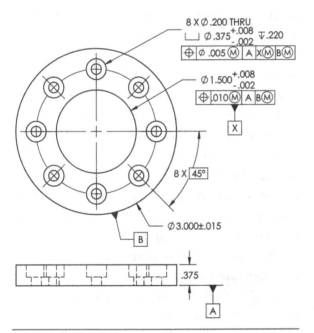

Figure 1-219. Positional tolerance at MMC

The center hole or datum feature X must be located with basic dimensions. The hole must meet the specified size and location requirements. The hole size may vary from 1.498 inch to 1.508 inch, with a positional tolerance that

is .010 inch at MMC to .020 inch at LMC. For this example, the hole measures 1.502, which is .004 larger than MMC, plus the .010 positional tolerance, equals a .014 positional tolerance. The pattern of eight holes must be produced in relationship to the axis of datum feature X. The feature control frame for the eight holes contains datum reference letter X with an MMC modifier. This modifier permits some pattern shift/rotation, an amount equal to the feature size departure from MMC. This amount of departure is also the amount the pattern may vary from true position. Figure 1-220 illustrates the available pattern shift/rotation tolerance. The illustration contains a gauge pin at virtual condition for datum feature X. This datum feature is controlled by Rule Four, datum/virtual condition. The gauge pin measures 1.488 in.

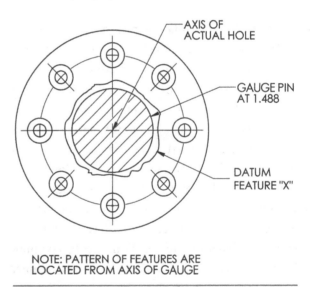

NOTE: PATTERN OF FEATURES ARE LOCATED FROM AXIS OF GAUGE

Figure 1-220. Pattern of features are located from axis of gauge

The pattern shift/rotation is achieved by the amount of clearance there is between the hole and gauge pin. The part may be shifted or rotated all around the gauge pin in relationship to the other datum features (part edges). The eight holes in the pattern establish a true position in relationship to the gauge axis. When the pattern orientation is established from the simulated datum axis, each of the eight features has an individual positional diametral tolerance of .005 inch at MMC. This tolerance permits each feature variation based on actual mating size. Figure 1-221 illustrates individual feature tolerance in relationship to the basic pattern orientation.

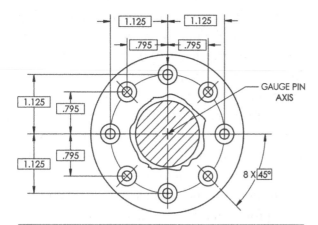

Figure 1-221. Feature tolerance in relationship to the basic pattern orientation

The MMC modifier permits the pattern some shift/rotation in relationship to the datum feature. Then each individual feature in the pattern is permitted variation from true position. The amount of variation is dependent on the actual mating size. For example, if the datum feature is produced at the virtual condition size, there is no pattern shift/rotation permitted. Likewise, if any or all of the eight features were produced at virtual condition, they would have to be located at their true position. Parts that have patterns of features related to another datum feature(s) may be verified with a coordinate measuring machine, paper gauging, or gauging. Setup time and accuracy required may determine the method of verification. Each of these methods provides advantages and disadvantages. The datum feature must be verified first with any method. Separate gauging is required to determine acceptable location. Then, from the actual datum feature axis, the pattern location is determined. Another gauge is required to verify pattern relationship to the datum feature axis.

Summary

Location tolerancing is the most advantageous part of GD&T. When location tolerances are properly specified, the rules, modifier principles datum references and feature control frames can be applied to their greatest advantages. It is through proper application and interpretation that parts are 100 percent interchangeable, production yields increase, tolerances are maximum, functional gauging techniques can be implemented and production costs are reduced, among many more advantages.

Location tolerances are the specified amount of variation from true position of another feature. True position is established by basic dimensions from specified datum features. The specified tolerance provides a cylindrical or non-cylindrical zone for the depth or length of the controlled feature. This zone is permitted to increase in size as the feature size varies from MMC. The actual feature axis or center line must lie within the zone. The axis or center line may take any form as long as it remains within the zone.

Position is the most frequently applied location tolerance. Concentricity should seldom be specified or it should be limited to features requiring dynamic balance. Position may be specified for nearly all features that require mating part assembly or interchangeability. This section should be studied to make certain the concepts are understood. Concentration on how the various concepts of GD&T are interrelated is important to the reader.

For full details and practical examples, refer to *Interpretation of Geometric Dimensioning and Tolerancing, Based on ASME Y14.5-2009*, Puncochar & Evans 2011.

End of excerpt of *Interpretation of Geometric Dimensioning and Tolerancing, Based on ASME Y14.5-2009*.

Part 5, GD&T Tolerances of Location, Study Questions

1. The geometric tolerances of location are _____ and _____.

2. Tolerances of location specify the amount of _____ permitted between features and datum features or between features.

3. Concentricity is always specified or implied _____.

4. The tolerance zone for position is either _____ or _____ as determined from the feature control frame.

5. True position is specified by _____ dimensions.

6. Datum feature references must be _____ on drawings.

7. For an internal feature, position tolerance at MMC _____ as the feature approaches LMC.

8. When a composite feature control frame is specified to locate a pattern of features in relationship to a datum reference frame, the first line of the feature control frame controls pattern _____.

9. Concentricity tolerance is always specified in relationship to another datum _____.

10. Position tolerance zones extend to the _____ or _____ of the controlled feature.

11. Datum features of size are considered at their _____ size when verifying patterns located from them.

12. Positional tolerances specified for features at _____ lend themselves to functional gauging verification.

13. Position tolerances specified at _____ provide the greatest advantages of GD&T.

14. Positional tolerances are specified for features of _____.

15. Location tolerancing is a _____ method of controlling part features.

Deburring Tools and Techniques

Objectives

1. Practice the safe use of files and other deburring tools.
2. Locate deburring requirements on engineering drawings.
3. Identify different methods for using files.
4. Select the proper tool for deburring of machined holes.
5. List other methods used for deburring machined parts.

Deburring Techniques

Deburring techniques are the methods used to remove sharp edges (burrs) left after machining of adjacent surfaces and is sometimes called *edge finishing*. Removal of these burrs is best done during the machining operation that created them by using a rotating chamfer tool, center drill, spot drill or countersink for holes, while the part is clamped. Due to the lack of access to the back side of the part, it will still require hand deburring. In some heavy production operations, robots are used to perform edge finishing. Deburring is critically important between operations to ensure positive part location and mar-free clamping of finished surfaces. The thickness of a burr will affect location accuracy if it is not removed prior to placement.

Deburring Safety

Handling of machined parts can be dangerous due to the sharp edges that are created at adjacent surfaces when metal is removed. Always use care when handling freshly machined parts. When using deburring tools, be sure to use them properly. Some basic rules for safely using deburring tools are:

- Never use a file without a handle.
- Do not use a file as a pry bar. Files are made of hardened steel but are brittle and could break, creating the potential for injury.
- Never use an air nozzle to clean a file; instead, use a file card.
- Use a bench vise to hold large parts.
- When using an edge finishing tool, be sure to move the tool in a direction away from your body to avoid jabbing yourself.

Deburring Requirements

Edge break specifications (Figure 1-222) are found in the tolerance block of the engineering drawing or sometimes in notes. This information should be followed to avoid scrapping a part by removing too much material from part edges.

UNLESS OTHERWISE SPECIFIED:		
DIMENSIONS ARE IN INCHES		
TOLERANCES:		
FRACTIONAL		
ANGULAR		
XX		
XXX		
BREAK ALL EDGES .015 MAX		
GD&T PER ASME Y14.5 2009		
MATERIAL:		
SURFACE FINISH:		
DO NOT SCALE DRAWING		

Figure 1-222. Edge break specifications

Hand Tools Used for Deburring

Files

The file type most commonly used in machine shops is the mill file (Figure 1-223) with a tapered shape, single-cut tooth pattern and a bastard coarseness designation. Other file shapes commonly used are flat, 3-square (triangle), half round and round in varying lengths.

Figure 1-223. Mill file

Note: Single-cut files are designed to cut in one direction during the forward stroke. Applying pressure on the back stroke will eventually wear the file's cutting edge and probably make the surface finish worse. There are basically three techniques used: flat, edge and draw filing. The figures below show the proper methods for each of these.

Flat filing is performed by laying the face of the file flat on the surface being worked. This is best performed before edge filing. Lay the file at an angle to the surface and push across the surface keeping even downward pressure on the file (see Figure 1-224 and 1-225).

Figure 1-224. Flat file start position

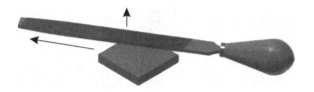

Figure 1-225. Flat file end position

Note: When the final surface finish is a priority, remember that flat-filing may leave unwanted marks on the surface.

Edge filing is used on the edges of the part. Position the file at approximately 45-degree angle to the two perpendicular surfaces while holding the part (see Figure 1-226). Make a smooth stroke across the entire length in one motion. It may take a couple of tries to break the edge depending on the material. Rotate the part so as to reverse the stroke direction and repeat the process. Carefully use a rag to remove dust shavings and check the results. Repeat this for all affected edges. (see Figure 1-227)

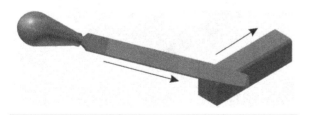

Figure 1-226. Edge filing start position

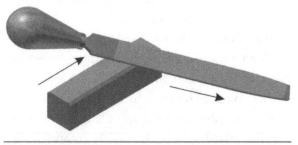

Figure 1-227. Edge filing end position

Draw filing is used to finish a surface, to make it flat and to create a nice surface finish. Place the file perpendicular to and flat on the surface being worked, hold the file with your thumbs on top of the file on either side of the work (see elliptical shapes on Figures 1-228 and 1-229). Make a stroke across the entire length of the part with even downward thumb pressure applied. In this case, the pressure can be applied on the back stroke as well. Always take care not to remove too much material as to affect dimensional tolerances.

Figure 1-228. Draw filing start position

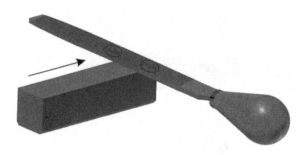

Figure 1-229. Draw filing end position

Swiss pattern and jewelers, commonly called *needle files*, are used for fine detailing work done by jewelers and tool and die makers. These are available in many shapes and cut styles and their small size (6¼ inches) make them especially useful for hard to reach features.

Proper File Care

A file should be cleaned after each use with a file card (Figure 1-230) to remove metal shavings from the file. Hold the file by the handle at an angle with the tip resting on a worktable (Figure 1-231). Use the wire-brush side to push the brushes across the teeth at the same angle to remove the shavings material.

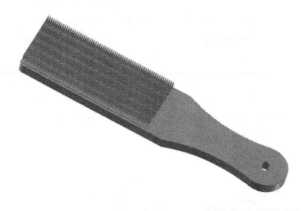

Figure 1-230. File card

Figure 1-232. Rout-A-Burr tool

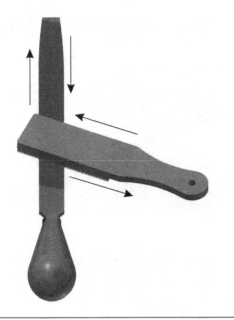

Figure 1-231. File cleaning motion and direction

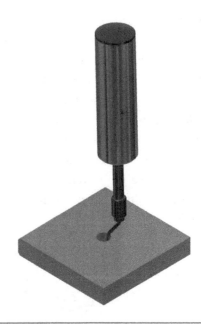

Figure 1-233. Rout-A-Burr hole deburring

Rout-A-Burr is a trade name for a hand deburring set that includes several different tips that fit into a handle that is held in the hand with the tip against the part edge and then scraped across or rotated around holes to remove the burr (Figures 1-232 and 1-233). Multiple tip configurations are available and easily exchanged including triangle-shaped knifelike edges that come to a point.

Another tool for deburring holes is countersinks installed into a handle (Figure 1-234) and used as shown in Figure 1-235.

Sometimes a simple secondary operation machine (like a benchtop drill press) is devoted to these types of operations where a countersink tool is mounted into the spindle chuck and the part is pressed against the slowly rotating tool.

Figure 1-234. Hand countersink

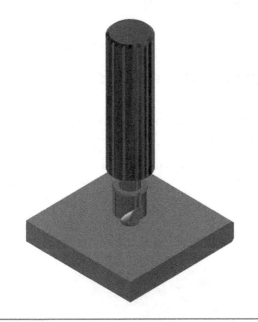

Figure 1-235. Hole deburring application

Other Deburring Methods

Vibratory tumbling machines are an excellent deburring device that range in size from bench top for small parts and quantities, to large tub type for large parts and capacities. These tumblers produce a light cutting action by shaking a part tub at a high rate to create vibration (sometimes variable vibration speeds are available). Special tumbling media of different shapes and sizes are used in the tub, then the parts are added and the machine is started. The parts and media rub against each other during this vibrating motion, thus removing sharp edges. This process can be done with or without water. The movement of the parts and tumbling media removes sharp edges from all types of geometric features.

Pneumatic die grinding tools are often used in Mold and Die work because of their high rotational speed capacities and a multitude of carbide burr and grinding stone tool options. These are available in many configurations (straight, right-angle and pencil) that enable accessing many hard to reach spots on parts that need finishing work.

Belt and disk sanders can be used but are usually limited to raw material preparation in precision machine shops. They are capable of aggressive metal removal and are widely used to prepare weldment parts.

Robots are being utilized for deburring and polishing in high production environments for large and complex components.

Summary

A few methods for deburring machined parts have been presented here. As stated before, the best practice is to make post process deburring unnecessary by combining the operation within the CNC machining program when possible. Always be on the lookout for and research new methods and never underestimate the adverse effect that poor edge finishing can have on the quality of your work.

Deburring Tools and Techniques, Study Questions

1. File handles are:
   a. Required for safe use
   b. Not necessary for safe use
   c. For convenience only
   d. A good alternative for a pry bar

2. Burrs are sharp edges created by machining. Name three tools that can be used for removing burrs.

3. Files should be cleaned with:
   a. A rag
   b. Pressurized air
   c. A file-card
   d. Alcohol

4. Edge break requirements are:
   a. Sometimes listed in notes on the engineering drawing
   b. Listed in the tolerance block of the engineering drawing
   c. Both a and b
   d. Determined by the shop lead

5. The best practice for removing burrs is:
   a. After machining has been completed
   b. By using a vibratory tumbler
   c. During the machining process with programmed tool paths and appropriate cutting tools
   d. With a knife

Deburring Tools and Techniques, Practical Exercises

Perform different filing techniques on machined parts.

Perform file cleaning.

Perform use of a Rout-A-Burr tool on edges and holes.

Perform use of a hand countersink to deburr holes.

Machined Part Inspection

Objectives

1. Learn common types of measuring tools used in a machining environment.
2. Demonstrate how to read a vernier caliper scale.
3. Demonstrate how to read a micrometer.
4. Demonstrate how to read a vernier micrometer.
5. Properly use a thread micrometer.
6. Properly use a depth micrometer.

Measuring Tools for Specific Applications

Measuring tools are used to ensure accuracy from the beginning of the manufacturing process to the final product. The following tools are commonly used in the machining environment. There are numerous other tools for specialized applications beyond the scope of what is covered in this section. It is strongly suggested that you research all measuring tools available and select what is best suited for your situation.

The following text and graphics have been provided by "Tools and Rules for Precision Measuring" bulletin #1211, L.S. Starrett Company.

Steel Rules and Related Tools

The rule is a basic measuring tool from which many other tools have been developed. Rules range in size from as small as one–quarter inch in length for measuring in grooves, recesses and keyways to as much as twelve feet in length for large work. Steel rules are graduated in the English (inch) or Metric system and sometimes scales for both systems are provided on a single rule. They can be graduated on each edge of both sides and even on the ends. English graduations are commonly as fine as one-hundredth (.010) inch in decimals or one sixty-fourth (¹⁄₆₄) inch in fractions. Metric graduations are usually as fine as one-half millimeter (0.50 mm). Starrett rules are graduated to agree with standards calibrated by the National Institute of Standards and Technology (NIST) (Figure 1-236).

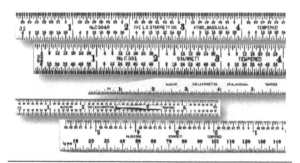

Figure 1-236. The steel rule is one of the basic measuring tools. Various types in english and metric are shown, *courtesy L.S. Starrett Company*

There are different types of steel rules. For instance, the 6–inch rule is the most convenient length for carrying. A spring tempered rule is desirable since it is both thin and flexible yet has ample stiffness to provide a straight measuring edge. Small rules are available with a tapered end for measuring in small holes, narrow slots, from shoulders, etc. The hook feature is available on many rules. It provides an accurate stop for setting calipers, dividers, etc., and it can also be used for taking measurements where it is not possible to be sure that the end of the rule is even with the edge of the work (Figures 1-237 and 1-238).

Figure 1-237. Starrett H604R–6 spring tempered, 6 inches long, *courtesy L.S. Starrett Company*

Figure 1-238. Starrett No. C331 150 mm long steel rule, *courtesy L.S. Starrett Company*

Vernier Slide Calipers

Vernier slide calipers are the oldest type and most accurate slide calipers. They require more skill to use than dial or electronic types and are described below in the Vernier Tools section.

Vernier Tools

The Vernier was invented by a French mathematician, Pierre Vernier (1580–1637). The Vernier caliper consists basically of a stationary bar and a movable Vernier slide assembly. The stationary rule is a hardened graduated bar with a fixed measuring jaw. The movable Vernier slide assembly combines a movable jaw, Vernier plate, clamp screws and adjusting nut. The Vernier slide assembly moves as a unit along the graduations of the bar to bring both jaws in contact with the work. Readings are taken in thousandths of an inch by reading the position of the Vernier plate in relation to the graduations on the stationary bar. Starrett Vernier gauges feature 50 divisions, with widely spaced, easy-to-read graduations allowing fast, accurate and simplified readings that can be read without a magnifying glass. The Vernier principle is applied to many Starrett tools such as Vernier Height Gauges, Vernier Depth Gauges, Vernier Protractors, Gear Tooth Vernier Calipers, etc. (see Figures 1-139 and 1-140).

How to Read Vernier Calipers (Inch)

The bar is graduated into twentieths of an inch (.050 inch). Every second division represents a tenth of an inch and is numbered (see Figure 1-241).

The Vernier plate is divided into fifty parts and numbered 0, 5, 10, 15, 20, 25…45, 50. The fifty (.050 inch). Every second division represents a tenth of an inch and is numbered (see Figure 1-142).

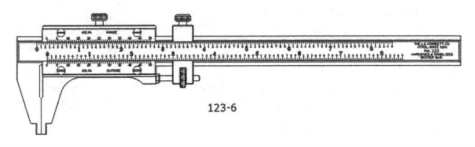

Figure 1-239.   Vernier caliper

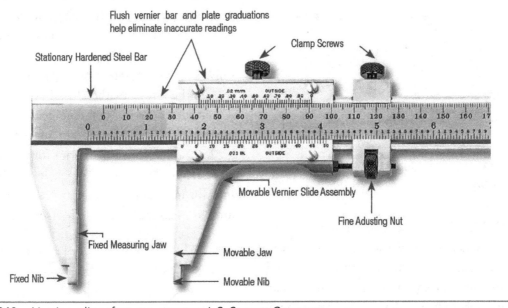

Figure 1-240.   Vernier caliper features, *courtesy L.S. Starrett Company*

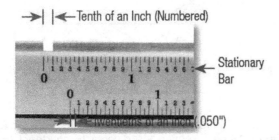

Figure 1-241.   Vernier graduations, *courtesy L.S. Starrett Company*

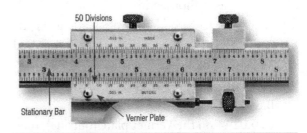

Figure 1-242.   Vernier stationary bar and plate, *courtesy L.S. Starrett Company*

The difference between the width of one of the fifty spaces on the Vernier plate and one of the forty–nine spaces on the bar is therefore 1/1000 of an inch (1/50 of 1/20). If the tool is set so that the 0 line on the Vernier plate coincides with the 0 line on the bar, the line to the right of the 0 on the Vernier plate will differ from the line to the right of the 0 on the bar by 1/1000; the second line by 2/1000 and so on. The difference will continue to increase 1/1000 of an inch for each division until the 50 on the Vernier plate coincides with the line 49 on the steel rule.

To read the tool, note how many inches, tenths (or .100) and twentieths (or .050) the mark on the Vernier plate is from the 0 mark on the bar. Then note the number of divisions on the Vernier plate from the 0 to a line which EXACTLY COINCIDES with a line on the bar (see Figure 1-243).

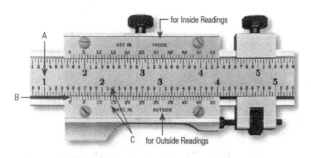

Figure 1-243. Vernier reading example, *courtesy L.S. Starrett Company*

EXAMPLE: In the above illustration for outside measurements the Vernier plate has been moved to the right one inch and four hundred and fifty thousandths (1.450") as shown on the bar, and the fourteenth line of the Vernier plate exactly coincides with a line, as indicated on the illustration above. Fourteen thousandths of an inch (.014") are, therefore, to be added to the reading on the bar and the total reading is one and four hundred and sixty–four thousandths inches (1.464").

    A. 1.000" on the bar

    B. 0.450" also on the bar, and

    C. 0.014" on the Vernier plate (outside), then

    D. 1.464" is your measurement

Vernier Gauge, Inside and Outside Measurements

If you are using a Starrett No. 123 (English) Vernier Gauge, the same procedure is used for obtaining inside measurements as for outside measurements, however the top row of measurements is to be used (see Figure 1-244).

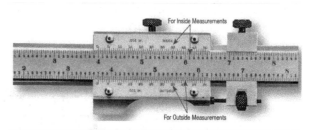

Figure 1-244. Vernier inside and outside measurements, *courtesy L.S. Starrett Company*

The top row of measurements is for inside work and the bottom row is for outside work. The bottom row on the bar and Vernier plate have inch graduations. Both these scales can be used for direct reading for outside dimensions only. For inside measurements, it is necessary to add the width of the closed contacts to your dimension to arrive at the correct and complete measurement. The minimum measurement "A" is .250" (6.35 mm for metric) for a Starrett No. 123E&M–6" Vernier Caliper. It is .300" (7.62 mm for metric) for a Starrett No.123E&M–12" and No. 123E&M–24" series (see Figure 1-245).

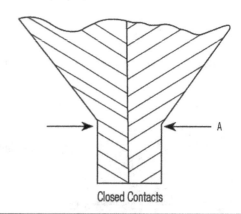

Figure 1-245. Vernier closed contacts, *courtesy L.S. Starrett Company*

Fine Adjustment

After bringing the measuring contacts to the work by sliding the movable jaw along the steel rule, tighten the fine adjustment clamp screw. Rotate the fine adjusting nut to bring the measuring contacts into final position against the work. Tighten the lock screw to clamp the sliding Vernier scale into position (see Figure 1-246).

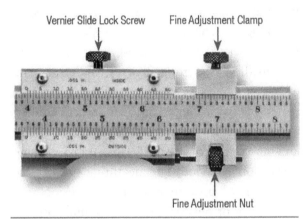

Figure 1-246. Vernier fine adjustment, *courtesy L.S. Starrett Company*

Vernier Height Gauges

Like the Vernier caliper, the Vernier height gauge consists of a stationary bar or beam and a movable slide. The graduated, hardened and ground beam is combined with a hardened, ground and lapped base. The slide Vernier assembly can be raised or lowered to any position along the bar and adjusted in thousandths of an inch by means of the Vernier slide fine adjusting knob. Set up as shown right in Figure 1-247 with hardened steel scriber clamped to the Vernier slide, the Vernier height gauge is used on a surface plate or machine table to mark off vertical distances and locate center distances. Other accessories include depth attachments, tungsten–carbide scribers, offset scribers, test indicators, electronic transducers and attachments that allow use with many dial indicators. Beam and Vernier plate graduations are identical with the outside scale of the Vernier caliper and readings as described above (see Figure 1-248).

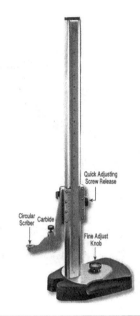

Figure 1-247. Vernier height gauge, *courtesy L.S. Starrett Company*

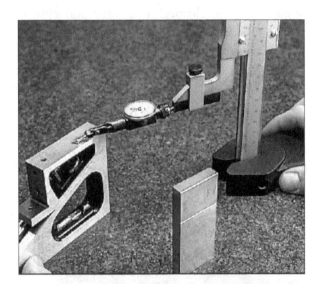

Figure 1-248. Last word® dial test indicator attached to vernier height gauge, *courtesy L.S. Starrett Company*

Granite Surface Plates

Every linear measurement depends on an accurate reference surface from which final dimensions are taken. Starrett Precision Granite Surface Plates provide this reference plane for work inspection and for work layout prior to machining. Their high degree of flatness, overall quality and workmanship also make them ideal bases for mounting

Figure 1-249. Surface plates and parallels, *courtesy L.S. Starrett Company*

sophisticated mechanical, electronic and optical gauging systems (see Figure 1-249).

Gauge Blocks

The international standards of length (the meter and international inch) are established in terms of light waves using the definition of the speed of light and atomic clocks (Figure 1-251).

Light waves, of course, cannot be handled like a micrometer or Vernier caliper, but they are used to establish the length of physical standards having accuracies in millionths of an inch (.000001"). These standards are called gauge blocks. Precision gauge blocks are the primary standards vital to dimensional quality control in the manufacture of interchangeable parts. These blocks are used for calibrating precision measuring tools and for setting numerous comparative type gauges used in incoming, production and final inspection areas. Gauge blocks provide the most economical, accurate method of setting dial test indicators and electronic gauges used in conjunction with surface plates for inspecting parts with exacting tolerances. Essentially, they consist of blocks of a hard and stable material with a flat, parallel gauging surface on each end. The measuring surfaces are ground and lapped to an overall dimension with a tolerance of plus or minus a few millionths of an inch. Gauge blocks may be stacked (or wrung) together to form accurate standards of practically any length (Figure 1-251).

Figure 1-251. Starrett–Webber gauge block set, *courtesy L.S. Starrett Company*

Slide Calipers

Slide calipers are essentially an extension of the steel rule that ensures greater accuracy by aligning the graduated scale with the edges or points to be measured. They add a pair of jaws (head)

Figure 1-250. Starrett–Webber gauge blocks, *courtesy L.S. Starrett Company*

to the rule—one jaw fixed at the end and the other movable along the scale. The slide is graduated to read inside or outside measurements. They have a knurled clamping screw to lock the slide to the desired setting. The configuration allows the tool to be operated with one hand, leaving the other hand free to hold the workpiece (Figure 1-252).

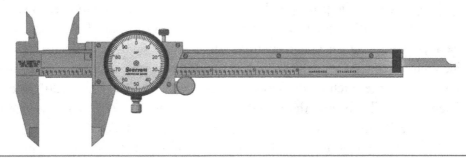

Figure 1-252. 120 6" Dial caliper, *courtesy L.S. Starrett Company*

Dial Slide Calipers

A most versatile and easy to read instrument, this four-way stainless steel dial caliper has knife edge contacts for inside and outside measurements, and a rod connected to its slide for obtaining depth dimensions. The rod contact is cut out to provide a nib for gauging small grooves and recesses. By placing the front end of the reverse side of the movable jaw against the edge of a work piece, parallel lines may be scribed against the front end of the fixed jaw. All readings are taken directly from the dial indicator and the bar, which have sharp clear dial graduations of .001" or 0.02mm – .100" or 2 mm in one revolution), and .100" or 1mm on the satin finished bar. Measurements may be made with one hand; a thumb roll being provided for fine adjustment. Knurled thumb screws lock the movable jaw and adjustable indicator dial at any setting. With the addition of a depth attachment, the dial caliper becomes a convenient and easy-to-use depth gauge. Dial calipers are available in 6", 9" and 12" ranges (150 mm, 225 mm and 300 mm) with depth capabilities (Figures 1-253 and 1-254).

Micrometers

The first micrometers originating in France were rather crude. Laroy S. Starrett (1836–1922), founder of the L.S. Starrett Company, is responsible for the many improvements that make it

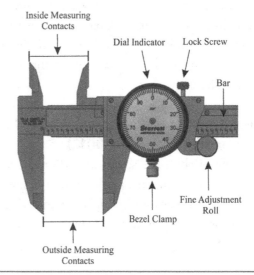

Figure 1-253. Dial caliper features, *courtesy L.S. Starrett Company*

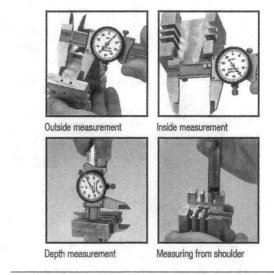

Figure 1-254. Dial caliper measurement techniques, *courtesy L.S. Starrett Company*

the modern precision measuring tool that we know today. In effect, a micrometer combines the double contact of a slide caliper with a precision screw adjustment which may be read with great accuracy. It operates on the principle that a screw accurately made with a pitch of forty threads to the inch will advance one–fortieth (.025) of an inch with each turn.

In the metric system the pitch of the screw is 0.5 mm. Each turn will advance the screw 0.5 mm, so it will take 50 turns to complete the normal range of 25 mm. As Figure 1-255 illustrates, the screw threads on the spindle revolve in a fixed nut concealed by a sleeve. On a micrometer caliper of one-inch capacity, the sleeve is marked longitudinally with 40 lines to the inch, corresponding with the number of threads on the spindle.

How to Read a Micrometer Graduated in Thousandths of an Inch (.001″)

Since the pitch of the screw thread on the spindle is $\frac{1}{40}$″ or 40 threads per inch in micrometers graduated to measure in inches, one complete revolution of the thimble advances the spindle face toward or away from the anvil face precisely $\frac{1}{40}$ or .025 of an inch. The reading line on the sleeve is divided into 40 equal parts by vertical lines that correspond to the number of threads on the spindle. Therefore, each vertical line designates $\frac{1}{40}$ or .025 of an inch and every fourth line, which is longer than the others, designates hundreds of thousandths. For example, the line marked "1" represents .100″, the line marked "2" represents .200″ and the line marked "3" represents .300″, etc.

The beveled edge of the thimble is divided into 25 equal parts with each line representing .001″ and every line numbered consecutively. Rotating the thimble from one of these lines to the next moves the spindle longitudinally $\frac{1}{25}$ of .025″ or .001″ of an inch; rotating two divisions represents .002″, etc. Twenty-five divisions indicate a complete revolution, .025″ or $\frac{1}{40}$″ of an inch. To read the micrometer in thousandths, multiply the number of vertical divisions visible on the sleeve by .025″, and to this add the number of thousandths indicated by the line on the thimble which coincides with the reading line on the sleeve.

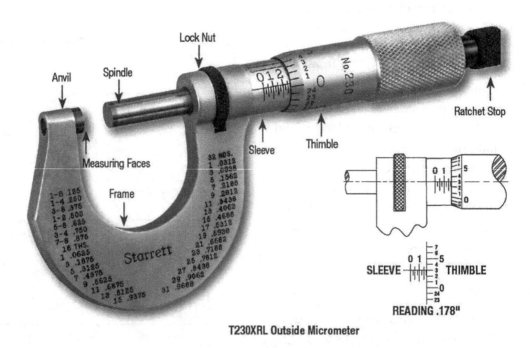

T230XRL Outside Micrometer

Figure 1-255. Outside micrometer features and reading; *courtesy L.S. Starrett Company*

T230XRL Outside Micrometer

EXAMPLE (refer to Figure 1-255):

The "1" line on the sleeve is visible, representing .... 0.100"

There are 3 additional lines visible, each representing .025" .... 3 × .025" = 0.075"

Line "3" on the thimble coincides with the reading line on the sleeve, each line representing .001" .... 3 × .001" = 0.003"

The micrometer reading is .... 0.178"

An easy way to remember is to think of the various units as if you were making change from a ten-dollar bill. Count the figures on the sleeve as dollars, the vertical lines on the sleeve as quarters and the divisions on the thimble as cents. Add up your change and put a decimal point, instead of a dollar sign, in front of the figures.

How to Read a Micrometer Graduated in Ten–Thousandths of an Inch (.0001″)

Once you have mastered the principle of the Vernier as previously explained, you will have no trouble reading a Vernier micrometer in ten-thousandths of an inch. The only difference is that on a Vernier micrometer, there are ten divisions marked on the sleeve occupying the same space as nine divisions on the beveled edge of the thimble. Therefore, the difference between the width of one of the ten spaces on the sleeve and one of the nine spaces on the thimble is one-tenth of a division on the thimble. Since the thimble is graduated to read in thousandths, one-tenth of a division would be one ten-thousandth. To make the reading, first read to thousandths as with a regular micrometer, then see which of the horizontal lines on the sleeve coincides with a line on the thimble. Add to the previous reading the number of ten-thousandths indicated by the line on the sleeve which exactly coincides with a line on the thimble (Figure 1-256).

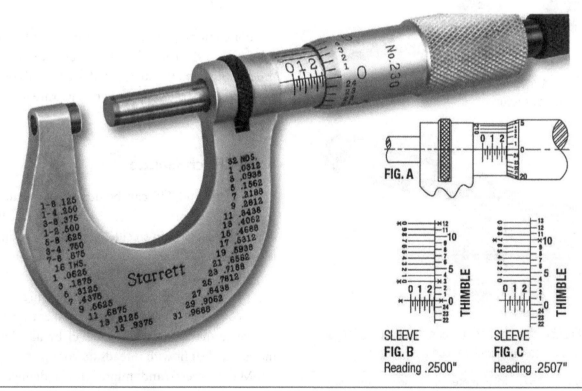

Figure 1-256. How to read a micrometer, *courtesy L.S. Starrett Company*

Quick Measurements

Micrometers are available with either a friction thimble or ratchet stop designed so that the spindle will not turn after more than a given amount of pressure is applied. This feature is of value when a number of measurements are made or when the measurements are made by more than one person with the same caliper. With the ratchet type, when the anvil and spindle are in proper contact with the work, the ratchet slips by the pawl and no further pressure is applied. The ratchet stop is incorporated in a small auxiliary knurled knob at the end of the thimble. The friction type mechanism is built into the thimble as a thimble friction where it reduces the required span of thumb and fingers and makes it easier to use the micrometer with one hand. A locknut is provided for retaining a reading (Figure 1-257).

Ratchet stop on micrometer ensures consistent gauging pressure.

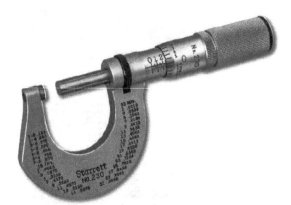

Friction thimble permits one–hand operation of micrometer and uniform contact pressure.

Figure 1-257. Micrometer quick measurements; courtesy L.S. Starrett Company

Measuring Screw Threads

To determine the number of threads per inch (TPI), the pitch of a few bolts, nuts, threaded holes or studs, simple measuring devices can be used. For example, when a count of threads per inch is desired, an ordinary steel rule can suffice. Simply line up the one-inch line graduation of a steel rule with the crest (point) of a thread (see Figure 1-258) and count the number of crests over that one-inch thread length. If that number is twelve over a one-inch thread length, then the TPI is twelve. If the overall thread length is shorter than an inch, count the crests over only a half inch, then multiply the number by two.

Figure 1-258. Micrometer quick measurements; courtesy L.S. Starrett Company

For example, if you counted six threads over a half inch, then 6 × 2 = 12. There are twelve threads per inch. If you counted five threads over a quarter inch, then there are twenty threads per inch, and so on.

Screw Pitch Gauge

Screw pitch and TPI can be determined readily with a screw pitch gauge (see Figure 1-259), which is a set of thin steel leaves. The individual leaf edges have "teeth" corresponding to a fixed thread pitch.

The leaves on Starrett screw pitch gauges are stamped to show the thread pitch number to easily determine the thread pitch and TPI by matching the correct leaf fit with threads on your part.

Major (crest) and minor (root) diameters of threaded parts can vary depending upon the

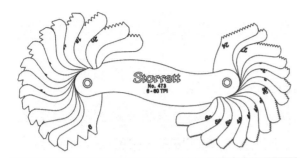

Figure 1-259. Screw pitch gauge, *courtesy L.S. Starrett Company*

sharpness or fullness (truncations) of the thread. Measurements are usually made at the pitch line to determine the pitch diameter. Therefore, pitch diameter is the diameter of a cylinder passing through the thread profile, so as to make the widths of thread ridges and widths of thread grooves equal.

Screw thread micrometers with a pointed spindle and V-shaped anvil is used to measure pitch diameters. The point of the spindle and the V-shaped anvil are designed so that contact is made on the side (flank) of the thread (Figures 1-260 and 1-261).

Figure 1-260. Measuring pitch diameter of screw thread with screw thread micrometer caliper. Reads directly in inches, *courtesy L.S. Starrett Company*

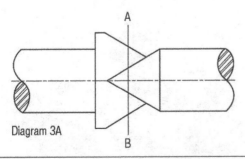

Figure 1-261. With anvil and spindle in position shown in sketch, line A–B corresponds to 0 reading, *courtesy L.S. Starrett Company*

How to Use a Depth Micrometer

A depth micrometer, as the name implies, was designed to measure the depth of holes, slots, recesses, keyways, etc. Available in both standard or digital readout (Figures 1-262, 1-263 and 1-264).

440, 445 Micrometer Depth Gauge

Figure 1-262. Depth micrometer, *courtesy L.S. Starrett Company*

446 Digital Micrometer Depth Gauge

Figure 1-263. Digital depth micrometer, *courtesy L.S. Starrett Company*

749 Electronic Depth Gauge with output

Figure 1-264. Electronic depth micrometer, *courtesy L.S. Starrett Company*

The tool consists of a hardened, ground and lapped base combined with a micrometer head. Measuring rods are inserted through a hole in the micrometer screw and brought to a positive seat by a knurled nut. The screw is precision ground and has a one-inch movement. The rods are furnished to measure in increments of one inch. Each rod protrudes through the base and moves as the thimble is rotated. The reading is taken exactly the same as with an outside micrometer, except that sleeve graduations run in the opposite direction. In obtaining a reading using a rod other than the 0–1", it is necessary to consider the additional rod length. For example, if the 1–2" rod is being used, one inch must be added to the reading on the sleeve and thimble. Before using the depth micrometer, be sure that base, end of rod and work are wiped clean and that the rod is properly seated in micrometer head. Hold base firmly against work as in Figure 1-265, and turn thimble until rod contacts bottom of slot or recess. Tighten locknut and remove tool from work to read measurement. Adjustment to compensate for wear is provided by an adjusting nut at the end of each rod. Should it become necessary to make an adjustment of a rod, back off the adjusting nut one-half turn before turning to new position, then check against a known standard such as a Webber gauge block.

Telescoping Gauges

Telescoping gauges are sometimes better than leg calipers for measuring internal diameters. The head of a telescoping gauge expands across the hole, is locked and then measured with a micrometer. Or, the gauge can be set to a standard and used to make shrink, close, or loose fits. Handles up to 12 inches are available (Figures 1-266 and 1-267).

Small hole gauges serve the same purpose for holes ranging from ⅛ to ½ inch. They are made with a split ball at the contact end which is expanded to get the measurement which is then transferred to a micrometer, duplicating as close as possible the "feel" of the gauge in the hole (Figures 1-268 and 1-269).

Radius gauges are useful for machining, inspection and layout. They come in radii from ¹⁄₆₄" to ½" and decimal sizes .010"–.500" as well as metric sizes 0.5 mm–15 mm. Each gauge has five different gauging surfaces for both convex and concave radii. A gauge holder is available for checking radii in hard-to-reach locations (Figures 1-270 and 1-271).

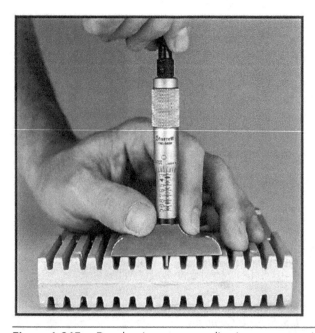
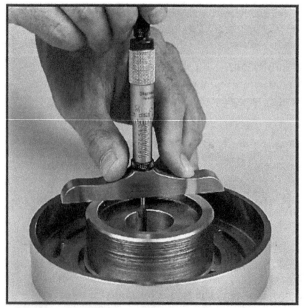

Figure 1-265.   Depth micrometer applications, *courtesy L.S. Starrett Company*

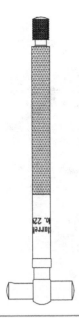

Figure 1-266. Telescoping gauge, *courtesy L.S. Starrett Company*

Figure 1-269. Small hole gauge in use, *courtesy L.S. Starrett Company*

Figure 1-267. Accurate measurements are ensured with telescoping gauge, *courtesy L.S. Starrett Company*

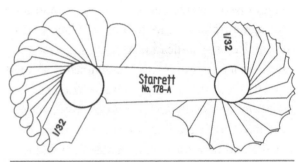

Figure 1-270. Radius gauge set, *courtesy L.S. Starrett Company*

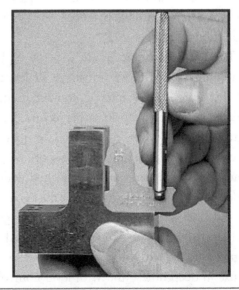

Figure 1-268. Small hole gauge, *courtesy L.S. Starrett Company*

Figure 1-271. Radius gauge in holder provides fast check of ¼″ radii milled on this workpiece, *courtesy L.S. Starrett Company*

Dial Indicators

One of the most widely used tools today in lay-out, inspection and quality control operations is the dial indicator. Specially designed with shock-less hardened stainless steel gear train and man-ufactured to fine watchmaking standards with jeweled bearings, the dial indicator has precisely finished gears, pinions and other working parts that make possible measurements from one-thousandth-fifty-millionths of an inch, depending on accuracy requirements. Any gear unit and any case assembly can be combined to give a com-plete dial indicator of the style desired. Dial faces are color-coded to avoid errors, white dial for English measurement and yellow face for metric. The contact point is attached to a spindle or rack with movement transmitted to a pinion and then through a train of gears to a hand which sweeps the dial of the indicator. A small movement of the contact is thus greatly magnified and read directly from the dial in thousandths or as close as 50 millionths of an inch. Long range indica-tors have direct reading count hands and a dou-ble dial. Graduations are available for reading in .001", .0005", .00025", .0001" and .00005" – with ranges from 12" down to .006"; also, in .01 mm, .002 mm and .001 mm with ranges up to 125 mm. Dials can have balanced or continuous graduations (Figure 1-272).

The regular contact point may be replaced with contacts of almost any shape or length to suit the work. These include contact points of extra length, special form, tapered and a roller contact for use on moving material. Many useful attachments are available to suit work require-ments. Dial indicators can be furnished with tol-erance hands, special dials, rubber dust guards to seal out dust and foreign matter; anti–magnetic mechanisms when the dial indicator is used near magnetic fields, with long stems up to 12 inches for use in deep holes; and with lever control for lifting the indicator spindle (Figure 1-273).

Super-precision dial indicators with grad-uations in 50 millionths (.00005) of an inch

Figure 1-272. Typical starrett dial indicators in the four standard American gauge design sizes, *courtesy L.S. Starrett Company*

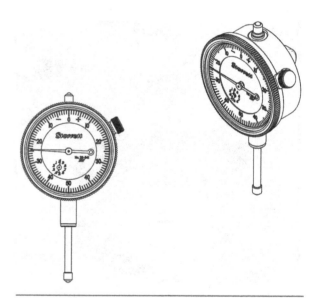

Figure 1-273. Travel dial indicator, *courtesy L.S. Starrett Company*

and accuracy to plus or minus 10 millionths (±.00001) are also available. These are used for applications requiring extreme precision such as in shop inspection to laboratory standards or in laboratory work. The long-range dial indi-cator with ranges of 2, 3, 4, or 5 inches and on up to 12 inches (and comparable metric ranges through 125 mm) makes possible all types of long-range gauging such as on jig and fixture work, for production measuring on machine tools or as precision stops. Count hands and double dials permit direct reading in thousandths of an inch.

Magnetic backs provide a quick and easy means of attaching any Starrett dial indicator to flat ferrous metal surfaces. A real time-saver for machine, jig and fixture setup (Figure 1-274).

Figure 1-275. Travel dial indicator features, *courtesy L.S. Starrett Company*

Figure 1-274. Dial indicator with magnetic back, *courtesy L.S. Starrett Company*

Dial Gauges. The principle of direct reading from a pointer and graduated dial provides both the accuracy and the speed of reading essential in many of today's inspection operations, and consequently, the dial indicator has been incorporated in all types of special and standard gauging equipment, as well as in many machine tools. Some gauges are direct reading and others serve as comparators showing plus or minus variations in size (Figure 1-275).

Dial Test Indicators

The dial test indicator is an all-purpose tool used in layout, inspection and on machine tools for truing

up work, checking runout, concentricity, straightness, surface alignment and for transferring measurements. They work with various attachments, including a tool post holder, adapt it to a wide range of applications, such as checking runout of a spindle turned on a lathe. Dovetail mount indicators such as the Starrett 708 and 709 Precision Dial Test Indicators with dovetail mounts can be positioned for easy and accurate readability. The versatility of the angled head, combined with the three dovetail mounts eliminates the need for having both vertical and horizontal style test indicators and work with existing test indicator accessories available with English and metric dials (Figures 1-276 and 1-277).

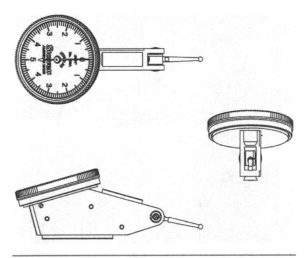

Figure 1-276. Dial test indicator, *courtesy L.S. Starrett Company*

Figure 1-277. Dial test indicators, *courtesy L.S. Starrett Company*

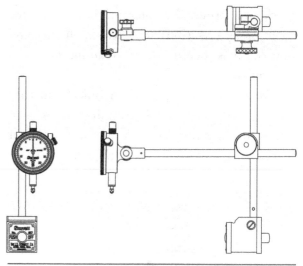

Figure 1-279. Dial indicator with magnetic base, *courtesy L.S. Starrett Company*

Magnetic Base Holders

A magnetic base indicator holder greatly increases the usefulness and application of the dial indicator in all shop setups, checking and inspection jobs. The time spent clamping an indicator to a machine is eliminated since its powerful, permanent magnet base holds to any flat or round steel or iron surface—horizontally, vertically, upside down or on shafts using a precision ground "V" on one face. A push button or lever turns the magnetic force on or off. They work with a variety of indicators and holding arrangements to suit the work (Figures 1-278 and 1-279).

Figure 1-278. Dial test indicator in use, *courtesy L.S. Starrett Company*

Dial Bore Gauges

Dial bore gauges are used for precision measuring of cylindrical bores. They quickly inspect hole diameters and will detect and measure any variation from a true bore, such as taper, out-of-roundness, bell mouth, hourglass or barrel shapes. Basically, the gauge consists of a contact head, an indicator housing and handle. An adjustable range screw plus two centralizing plungers give the gauge 3-point contact to insure true alignment. The gauging contact actuates a dial indicator for comparative readings in half-thousandths (.0005") or ten-thousandths (.0001"). Models are available with ranges from 1-½–12-⅛" diameter and 6–7" bore depths. For smaller diameters, a series of split-ball type bore gauges are available that will measure holes ranging from .107"–1.565" diameters and 13/16–5" bore depths in .0001" increments (Figures 1-280 and 1-281).

Chamfer, Countersink and Hole Gauges

These gauges provide fast, accurate, easy-to-read measurements with one hand operation. A check gauge stand is also available.

Chamfer gauges measure any chamfered hole that has an included angle equal to or less than

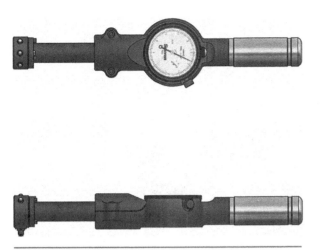

Figure 1-280. Dial bore gauge, *courtesy L.S. Starrett Company*

Figure 1-281. Dial bore gauge in use, *courtesy L.S. Starrett Company*

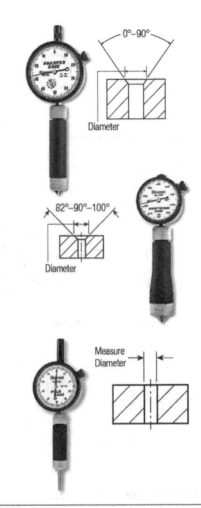

Figure 1-282. Chamfer, countersink and hole gauges, *courtesy L.S. Starrett Company*

the angle on the dial face. They are available in 0°– 90° and 90°–127° angles in both I.D. and O.D. models with a range of 0–⅜" and 1–2". These gauges need no setting master.

Countersink gauges directly read countersinks and chamfers in .002 inch diameter increments. This series of gauges comes in 80°, 90° and 100° angles in a range of .020–.078". A "set *master*" ring is included with each gauge. Indicator reading is "held" until the reset button is activated, giving you added control.

Hole gauges will check hole diameters to .001" English and .02mm Metric. The English range is .010–.330". Metric range is .25 mm–8.35 mm.

An optional set master ring is available. Metric dial faces are yellow (Figure 1-182).

Wireless Data Collection

Wireless data collection systems consist of three primary elements: miniature radios (called *end nodes*) that are attached to the data output ports of electronic tools, a gateway that connects to a PC, and routers that extend the system's range and make the radio network more robust. DataSure® gathers data from electronic measuring tools and delivers it to a PC. The data can then be analyzed by your statistical process control (SPC) application.

Why go wireless? A wireless data collection system can significantly reduce human error in

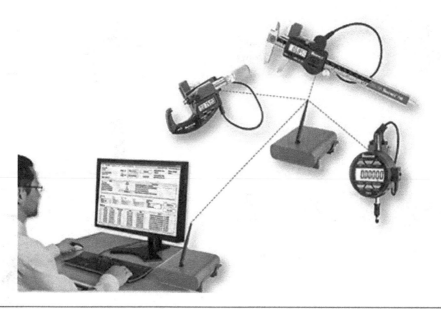

Figure 1-283.   Data collection, *courtesy L.S. Starrett Company*

data recording. It removes wiring-related placement, installation, safety and cost issues. It also makes it easier for you to bring a precision measuring tool to the work, rather than having to bring work to the measuring tool (Figure 1-283).

SPC Software for Data Collection

Starrett 728-3 Shop Floor Plus® II Software is an advanced Windows-based program that is fast, versatile and easy to use for quality control and SPC/QC applications. Data can be entered from any electronic measuring tool with output. The software allows data manipulation for analysis and dynamic data exchange for real time data input.

- Read directly from Excel files to immediately begin to analyze with control charts.
- A variety of gauges can be used to add data to the spreadsheet using our Real Time Gauge.
- Data is entered directly into SPC-PLUS II's spreadsheet which has many functions for calculating user-defined parameters.

Shop Floor Pro™ Software provides powerful data acquisition and analysis features in an intuitive, simple-to-use interface.

- Easy to learn real-time data collection and control charting software package.
- Operates in a stand-alone or networked environment.
- Allows for data entry directly from gauges or through the keyboard.
- Pre-defined Work Sets can be created for Work Centers, part numbers, etc. Each Work Set opens in a single step and supports up to 20 control charts.
- Control charts can show different characteristics for the same part or characteristics from multiple parts.
- Chart types include X-Bar/Range, X-Bar Sigma, Individual-X/Moving Range and Run Charts. User definable subgroup sizes range from 1–20.

End of excerpt from "Tools and Rules for Precision Measuring."

The tools listed above are used to measure specific types of features identified on engineering drawings to confirm dimensional accuracy. These tools represent the core group of inspection tools used, but there are many more tools available beyond the scope of this text. Always research the best tool for the job at hand. Generally speaking, selection of the tool used is dependent on the type of and tolerance of the feature. Use a tool capable of measuring ten times more accurate than the design tolerance requires. For example, if the tolerance listed for the outside diameter of a shaft is .001", then a one-ten thousandths O.D. Vernier micrometer should be used.

Final part quality is verified by measurement of 100 percent inspection of every dimension and recorded on a Quality Control Check sheet after the part is completed. This is necessary for the first part but it is not feasible for every part thereafter. This is where Statistical Process Control becomes effective (Figures 1-284 and 1-285).

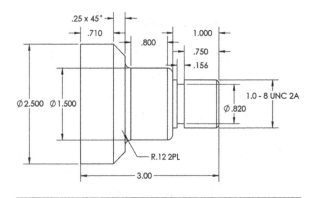

Figure 1-284. Machined part example

Turning Center
Quality Control Check Sheet

Date: Today			Checked By: You
Part Name: Turning Center Project		Part Number: 1234	
Blueprint Dimension	Tolerance	Actual Dimension	Comments
Ø 2.50	±.010		
Ø 1.50	±.010		
.75	±.010		
.25	±.010		
45°	±.5°		
R.1 2PL	±.015		
1.0 - 8 Thread	.9980/.9830		
1.0 - 8 Thread Pitch Ø	.9168/.9100		
.05	±.010		
45°	±.5°		
.75	±.010		
.156	±.005		
.09	±.010		
3.00	±.010		

Figure 1-285. Quality control check sheet

Statistical Process Control

In manufacturing production environments, the process of monitoring part specifications by recording inspection data on charts used for the analysis of trends and that enable more control of the process is called statistical process control (SPC). The results of a manufacturing process that is in control, are the elimination of waste caused by scrap and rework, as well as higher efficiency and less need for final inspection. When a dimensional value, identified by the engineering drawing specification, is outside tolerance, then the part is nonconforming and should be rejected.

Going forward in the production process, inspection plans are developed by identifying control dimensions or those most critical to overall part quality and customer satisfaction. The identified dimensions are listed on a control chart and then measurements are taken and recorded on the chart based on a sampling plan. The most common types of SPC charts used in a machining environment are X-Bar and R charts. The ultimate goal is process stability which indicates that the process is *in control* as opposed to an unstable process that is *out of control*. By analyzing the data on these charts, trends can be detected and corrective action taken before parts are rejected. The software and data connected tools listed above and others like them automate the data collection and charting process to simplify and store data for decision-making and analysis that improves overall production quality. For a more detailed study on the subject of statistics for quality control, check out "Statistics for Quality Control" at https://books.industrial-press.com/statistics-for-quality-control.html.

Summary

In this section, many common machined part inspection tools have been introduced. There are many more inspection measurement options available beyond the scope of this text including coordinate measuring machines, comparators, vision systems and laser scanners, to name a few. Final part quality is dependent upon a sound manufacturing process that includes proper tooling for machining and tools for inspection that confirm dimensional accuracy.

Machined Part Inspection, Study Questions

1. Give the micrometer reading for the following graphic (Figure 1-286). _____

Figure 1-286. Machined part inspection, problem 1

2. Give the micrometer reading for the following graphic (Figure 1-287). _____

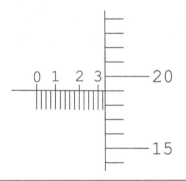

Figure 1-287. Machined part inspection, problem 2

3. Give the micrometer reading for the following graphic (Figure 1-288). _____

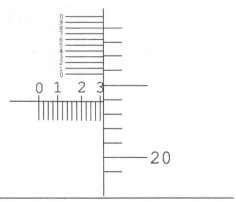

Figure 1-288. Machined part inspection, problem 3

4. Telescoping gauges are available in sets that allow inside diameter measurements up to 6 inches. What other precision measuring tool is required to measure diameters with telescoping gauges?

   a. Bore gauge
   b. OD micrometer that covers the hole size range
   c. Vernier slides
   d. Slide calipers

5. Height gauges are used in conjunction with:

   a. Test indicators
   b. Surface plates
   c. Gauge blocks
   d. All of the above

6. A magnetic base is used in conjunction with:

   a. Travel indicators
   b. Test indicators
   c. Surface plate
   d. Either a or b

7. A depth micrometer reading of over two inches requires:

   a. Gauge block
   b. A 2–3, inch depth micrometer
   c. Changeout to the 2 inch–3 inch rod length
   d. Height gauge

8. What tool is used to check the pitch diameter on external threads?

   a. Screw pitch gauge
   b. OD micrometer
   c. ID micrometer
   d. Thread micrometer

9. Which of the following is a Statistical Process Control (SPC) chart?

   a. Radian chart
   b. X-bar and R control chart
   c. Sampling plan
   d. Quality Control Check sheet

10. Which of the following information is found on an SPC X-bar and R control chart?

   a. Upper control limit
   b. Lower control limit
   c. Mean
   d. All of the above

English Gage Blocks Set									
Increment	Size in Inches								
0.0001	0.1001	0.1002	0.1003	0.1004	0.1005	0.1006	0.1007	0.1008	0.1009
0.001	0.101	0.102	0.103	0.104	0.105	0.106	0.107	0.108	0.109
	0.110	0.111	0.112	0.113	0.114	0.115	0.116	0.117	0.118	0.119
	0.120	0.121	0.122	0.123	0.124	0.125	0.126	0.127	0.128	0.129
	0.130	0.131	0.132	0.133	0.134	0.135	0.136	0.137	0.138	0.139
	0.140	0.141	0.142	0.143	0.144	0.145	0.146	0.147	0.148	0.149
0.050	0.050	0.100	0.150	0.200	0.250	0.300	0.350	0.400	0.450
	0.500	0.550	0.600	0.650	0.700	0.750	0.800	0.850	0.900	0.950
1.000	1.000	2.000	3.000	4.000					

Figure 1-289. Gauge block set

11. Use the gauge block set in Figure 1-289 to select the least number of gauge blocks needed to build a stack needed to check the measurements (Figure 1-290).

Measurement	1.000 inch Increment	0.050 inch Increment	0.001 inch Increment	0.0001 inch Increment	Number of Blocks Needed
5.4197					
2.9081					

Figure 1-290. Gauge block set problems

Basic CNC Program Structure and Format

Objectives

1. Define the main components of a CNC program.
2. Understand the program block construction.
3. Identify and use the appropriate letter address.
4. Define common G-codes.
5. Define common M-codes used in CNC programs.

CNC Programming Defined

CNC programming is a method of defining machine tool movements through the application of numbers and corresponding coded letter symbols. As shown in the list below, all phases of manufacturing production are considered in CNC programming, beginning with the engineering drawing or blueprint and ending with the final product:

- Part model, engineering drawing or blueprint
- Work-holding requirements
- Cutting tool selection
- CNC part program creation
- CNC part program tool path verification and simulation
- Machine setup including measuring of tool and work offsets
- Program testing by dry run
- Automatic operation or CNC machining

Begin all programming by closely evaluating the part model, engineering drawing or blueprint. Pay special attention to dimensional tolerances assigned for particular features, cutting tool selection and the choice of a machine. Next, select the machining process. The machining process refers to the selection of work-holding fixtures and determination of the operation sequence. The next step is selection of the appropriate cutting tools and the sequence for their application. Before writing the part program, the spindle speeds and feedrates for each tool need to be calculated based on the tool and workpiece materials. Calculating spindle RPM (r/min) and feedrates were covered in detail in Cutting Tool Selection and Machining Mathematics sections (review if necessary). Always pay special attention to the specific tool movements necessary to complete the finished part geometry, including non-cutting movements that will be at rapid traverse speeds. Select each individual tool and note them in the program manuscript. Also, note any miscellaneous functions for each tool, such as flood coolant or spindle direction. Then, once the program is written, transfer it to the machine through an input medium like one of the following: ethernet, USB, or RS-232 interface.

Initiate the machining by preparing the machine for use, commonly called *setup*. For example, measure and input workpiece zero and tool length offsets into CNC memory offset registers. When the CNC program is created from a CAM program, toolpath simulation should be performed prior to post processing. Perform a graphical simulation of the programmed tool path on the CNC controller if equipped. This step enables the machinist or setup person to verify that the program has no syntax errors and to visually inspect the tool path movements. Then, if all looks well, machine the first part with increased confidence.

After program execution is completed, a thorough dimensional inspection is needed to compare and confirm dimensions of the final product to match those on the part model, engineering drawing or blueprint. Evaluate and correct any differences between the actual dimensions and the dimensions on the drawing by inserting wear offset values into the offset register of the machine. In this manner, you can obtain the correct dimensions of consecutively machined parts.

Coordinate Input Format

CNC machines allow input values of inches (specified by the command G20), millimeters (specified by the command G21), and degrees using a decimal point with significant zeros in front of (leading) or at the end (trailing) of the values. As a general rule, use decimal point entry when programming. When using inch programming, values can be specified by:

- Programming with a decimal point:
  - Input needed for one inch is: 1. or 1.0
  - Input needed for one and one quarter inch is: 1.250 or 1.25

- Input needed for one sixteenth inch is: 0.0625 or .0625
- Input needed for an angular value of ninety degrees is: A, B or C (axis designator) 90.

Program Format

The language described in this book is used for controlling machine tools and is known informally as "G-Code." This language is used worldwide and is reasonably consistent. The standard by which it is governed was established by the Electronics Industries Association and the International Standards Organization, called EIA/ISO for short. Because of this standardization, a program created for a particular part on one machine may be used on other similar machines with minimal changes required.

Each program is a set of instructions that controls the tool path. The program is made up from blocks of information separated by the semicolon symbol (;). This symbol is defined as the end of the block (EOB) character. Each block contains one or more program words. For example, see Table 1-1.

Each word contains an address, followed by specific data. For example, see Table 1-2.

Chart 1-11 is a list for all of the letter addresses that are applicable in programming, along with brief explanations for each. Chart 1-12 lists symbols commonly used in CNC programs.

Word	Word	Word	Word	Word
N02	G01	X3.5	Y4.728	F8.0

Table 1-1. Block with Program Words

Address	Data	Address	Data	Address	Data
N	02	G	01	X	3.5

Table 1-2. Address and Data Block

Basic Preparatory Functions "G-Codes" Definition

G-codes are officially titled, preparatory functions. They identify a specific functional activity when used in combination with letter address codes and axis designations that the machine will execute. Preparatory functions are programmed with the letter address G, normally followed by two digits, to establish the mode of operation in which the tool moves. One or more G-codes are listed at the beginning of the program block to set the desired function. There are two types of G-codes: modal and non-modal (one-shot). Modal commands remain in effect, in multiple blocks, until they are replaced by another G-code command from the same group. There are several different groups of G-codes as indicated in column 2 of Chart 2-3 and Chart 3-3. One code from each group may be specified in an individual block. If two codes from the same group are entered in the same block, the first will be ignored by the control and the second will be executed. Those G-codes that are active upon startup (default) of the machine are indicated by an asterisk (*) in the chart.

A *safety block* is commonly placed in the first line of the program and just after tool changes where cancellation codes are used to ensure all G-codes that have been in effect in prior program

CHARACTER	MEANING
A	Additional rotary axis parallel and around the X axis
B	Additional rotary axis parallel and around the Y axis
C	Additional rotary axis parallel and around the Z axis
D	Tool radius offset number (Turning) Depth of cut for multiple repetitive cycles
E	User macro character
F	Feed rate (Turning) Precise designation of thread lead
G	Preparatory function
H	Tool length offset number
I	Incremental X coordinate of circle center (Turning) parameter of fixed cycle
J	Incremental Y coordinate of circle center
K	Incremental Z coordinate of circle center (Turning) parameter of fixed cycle
L	Number of repetitions (subprogram, hole pattern) Fixed offset group number
M	Miscellaneous function
N	Sequence or block number
O	Program number
P	Dwell time, program number, and sequence number designation in subprogram (Turning) Sequence number start for multiple repetitive cycles
Q	Depth of cut, shift of canned cycles (Turning) Sequence number end for multiple repetitive cycles
R	Point R for canned cycles, as a reference return value Radius designation of a circle arc Angular displacement value for coordinate system rotation
S	Spindle-speed function
T	Tool function
U	Additional linear axis parallel to X axis
V	Additional linear axis parallel to Y axis
W	Additional linear axis parallel to Z axis
X	X coordinate
Y	Y coordinate
Z	Z coordinate

Chart 1-11.   Programming Letter Address

Common Symbols Used in Programs	
SYMBOL	**MEANING**
-	Minus sign, used for negative values
/	Slash, used for block skip function
%	Percent sign, necessary at program beginning and end for communications only
( )	Parentheses, used for comments within programs
:	Colon, designation of program number
;	Semicolon, end-of-block character
.	Decimal point, designation of fractional portion of a number

Chart 1-12. Programming Symbols

sections are cleared. On CNC mills, this block typically contains (G40) cutter compensation cancel, (G49) tool length compensation cancel, (G80) canned cycle cancel and (G17) XY plane selection. These cancellations are important because of modal commands that stay in effect until either cancelled or replaced by a command from the same group. By inserting the safety block, there is no chance of modal commands remaining active.

It is important to note that if the measurement system is changed (for example, from the G20-inch to the G21 metric system) then G21 will be in effect at the next startup of the machine or until a program call of G20 is executed. On machines sold in the United States, the parameters are set to default to the G20-inch system upon startup. When programming, always consult the programming manual for the specific control prior to ensure proper selection and use. Complete lists of G-codes specific to CNC Milling or CNC Turning will be included in Chapters 2 and 3, respectively.

Basic Miscellaneous Functions "M-Codes" Definition

M-codes are miscellaneous functions usually listed at the end of the program block to activate

functions to turn on and off coolant, set spindle direction and to execute tool change commands, etc. Complete lists of M-codes specific to CNC Milling or CNC Turning will be included in Chapters 2 and 3, respectively.

The code consists of the letter M typically followed by two digits. Normally, one block will contain only one M-code function; however, up to three M-Codes may be in a block depending upon parameter settings. Most of the common M-codes are listed in charts in sections for CNC Lathes and CNC Mills. Many machine tool builders assign other codes for specific purposes relative to their equipment. Always consult the manufacturer manuals specific to the machine in use for pertinent M-codes.

Summary

All CNC programs are a combination of codes, letters and symbols that command the machine tool to execute movements that are needed to drive tooling to cut desired feature shapes from raw materials. Understanding the basic structure of CNC programs is the foundation required for creating successful CNC programs.

Basic CNC Program Structure and Format, Study Questions

1. CNC programming is:

   a. A combination of letters, numbers, codes and symbols
   b. A collection of coded instructions the machine tool will execute
   c. Exclusively done using Computer Aided Manufacturing (CAM)
   d. Both a and b

2. CNC programming requires which of the following:

   a. Knowledge of cutting tools and proper selection for their use
   b. A thorough understanding of part features communicated on engineering drawings
   c. Understanding of G- and M-codes
   d. All of the above

3. What does the acronym EIA stand for?

   a. Engineering Information Association
   b. Electronics Industries Association
   c. Engineering Institute of Accountability
   d. Electronic Inventory Association

4. A CNC program block is:

   a. A single program word
   b. A line or sequence number
   c. An entire single line of CNC program code
   d. The End of Block character

5. G-codes are:

   a. Preparatory functions
   b. Generic instructions
   c. Rapid traverse commands
   d. Miscellaneous commands

2

Setup, Operation and Programming of CNC Mills

CNC Mill Setup and Operation

Objectives

1. Identify and use common CNC mill operator panel functions.
2. Identify and use common machine control panel functions for CNC mills.
3. Identify work-holding components and related accessories.
4. Perform common setup and operation functions at the CNC mill machine control.
5. Use the control to input setup data including work and tool offsets.
6. Use the controller EDIT mode to modify part programs.
7. Identify and solve common problem situations encountered during setup and operation of CNC mills.

Machine Safety for CNC Mills
Machine Guarding

CNC machines today are provided with enclosures that envelope the worktable, chuck and moving parts to mitigate the potential for flying chips and coolant from hitting the operator. Sliding doors are fitted with safety interlocks to prevent opening during cutting cycles. These doors are equipped with windows so the operator can observe cutting activities. The windows are made of glass or Plexiglas and are not a guarantee that flying objects will not penetrate them. Never operate any machine tool without proper guards in place and do not alter door safety interlocks!

Every CNC machine tool should have a power-disconnect mounted on a wall near the back of the machine to remove power from the machine and enable Lockout/Tagout during maintenance activities. The first step in machine startup is to ensure this disconnect is in the ON condition.

CNC Mill Operator Control Panel

Every CNC machine tool has an operation control panel (operator pendant) that provides the interface for the operators and programmers with the machine tool, referred to as the human machine interface (HMI). By using the operator panel, we can physically manipulate the working components of the machine to do what we need; the control panel is where the program data are entered and stored. A thorough understanding of each is necessary for successful CNC machine use. Let us first study an example of an operation control pendant.

Operator Control Pendant Features

The following descriptions are for the machine models shown in Chapter 1, Figures 1-13 and 1-14 and represent the configuration for a common operator panel. Some differences will exist for each manufacturer's operator panel, but the functions and appearance are generally similar. Figure 2-1 shows the operator pendant for a Haas three-axis CNC mill. You should always consult the applicable manufacturer manual for detailed descriptions that match your specific needs.

POWER ON and POWER OFF

Located in the upper left-hand corner of the control in (Figure 2-1), the POWER ON and POWER OFF buttons (Figure 2-2) are used to activate/deactivate power to the control. Press these buttons to turn CNC power ON and OFF. The ON button is green. The OFF button is red. Below and to the right of these buttons is the operator control pendant keyboard, which is described later in in this chapter.

Note: The control is always turned ON after turning on the MAIN POWER switch, which is located on the door of the control system, typically at the back or side of the machine. The control is always turned OFF before the MAIN POWER switch is turned OFF.

Emergency Stop

The EMERGENCY STOP (E-Stop) is the large, red, mushroom-shaped button used to stop machine function when an emergency situation occurs. Examples of such situations are:

- Overloading of the machine spindle
- The part being machined has come loose
- Incorrect data in the program or work/ tool offsets have caused a collision (crash) between the tool work-holding and/or the workpiece

When this button is pressed, all program-commanded feedrates, spindle revolutions and coolant flow are halted immediately. To recover from an E-Stop condition, you must reset the

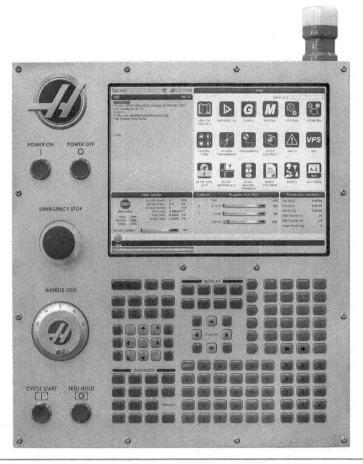

Figure 2-1. Operator pendant, *courtesy Haas CNC Automation*

Figure 2-2. Power on and power off symbols

Figure 2-3. Cycle start symbol

program controller and, in some cases, zero return/home the machine axes. To reset the E-Stop button, turn it clockwise. It should pop out of the depressed condition. Check the monitor for any alarm signals and take note of the alarm number and description; then eliminate the cause that forced the use of the emergency stop button. Press the RESET button to clear all pending commands; ensure no interference conditions are present between the work and tool-holding equipment and then press the HOME G28 button. All of the axes will return to the machine home position and normal operation may continue.

Handle/Jog

Pressing the HANDLE/JOG button on the control pendant keyboard activates the HANDLE/JOG mode and enables manual control of axis movements using the handle for a selected axis: X, Y, or Z; or for rotational axes A, B, or C, by use of the handle. This handle is also known as the manual pulse generator (MPG).

Cycle Start

The CYCLE START button (Figure 2-3) is used to start automatic operation. Use this button in order to begin the execution of a program from memory. When the CYCLE START button is pressed, the beacon light located on top of the control turns on and is green. The active program will be executed to the end.

Feed Hold

Figure 2-4.   Feed hold symbol

Pressing the FEED HOLD button (Figure 2-4) during automatic operation will halt all feed movements of the machine. It will not stop the spindle RPM or affect the execution of tool changes on some machines. When the FEED HOLD button is pressed, the beacon light located on top of the control will flash green until CYCLE START is pressed again. This button is used when minor problems are encountered, such as coolant flow direction or when checking the distance-to-go position during setup. When the problem is remedied or position is confirmed, press CYCLE START again to resume automatic cycle operation. It is not recommended using this button to interrupt a cut because the spindle does not stop; therefore, damage to the tool or part may occur. When pressed during the execution of the tapping, FEED HOLD will take effect after the tap is withdrawn. If the tap breaks during the tapping cycle, the only way to stop the machine is by pressing the RESET button on the controller or the EMERGENCY STOP button.

Universal Serial Bus, USB Port

Figure 2-5.   Universal serial bus, USB port symbol

The Universal Serial Bus (USB) port is located on the right side of the operator pendant and includes a screw-on cover (Figure 2-5). It is used to transfer programs to and from the control.

Memory Lock

Figure 2-6.   Memory lock symbol

When the MEMORY LOCK key switch (Figure 2-6) is in the ON condition (horizontal), it prohibits program changes to be made. This locked condition does not affect work or tool offset adjustments. Some shops set this condition to ON, remove this key, and allow only the programmer or setup person access to the key. This is especially true in larger shops with multiple shifts and many workers. Some quality programs like AS9100 require that CNC program integrity be insured by locking out access to editing.

Setup Mode

Figure 2-7.   Setup mode symbol

When the SETEUP MODE key switch (Figure 2-7) is in the horizontal (unlocked) position it allows opening of the doors and access to the work envelope during setup procedures. The spindle RPM is limited to 750 RPM with the doors open in this condition. When the switch is in the vertical (locked) position the doors cannot be opened during operation.

Second Home

Figure 2-8.   Second home symbol

The SECOND HOME button (Figure 2-8) is a machine option; when included, it allows all axes to be positioned at rapid traverse to position coordinates associated with G154 P20.

Servo Auto Door Override

Figure 2-9. Servo-auto door override symbol

If the machine tool is equipped with this automation feature the Servo Auto Door Override switch (Figure 2-9) is used to open or close the doors.

Work Light

Figure 2-10. Work light symbol

These toggle switches (Figure 2-10) enable activation of lights that illuminate the work envelope. The machine can be fitted with high intensity lamps which further aid viewing within the work envelope.

Beacon Light

The beacon light is attached to the top of the control and has the following illumination functions: when the light is off, this means the machine is in the idle state; when the light is on and green, the machine is running under automatic cycle; when the light is flashing green, the machine is stopped but remains in the ready state and requires operator intervention to continue; when the light is flashing red, a machine fault has occurred or an emergency stop condition exists; and lastly, when the light is flashing

yellow it indicates that the programmed tool life has expired.

CNC Mill Operator Control Pendant, Keyboard Descriptions

Most of the operator keyboard buttons are the same for mill and lathe; however, there are some differences. The following keyboard buttons are not available for the lathe: Tool Offset Measure, Tool Release, Part Zero Set, ATC FWD, ATC REV, CHIP FWD, +B/+A/C, CLNT UP, CHIP STOP, JOG LOCK, CLNT DOWN, CHIP REV, -A/C/-B, AUX CLNT; however, the alternates for these buttons will be fully outlined in Chapter 3.

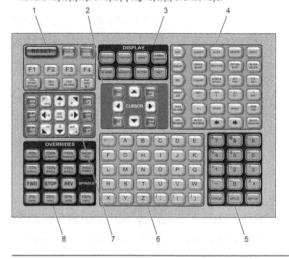

Mill Keyboard: [1] Function Keys, [2] Cursor Keys, [3] Display Keys, [4] Mode Keys, [5] Numeric Keys, [6] Alpha Keys, [7] Jog Keys, [8] Override Keys.

Figure 2-11. Operator control pendant, *courtesy Haas CNC Automation*

The following sections describe the various operator control buttons (Figure 2-11): Function Keys-1, Cursor Keys-2, Display Keys-3, Mode Keys-4, Numeric Keys-5, Alpha Keys-6, Jog Keys-7 and Override Keys-8.

Function Keys

RESET. Pressing the RESET button resets or cancels an alarm; it can also be used to cancel an automatic operation. An alarm should only be cancelled if its cause has been eliminated. When the RESET button is pressed during automatic operation, all program-commanded axis feeds and spindle revolutions are cancelled. The program will return to its starting block when this button is pressed.

POWER UP. This button is used after the POWER ON button is pressed to start up the machine. Pressing this button initializes and positions all of the axes of the machine to the machine zero location and places tool #1 in the spindle. (On legacy controls this button was called POWER UP/RESTART.)

RECOVER. This button is used to recover from automatic tool changer malfunctions. When pressed, the operator is prompted on the display to perform the steps necessary to recover.

F1–F4. These function buttons are used in conjunction with other actions depending on what tab is active (e.g., when inputting data for offsets).

TOOL OFFSET MEASURE. This button is pressed to register tool offset into the register during tool length measurement.

NEXT TOOL. During the tool measurement process, this button initiates placing the next tool into the spindle from the ATC when pressed.

TOOL RELEASE. This button is used to release the tool in the spindle during setup procedures. One of the MDI, ZERO RETURN or HANDLE/JOG modes must be active. Note: You must place your hand around the tool being removed to avoid having the tool drop. There is also a button on the spindle housing that does the same

thing and this proximity makes it easier. Some shops Lock Out (Setting 76) this function because of the potential of accidentally dropping tools onto the worktable.

PART ZERO SET. This button is pressed to register the position of the axes in order to register the work offset (Part Zero) into the offset page. Be sure to compensate for the radius of the measuring tool after setting this value. More detail will be given in the procedures section later.

Cursor Keys

HOME. When this button is pressed, while using the EDIT mode, the cursor position is moved to the program beginning (the first block of the program).

UP, RIGHT, LEFT, DOWN. These buttons are the direction arrows that determine which direction the cursor is moved with the active program while in the EDIT mode. Each press of the UP or DOWN arrow moves the cursor one block at a time. When using the RIGHT or LEFT buttons the cursor is moved one program word at a time.

PAGE UP or PAGE DOWN. When pressed, while using the EDIT mode, the cursor position is moved either up or down one full viewing page. This is an effective method to advance through the program in either direction based on the number of lines the screen can display. The last block of a given page becomes the first block of the next page. The page keys allow for scrolling through long programs more effectively.

END. When this button is pressed, while using the EDIT mode, the cursor is moved to the program end (the last block of the program).

Display Keys

Here we describe the function of buttons related to whichever area of the display is active when they are pressed while the machine is in MEMORY mode (Figure 2-12).

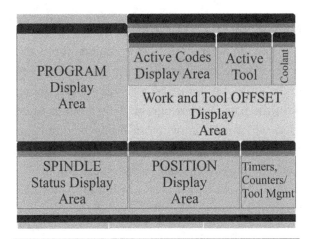

Figure 2-12. Main display screen areas

PROGRAM. Pressing this button activates the program pane of the display screen. The area of the screen changes to white when it is activated, and data in this pane can only be adjusted when it is active.

POSITION. This button, when pressed, selects which position display is present in the POSITION display area.

OFFSET. This button, when pressed, selects either the tool or work offset to be displayed.

CURRENT COMMANDS. When this button is pressed, it changes the display to include Timers, Active Codes, Advanced Tool Management information as well as Tool Table and Pallet information on machines so equipped.

ALARMS. Pressing this button displays a window with detailed information, related to an alarm that has occurred in the control.

DIAGNOSTIC. This button is used by authorized service technicians to access diagnostic information for maintenance and repair.

SETTING. When this button is pressed, it allows access to user settings. *Note:* These settings should only be adjusted by a setup person or lead machinist who fully understands the consequences of any changes made.

HELP. Pressing the HELP button gives the operator access to an assortment of help screens on how to operate the machine functions such as G-codes and M-codes.

Mode Keys

Descriptions of the various mode key functions follows.

EDIT

Pressing this button selects the program edit mode. The EDIT mode enables you to enter the part program to control memory via the numeric and alpha keys, enter any changes to the program and transfer program data via communication interfaces (USB, or ethernet) to or from an offline storage device. The Haas Visual Programming System (VPS) is accessed from this mode also.

INSERT. Use this button to input desired data, entered into the input line or from the clipboard, into the program. The data will be placed after the position of the cursor in the program.

ALTER. This button is used to replace an entire program word.

DELETE. This button is used to delete a single item from where the cursor is placed or an entire program block, if selected.

UNDO. The UNDO button allows up to 40 changes and deselects a highlighted block.

MEMORY

The following buttons are related to the automatic operation of the machine. Activating one of these buttons has an effect on the operation. When the Memory mode is active, CNC programs stored in memory are executed when the CYCLE START button is pressed.

SINGLE BLOCK. When this button is pressed, the execution of a SINGLE BLOCK of information is executed. Each time the CYCLE START button is pressed, only one block of information will be executed. This switch can also be used if you intend to check the initial performance of a new program on the machine or when the momentary interruption of a machine's work is necessary.

GRAPHICS. When this button is pressed, access is given to the graphics display used for program simulation.

OPTION STOP. When this button is pressed, the OPTIONAL STOP mode is active. The OPTIONAL STOP function interrupts the automatic cycle of the machine if the program word M01 appears in the program. Quite often, function M01 is placed in the program after the work of a particular tool is completed or before a tool change. This enables you to perform chip removal or a routine measurement directly on the machine and, if necessary, adjust, and then rerun the same tool to correct inaccuracies.

BLOCK DELETE. When this button is pressed and is active simultaneously in the memory mode, the controller skips execution of the program blocks that are preceded by the slash (/) symbol and that end with the end of block (;) character. If the BLOCK DELETE is not active, all blocks, regardless of the symbol (/), will be executed.

MDI—Manual Data Input

Pressing the MDI mode button enables automatic control of the machine, using information entered in the form of program blocks without interfering with the basic part program. This mode is often used when machining work-holding equipment such as soft jaws and during setup. It corresponds to single moves (milling surfaces, drilling holes), descriptions of which need not but may be saved to memory storage.

COOLANT. Pressing this button controls coolant flow during manual or automatic operation.

HANDLE SCROLL. By activating this button, the HANDLE/MPG can be used to scroll through menus while the controller is in the HANDLE/JOG mode.

ATC FWD. When this button is pressed the ATC will rotate forward (number sequence) to the next tool in the magazine and place it into the spindle. The control must be in the MDI mode.

ATC REV. When this button is pressed, the ATC will rotate reverse (number sequence) to the previous tool in the magazine and place it into the spindle. The control must be in the MDI mode.

HANDLE/JOG

When the control is in the HANDLE/JOG mode, the increment or feedrate amount is determined by which subsequent button is active. The desired axis button (and direction, positive or negative) must be pressed prior to movement when using JOG. Conversely, the direction of movement for the HANDLE is as marked on the control pendant. *Note:* For all HANDLE movements, regardless of the axis selected, negative equals counterclockwise (CCW) and positive equals clockwise (CW) rotation. One full revolution of the HANDLE (360 degrees) corresponds to 100 units on the scale.

.0001/1. This button causes movement of the selected axis one ten thousandth of an inch per click when using the HANDLE, or one hundred thousandths (.1) of an inch per minute (IPM) when using the JOG button. Press and hold the axis direction needed to execute movement to JOG. This button is used when you are precisely dialing-in the part zero of the workpiece or when you are determining the tool length offset.

.001/1. This button causes movement of the selected axis one thousandth of an inch per click on the dial when using the HANDLE, or one (1.0") inch per minute (IPM) when using the JOG button. Press and hold the axis direction needed to execute movement to JOG.

.01.10. This button causes movement of the selected axis ten thousandths of an inch per click on the dial when using the HANDLE, or one (10.0") inches per minute (IPM) when using the JOG button. Press and hold the axis direction needed to execute movement to JOG.

.1/100. This button causes movement of the selected axis one hundred thousandths of an inch per click on the dial when using the HANDLE, or one (100.0") inches per minute (IPM) when using the JOG button. Press and hold the axis direction needed to execute movement to JOG. **CAUTION:** If the handle is rotated quickly while the magnitude is set at .1/100., the tool will move at a rapid feedrate and a crash could occur!

Zero Return

This mode displays axis position in the following categories: Operator; Work (G54); Machine and Distance-To-Go. The tab selected determines which is displayed.

ALL. Pressing of this button returns all axes to machine zero in the same way that POWER UP does at machine startup, excluding the tool change.

ORIGIN. When this button is pressed while in the HANDLE/JOG mode, this button sets the selected axis data to zero. This button is often used in conjunction with the Operator Position display to track incremental movements somewhat like a digital readout (DRO). Steps to perform this function are:

1. HANDLE/JOG mode active.
2. Operator Position Display active.
3. Select desired axis with the respective jog key.
4. Press ORIGIN.
5. Repeat for each axis, as needed.

Another function in the HANDLE/JOG mode and with the Offset Display active, is that when ORIGIN is pressed, a dialog will display with options to clear data from the offset page by cell, column, row, work offsets and even clear all tables. Obviously, care should be taken not to erase data unintentionally.

SINGLE. By using this button, a single axis can be returned to machine zero position by first pressing the desired axis key and then the SINGLE AXIS button.

HOME G28. Pressing the HOME G28 button causes the machine to return to the machine zero position for each axis in relation to the machine coordinate system at the rapid traverse rate (maximum federate). The Z-axis is moved first then the X and Y axes are moved simultaneously.

List Program

The LIST PROGRAM button is used to access the program file memory, load and save files to memory or offline locations and to monitor storage capacity.

SELECT PROGRAM. When the cursor is used to navigate to a specific file, this button is used to select the highlighted program making it active.

BACK ARROW. This button allows the user to return to a previous display screen and operates in a similar way as the Back button on a web browser.

FORWARD ARROW. This button allows the user to return to a previous display screen and operates in a similar way as the Forward button on a web browser (functions only after the Back Arrow has been used).

ERASE PROGRAM. Pressing the Erase Program button will delete the highlighted program from the List Program memory. It can also be used to delete lines of code that are entered into the MDI program screen.

Numeric Keys

This keypad of numbers and symbol characters is used to input data while writing or editing

programs at the control. These keys are also used to enter numerical data and offsets into memory. Many of the keys are used in conjunction with other keys. Because of limited space on the control panel, some keys have two characters on them. When the letter or symbol indicated in the upper left corner of the key is needed, the operator first presses the SHIFT key, which switches the key to that character. This sequence must be followed each time an alternate letter is needed. The SHIFT key functions the same way as its equivalent on a computer keyboard. When the SHIFT key is pressed, the desired (second) character on the key may be entered.

CANCEL. The CANCEL key is used while inputting data in the MDI mode. It is essentially a destructive backspace key that can be used to correct an erroneous entry. Press this key to delete the last character or symbol input to the key input buffer.

SPACE. Use this key to add a space between characters.

ENTER. Use the ENTER key to accept display prompts and to input data such as offsets. After the data are entered via the keypad, the ENTER key is pressed. The incremental value of the data is entered into the offset register.

Alpha Keys

This keypad of letters and symbol characters is used to input data while writing or editing programs at the control. As with the numeric keypad, many of the keys are used in conjunction with other keys. The standard alpha keypad includes three second function keys that are accessed by pressing the SHIFT key on the alpha keypad.

Jog Keys

These buttons are used to select the manual feed and axis direction. Pressing these buttons executes movement along the selected axis in the positive or negative direction relative to the machine coordinate system. The same is true for each of the linear and rotational axis buttons. As long as the button is held, the axis will move at a feedrate determined by the HANDLE/JOG setting until released.

CHIP FWD. Pressing this button activates the chip removal auger (if equipped) in the forward direction.

+B/+A/C. These buttons allow the user to move the selected rotary axis (if equipped) manually in the positive direction at a specific feedrate as determined by the HANDLE/JOG increment chosen. To access +B, the SHIFT key must be pressed first.

+Z. This button allows the user to move the Z-axis manually in the positive direction at a specific feedrate as determined by the HANDLE/JOG increment chosen.

−Y. This button allows the user to move the Y-axis manually in the negative direction at a specific feedrate as determined by the HANDLE/JOG increment chosen.

CLNT UP. This button allows the programmable coolant nozzle to be moved up to a preset position (if equipped).

CHIP STOP. Pressing this button stops the chip removal auger (if equipped).

+X. This button allows the user to move the X-axis manually in the positive direction at a specific feedrate as determined by the HANDLE/JOG increment chosen.

JOG LOCK. When JOG LOCK button is pressed and then the axis jog direction is selected movement will continue at the specified feedrate determined by the

HANDLE/JOG increment chosen until the button is pressed again. CAUTION: A collision could result.

−X. This button allows the user to move the X-axis manually in the negative direction at a specific feedrate as determined by the HANDLE/JOG increment chosen.

CLNT DOWN. This button allows the Programmable Coolant nozzle to be moved down to a preset position (if equipped).

CHIP REV. Pressing this button activates the chip removal auger (if equipped) in the reverse direction (sometimes needed to clear a jam).

+Y. This button allows the user to move the Y-axis manually in the positive direction at a specific feedrate as determined by the HANDLE/JOG increment chosen.

−Z. This button allows the user to move the Z-axis manually in the negative direction at a specific feedrate as determined by the HANDLE/JOG increment chosen.

−A/C/−B. These buttons allow the user to move the selected rotary axis (if equipped) manually in the negative direction at a specific feedrate as determined by the HANDLE/JOG increment chosen. To access +B, the SHIFT key must be pressed first.

AUX CLNT. Use this button to turn ON/OFF the Through-Spindle Coolant function (if equipped).

Override Keys

−10% FEEDRATE. Pressing this button decreases the programmed feedrate in 10 percent increments each time it is pressed. Zero percent of the programmed feedrate (stopped) is the minimum result possible.

100% FEEDRATE. This button is pressed to reestablish the spindle feedrate to 100 percent of the programmed value.

+10% FEEDRATE. Pressing this button increases the programmed feedrate in 10 percent increments each time it is pressed. Nine hundred ninety-nine percent of the programmed feedrate is the maximum result possible (up to the machine maximum specification only).

HANDLE FEED. By pressing this button, adjustment control of the programmed feedrate in one percent increments is completed by using the MPG.

Note: On legacy controls there was a Dry Run button. When activated during automatic cycle, all of the rapid and work feeds are changed to the maximum feed (rapid traverse) instead of the programmed feed. This feature was used to check a new program on the machine without any work actually being performed by the tool. This is particularly useful on programs with long cycle times to progress through the program more quickly. *Note:* Use caution when using this function. It is NOT intended for metal cutting. Using the Feedrate Override maximum percentage value accomplishes the same thing.

Spindle

The spindle buttons are used during manual machine operation and for setup functions. The following descriptions explain their specific functions.

−10% SPINDLE. Pressing this button decreases the spindle RPM in 10 percent increments each time it is pressed. Zero percent of the programmed spindle RPM is the minimum result possible or zero RPM.

100% SPINDLE. This button is pressed to reestablish the spindle RPM to 100 percent of the programmed value.

+10% SPINDLE. Pressing this button increases the spindle RPM in 10 percent increments each time it is pressed. Nine hundred ninety-nine percent of the programmed spindle RPM is the maximum result possible (up to the machine maximum specification only).

HANDLE SPINDLE. By pressing this button, adjustment control of the programmed spindle RPM in 1 percent increments is completed by using the MPG.

FWD. By pressing the Spindle FWD button while in one of the operation modes MEM, MDI or HANDLE/JOG the spindle will start rotation in the CW direction. The spindle RPM is adjusted by using the Spindle Override % buttons. When set at 100 percent, the spindle will rotate the last RPM commanded in the program. The spindle command is retained and will restart upon pressing the FWD button. When the machine is first started, there has been no value established for the RPM. Therefore, if one of these buttons is pressed while in the modes listed above, the spindle will not start. An RPM must be input via MDI or by activating the program with the CYCLE START button. From that point on, as long as the machine is not turned off, the RPM will be activated at the last commanded value when one of these buttons is pressed.

STOP. Pressing the STOP button stops spindle motor rotation while in one of the operation modes listed above. However, pressing this button will NOT stop the spindle while in any of the Automatic execution modes.

REV. Pressing the REV button functions in exactly the same way as pressing the FWD button except that the spindle will start rotation in the CCW direction.

Rapid Overrides

The rapid feed override percentage can be adjusted to control the rapid feedrate. These buttons are used to reduce the programmed rapid feedrate (G00). When override buttons are used, they are incremented in percentage steps as follows:

5% RAPID. If the 5% button is pressed all of the programmed rapid feedrate movements will be reduced to 5 percent of the maximum that the machine can generate. It is advised to use this override when running a program for the first time.

25% RAPID. If the 25% button is pressed, all of the programmed rapid feedrate movements will be reduced to 25 percent of the maximum that the machine can generate.

50% RAPID. If the 50% button is pressed, all of the programmed rapid feedrate movements will be reduced to 50 percent of the maximum that the machine can generate.

100% RAPID. If the 100% button is pressed, all of the programmed rapid feedrate movements will be at the maximum the machine can generate.

Control Panel Functions

The control panel described here is quite typical of the control panels used on CNC machines. The control panel switches and buttons may be distributed differently on the panel for each individual machine; however, the purpose and function of each switch and button remains the same. Some control panels are equipped with

additional buttons or switches not shown here. Definitions and applications of these buttons or switches can be found in the manufacturer operation manuals for specific machines. The operator pendant is located at the front of the machine and is equipped with a display and with various buttons and switches, as illustrated in Figure 2-1. Today, machine tools are equipped with ethernet connectivity to allow access to your company's network, thereby offering unlimited part program and tool file storage with easy access via Windows-based operating systems. For this controller, the USB port is on the right side. Ethernet and USB are used as a file storage medium and transfer.

Display

The display screen is where all the program characters and data are shown. This display has a 15-inch diagonal viewing area. The Haas New Generation Control displays are also color enhanced. When a specific display area is active, it will have a white background. The area must be active in order to access the data to make changes.

The general screen display layout is shown below in Figure 2-13. Brief details are given for the individual display areas.

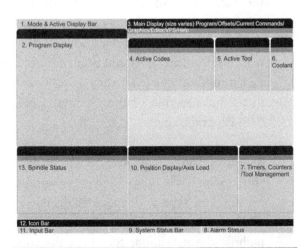

Figure 2-13. Main display area descriptions

Note: The operator should always consult the manufacturer manual for more specific detailed instructions on the display readouts.

1. Mode and Active Display Bar. There are three different modes that will be displayed: Setup, Edit and Operation. The mode that is currently active is displayed in the upper left of the screen (Item 1. Figure 2-13). For example, when the ZERO RETURN or HANDLE/JOG buttons are pressed, the active display bar will display SETUP, ZERO RETURN or JOG, respectively.

2. Program Display. The active program is displayed here. When the program exceeds the display area capacity, the Page keys can be used to progress to the part of the program desired.

3. Main Display. This area of the display accesses multiple functions and the size fields changed based on which data are being accessed: Program, Offsets, Current Commands, Settings, Graphics, Editor and Help.

4. Active Codes. This area of the display lists all codes that are currently active in the CNC control based on the default start-up values or those commanded in the program as it progresses. This display is for information only and is not editable. Learn more about what each code means in the next section of this chapter (CNC Mill Programming).

5. Active Tool. This area of the display shows which tool is mounted in the spindle by number, type and image is displayed of the tool. Other items displayed here include the maximum tool load the tool has experienced, percentage of tool life remaining (if programmed) and the next tool and its pocket number.

6. Coolant. When the control is in the Operation MEM mode, this display indicates if the coolant is ON or OFF, the position of the coolant nozzles (if equipped) and the coolant tank level. This display is only available on Haas "New Generation Controls".

7. Timers, Counters/Tool Management. This display is used to show cycle times, parts counters and tool management data. There are other counters that are covered in detail in the Operator Manual for the machine.

8. Alarm Status. When an alarm is triggered, there will be a (brief) description and a number for the alarm highlighted in red. The machine will not function until the problem is remedied. To learn more detailed information about the alarm, press the ALARM Display button on the control. After correcting the cause of the alarm, press the RESET button on the control to clear it. Refer to the Operators Manual for the machine that contains a list of alarms by number, the possible causes and corrective action recommendations.

9. System Status Bar. This area on the screen displays messages for the operator.

10. Position Display/Axis Load. This display has multiple options for showing the actual location of the machine axes within the work envelope. By pressing the POSITION button on the display one to five times, each of the following will be displayed:

WORK OFFSET (G54–G59, etc.). *This position display lists the machine axes* *coordinate position in relation to the measured work offset that is active based in the CNC program being used or the default at startup (G54).*

DIST TO GO (Distance to go). *This position display lists the remaining distance each axis must travel to reach the programmed value. This display is especially useful during the first-run of a new program setup.*

MACHINE. *This position display lists the axes in relation to machine home (extent of positive axis travels in most cases). For example, at startup of the machine, when the POWER UP button is pressed, the machined is moved to this location. If the POSITION button is pressed until the MACHINE item is displayed, then all of the axes will read zero.*

OPERATOR. *This position display is used to show distance traveled from one point to the next and can be set to zero (see ZERO RETURN, ORIGIN button descriptions for details). This display can be useful during manual operations similar to a DRO.*

ALL. *This position display lists all four of the displays listed above at the same time.*

11. Input Bar. This area is where data entered at the control via the alpha and numeric keypads is temporarily held until the ENTER or F1 key is pressed (for program or tool data). On some controls, this is called the input buffer.

12. Spindle Status. This display area shows information related to the spindle. It includes STOP or rotation direction, override percentages, color-coded load conditions and active speed and feed data.

CNC Mill Operation and Setup Procedures

The following explanations are for operations considered routine for users of CNC machine tools and are given in their sequence of use. Please note that these procedures are specific to the Haas New Generation Controller depicted in Figure 2-1. The procedures for another type of control may be similar. Be sure to consult the manufacturer manuals specific to your machine tool operation and control panel.

Process Planning Documents

The information given on the engineering drawing or blueprint will include the material, overall shape and the dimensions for part features (Figure 1-91). The part geometry determines the type of machine (mill or lathe) to be used to produce the part. By studying the engineering drawing or blueprint, material and operations (drilling, milling, boring, etc.) can be identified. The tools and work-holding method can also be determined. Occasionally, the geometry will require multiple machines to manufacture the part; thus, additional operations will be necessary.

Operation Sheet

The purpose of this planning document (sometimes called a traveler) is to identify the correct order for operations to be performed and the machine/s to be used. The operation sheet is particularly useful in a production environment when many identical parts are machined. The operation sheet is similar to directions or a how-to approach. The process needed to manufacture the finished part has been decided in advance and is documented for future use.

When small batches of parts are to be made, there may not be an operation sheet. It is the machinist's responsibility to study the engineering drawing or blueprint and decide the necessary steps and machines needed to manufacture the part. An operation sheet can aid in the decision-making process (Chart 2-1). This document is usually provided by the manufacturing engineering department.

Date: Today	Prepared By: You
Part Name: CNC Mill Project	Part Number: 1235
Order Quantity: 100	Sheet 1 of 1
Material: 6061 T6 Aluminum	
Raw Stock Size: 4.5" x 1.25" bar stock	

Operation Number	Machine Used	Operation Description	Time Standard
1	Saw	Cut the Bar Stock to 3.0625" lengths.	3 minutes each
2	Manual Mill	Machine Blanks to 4.25" x 3.75" finished size.	20 minutes
3	CNC Lathe	Machine complete to dimensions.	10 minutes each
4	Bench	Deburr as Needed.	5 minutes each
5	Quality Control	Final Quality Control Inspection.	15 min. each

Chart 2-1. Process planning operation sheet

CNC Setup Sheet

The CNC setup sheet is the document that informs the machinist what tools are to be used and any specific information related to setup and cutting tools. For example, depending on the part geometry features, it may be necessary to have a certain amount of tool projection/extension for a drill to be able to completely machine through the part without the tool holder contacting the part or work-holding system. CNC Setup Sheets consist of: the program number, a graphical representation of the machined part, the location of workpiece coordinate (WCS) origins (part zero), and each individual cutting tool used and the tool number (Chart 2-2).

In Chapter 4 of this text, you will be introduced to CNC Programming with CAD/CAM, and how you can develop and output CNC setup sheets directly from the CAD/CAM programs. Companies today are going to "paperless"

factories, wherein these documents will be available and viewed on the intranet in electronic form.

Determining the Work-Holding Method

Many tool and work-holding methods used on manual machines are also used on CNC machines. The machines themselves differ in their method of control, but otherwise they are very similar. The shape or geometry of the workpiece affects the metal-cutting operation and plays a large part in determining the type of work-holding method that must be used. The clamping method must hold the workpiece securely, be rigid, and minimize the possibility of any flex or movement of the part.

Of the typical work-holding methods used in milling operations—the vise, step block and strap clamps, and a universal 3-jaw chuck—the

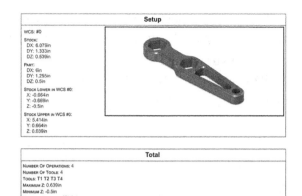

Setup

WCS: #0

STOCK:
DX: 6.079in
DY: 1.333in
DZ: 0.539in

PART:
DX: 6in
DY: 1.255in
DZ: 0.5in

STOCK LOWER IN WCS #0:
X: -0.664in
Y: -0.669in
Z: -0.5in

STOCK UPPER IN WCS #0:
X: 5.414in
Y: 0.664in
Z: 0.039in

Total

NUMBER OF OPERATIONS: 4
NUMBER OF TOOLS: 4
TOOLS: T1 T2 T3 T4
MAXIMUM Z: 0.639in
MINIMUM Z: -0.5in
MAXIMUM FEEDRATE: 65in/min
MAXIMUM SPINDLE SPEED: 10000rpm
CUTTING DISTANCE: 74.178in
RAPID DISTANCE: 10.461in
ESTIMATED CYCLE TIME: 2m:15s

Tools

T1 D1 L1
TYPE: face mill
DIAMETER: 2in
LENGTH: 1.563in
FLUTES: 5
DESCRIPTION: 2" Face Mill
VENDOR: Maritool
PRODUCT: MSAP16-D050A05R-25.4
MINIMUM Z: 0in
MAXIMUM FEED: 65in/min
MAXIMUM SPINDLE SPEED: 6500rpm
CUTTING DISTANCE: 7.843in
RAPID DISTANCE: 0.824in
ESTIMATED CYCLE TIME: 7s (5.4%)
HOLDER: Maritool CAT40-FMA1.0-1.5M
VENDOR: Maritool
PRODUCT: CAT40-FMA1.0-1.5M

T2 D2 L2
TYPE: center drill
DIAMETER: 0.437in
TIP ANGLE: 118°
LENGTH: 0.604in
FLUTES: 2
DESCRIPTION: #5 Center Drill
MINIMUM Z: -0.5in
MAXIMUM FEED: 29in/min
MAXIMUM SPINDLE SPEED: 6500rpm
CUTTING DISTANCE: 0.7in
RAPID DISTANCE: 1.579in
ESTIMATED CYCLE TIME: 1s (1.1%)
HOLDER: Maritool CAT40-ER32-2.35
VENDOR: Maritool
PRODUCT: CAT40-ER32-2.35

T3 D3 L3
TYPE: drill
DIAMETER: 0.375in
TIP ANGLE: 118°
LENGTH: 3.85in
FLUTES: 1
DESCRIPTION: 3/8
MINIMUM Z: -0.5in
MAXIMUM FEED: 29in/min
MAXIMUM SPINDLE SPEED: 6500rpm
CUTTING DISTANCE: 0.7in
RAPID DISTANCE: 1.579in
ESTIMATED CYCLE TIME: 1s (1.1%)
HOLDER: Maritool CAT40-APU13 Drill Chuck
VENDOR: Maritool
PRODUCT: CAT40-APU13

T4 D4 L4
TYPE: flat end mill
DIAMETER: 0.25in
LENGTH: 0.85in
FLUTES: 3
DESCRIPTION: 1/4" Flat Endmill
MINIMUM Z: -0.465in
MAXIMUM FEED: 60in/min
MAXIMUM SPINDLE SPEED: 10000rpm
CUTTING DISTANCE: 64.935in
RAPID DISTANCE: 6.48in
ESTIMATED CYCLE TIME: 1m:5s (48.1%)
HOLDER: Maritool CAT40-ER32-2.35
VENDOR: Maritool
PRODUCT: CAT40-ER32-2.35

Operations

Operation 1/4
DESCRIPTION: Face1
STRATEGY: Facing
WCS: #0
TOLERANCE: 0in
MAXIMUM STEPOVER: 1.9in
MAXIMUM Z: 0.63in
MINIMUM Z: 0in
MAXIMUM SPINDLE SPEED: 6500rpm
MAXIMUM FEEDRATE: 65in/min
CUTTING DISTANCE: 7.843in
RAPID DISTANCE: 0.824in
ESTIMATED CYCLE TIME: 7s (5.4%)
COOLANT: Flood
T1 D1 L1
TYPE: face mill
DIAMETER: 2in
LENGTH: 1.563in
FLUTES: 5
DESCRIPTION: 2" Face Mill
VENDOR: Maritool
PRODUCT: MSAP16-D050A05R-25.4

Operation 2/4
DESCRIPTION: Drill3
STRATEGY: Drilling
WCS: #0
TOLERANCE: 0in
MAXIMUM Z: 0.639in
MINIMUM Z: -0.5in
MAXIMUM SPINDLE SPEED: 2620rpm
MAXIMUM FEEDRATE: 29in/min
CUTTING DISTANCE: 0.7in
RAPID DISTANCE: 1.579in
ESTIMATED CYCLE TIME: 1s (1.1%)
COOLANT: Flood
T2 D2 L2
TYPE: center drill
DIAMETER: 0.437in
TIP ANGLE: 118°
LENGTH: 0.604in
FLUTES: 2
DESCRIPTION: #5 Center Drill

Operation 3/4
DESCRIPTION: Drill4
STRATEGY: Drilling
WCS: #0
TOLERANCE: 0in
MAXIMUM Z: 0.639in
MINIMUM Z: -0.5in
MAXIMUM SPINDLE SPEED: 3060rpm
MAXIMUM FEEDRATE: 29in/min
CUTTING DISTANCE: 0.7in
RAPID DISTANCE: 1.579in
ESTIMATED CYCLE TIME: 1s (1.1%)
COOLANT: Flood
T3 D3 L3
TYPE: drill
DIAMETER: 0.375in
TIP ANGLE: 118°
LENGTH: 3.85in
FLUTES: 1
DESCRIPTION: 3/8

Operation 4/4
DESCRIPTION: 2D Pocket1
STRATEGY: Pocket 2D
WCS: #0
TOLERANCE: 0.004in
STOCK TO LEAVE: 0.02in
MAXIMUM STEPOVER: 0.238in
MAXIMUM Z: 0.639in
MINIMUM Z: -0.465in
MAXIMUM SPINDLE SPEED: 10000rpm
MAXIMUM FEEDRATE: 60in/min
CUTTING DISTANCE: 64.935in
RAPID DISTANCE: 6.48in
ESTIMATED CYCLE TIME: 1m:5s (48.1%)
COOLANT: Flood
T4 D4 L4
TYPE: flat end mill
DIAMETER: 0.25in
LENGTH: 0.85in
FLUTES: 3
DESCRIPTION: 1/4" Flat Endmill

Generated by Fusion 360 CAM 2.0.5119   1

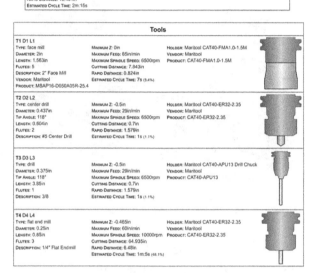

Chart 2-2. CNC milling setup sheet, *courtesy Autodesk Fusion 360*

most common device is the machining vice, as shown in Figure 2-14. Precision ground parallels are used with the vise to hold the part at a height

that allows ample clamping and clearance for tools to penetrate the workpiece. Remember to clamp as much in the vise as possible to ensure rigidity. A common parallel set (Figure 2-15) includes a range from 1.0-inch tall up to 2.125 inches in .125-inch increments. Before installing parallels, confirm they are both the same size and immaculately clean.

Positioning of the part in the vise is critical. When possible, center your part in the vise. If the part is small and requires clamping on one end of the vise, it may be necessary to insert an exact-size part at the opposite side of the vise to equalize clamping pressure. Never extend a large portion of the material outside the vise jaw.

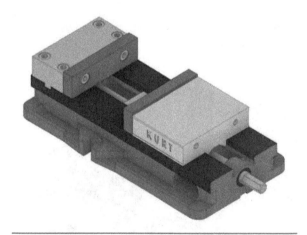

Figure 2-14. Kurt D810 vise, *courtesy Kurt Manufacturing*

Figure 2-15. Kurt parallel set, *courtesy Kurt Manufacturing*

Another accessory used in conjunction with the vise is a work stop. Figure 2-16 shows a *groove stop* and Figure 2-17 shows an *adjustable stop*. Both of these options allow for consistent location of parts within the vise for production runs.

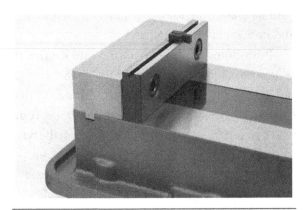

Figure 2-16. Kurt groove stop, *courtesy Kurt Manufacturing*

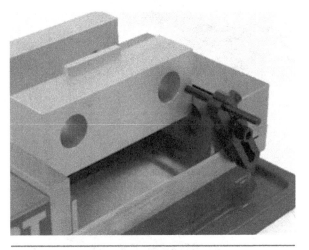

Figure 2-17. Kurt adjustable stop, *courtesy Kurt Manufacturing*

In Figure 2-17, special vise jaws are being readied for machining that will allow clamping of a part associated with the features cut into them. These are called *soft jaws* because they are machined into a mating shape to match the part to be machined. To machine a cylindrical part, soft jaws can be used after first machining the jaws to the exact diameter of the surface to be clamped.

Note: When machining soft jaws, the vise must be clamped. To do this, insert a piece of raw bar stock or something slightly smaller than the diameter to be machined. This same concept is used for soft jaws on a lathe.

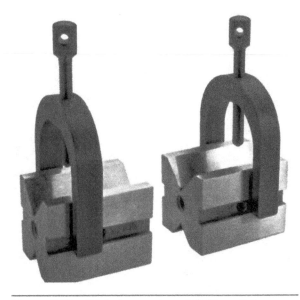

Figure 2-18. V-blocks. *courtesy, L.S. Starrett Company*

Precision V-blocks (Figure 2-18) can be inserted in the vise when clamping of a cylindrical part is required. In most cases, when this is necessary, the clamps are removed and the V-blocks are placed on the movable jaw side creating three points of contact for clamping.

Sometimes, when the workpiece is too large for a vise or other fixture, the part can be clamped directly to the worktable using the components from a *hold down clamp kit* as shown in Figure 2-19.

When using step blocks, T-nuts and strap clamps to hold a part, be sure the strap clamp is parallel with the table or part surface and possibly a little higher on the non-part side. Also, place the hold down stud as close to the material being clamped as possible, in order to transfer the maximum clamping pressure to the material. Notice that the strap clamps are parallel with the table in Figure 2-20. For successful results using this method, proper alignment as shown is essential. **CAUTION:** The machine table can be damaged if tool paths are programmed to exceed the thickness of the clamped part.

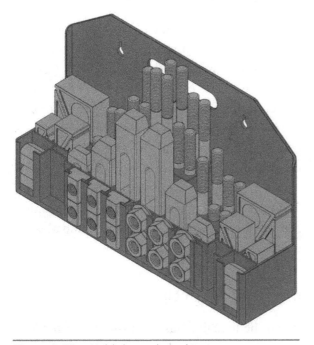

Figure 2-19. Hold down clamp kit, *courtesy Autodesk Fusion 360*

Figure 2-20. Proper clamping setup using step blocks and strap clamps, *courtesy Autodesk Fusion 360*

If the part shape characteristics won't allow clamping in a vise, a fixture can be used. In Figure 2-21, a fixture includes locating pins and clamping screws. The part is placed into the fixture in contact with the primary, secondary and tertiary datum planes and is then clamped. The fixture can be held in a vise or clamped to the worktable.

When clamping on a cylindrical workpiece surface is required, a 3-jaw universal chuck can be mounted to the machine table. It may be

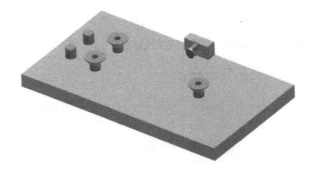

Figure 2-21. Part clamping using a fixture

necessary to use a chuck-clamping plate to mate the chuck to the table where the plate matches a pattern of holes in the chuck for fastening the plate to the chuck. Then the plate is clamped to the table.

Mounting Work-Holding Devices

For CNC mills, the work-holding device is mounted directly to the machine table. Some features of the device must be aligned to be parallel to an axis. When a vise is used, this alignment is to the fixed-jaw (non-movable).

If strap clamps and step blocks are used, follow appropriate mounting practices (see Figure 2-20); alignment to a machined edge is required. If a fixture is used, there are alignment features designed into the fixture that can be used to verify alignment. It cannot be over-emphasized that the clamping surfaces of the vise or fixture and the table must be clean and free of burrs. Even small metal chips under the clamping fixture can cause inaccurate results. Prior to fixture clamping, a honing stone should be used on the table to lightly hone the locating surface. Doing this ensures that the locating surface does not have any burrs or dings that would affect parallelism. Before installing the vise, put a light coat of Starrett M1 lubricant oil to keep the table and vise from developing rust and corrosion.

Routine CNC Milling Machine Procedures

Machine Startup

Complete the following steps to start the machine:

1. Confirm that the power disconnect to the machine is ON.
2. On the machine electrical enclosure, turn on the main power switch.
3. At the machine control pendant, press the POWER ON button.

   As the machine computer starts, a sequence of warnings will display on the screen. Heed these warnings at all times. When the boot-up finishes, the display will be active and an alarm will be displayed (highlighted red) in the lower right-hand corner (alarm status field) that reads "102 Servos Turned Off."

4. Check the E-Stop button to see if it has been pressed and release it by turning it clockwise.

5. Press the RESET button to clear the Alarm status.
6. With the machine doors closed, press the POWER UP button.

The machine will begin moving the Z-axis to its maximum positive location and then the X and Y-axes will move simultaneously to their maximum positive locations which is *machine home*. At this point, the machine is ready for use.

Once power-up is complete, the display areas (Figure 2-13) will read: 1. Operation: MEM-Program; 2. The active Program will be displayed; 3. Program Tool Offsets; 4. Active Codes; 5. Active Tool; 6. Coolant; and 10. The Position readout display will be on. Machine and all axes will be zero (Figure 2-22).

Machine Warm Up

When the machine has been turned off for longer than overnight, the machine is considered cold. It is a good idea to run the spindle for sev-

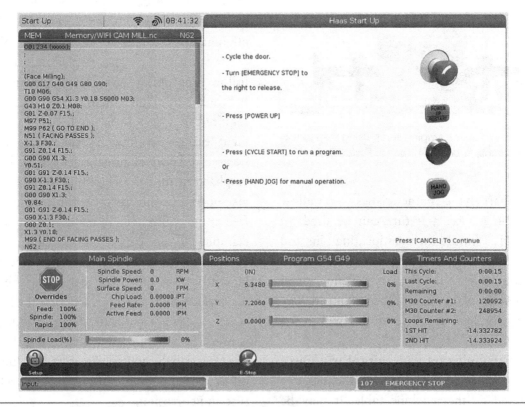

Figure 2-22. Haas CNC mill startup display, *courtesy Haas Automation*

eral minutes prior to starting automatic operation. This can be accomplished by using the MDI mode to program an RPM. Use the alpha and numeric keypads to key in: S500M3. Press the ENTER key and then the CYCLE START button. The spindle will activate at 500 RPM in the clockwise direction. Allow the spindle to run for several minutes. Alternatively, use the Spindle Warmup program that is resident on all Haas controllers. To do this, follow these steps:

1. Press the LIST PROGRAM button and use the cursor key to highlight the Spindle Warmup program from the list.
2. Press the SELECT PROGRAM button.
3. Press the MEMORY button. Verify that the program is listed in the Program Display area of the screen.
4. Press CYCLE START. Allow the program to run. *Note:* The maximum spindle RPM for the machine specification is the maximum it will run, even if a larger programmed amount is given.

Work-Holding Device Alignment Methods

The vise must be installed so that the clamping surfaces (fixed-jaw) are parallel to the X- or Y-axes of machine travel. This is done by using a dial indicator. The following steps make the procedure quick and effective.

Note: Some machining vises have alignment features in the bottom that can be used to quickly locate them off of the T-slot groove. These are in the form of a lock key or sine pin that is installed into mating features on the bottom of the vise. Check with the manufacturer to order them if not included (Figure 2-23).

Note: Always check the alignment with an indicator and make adjustments as needed.

After the prep described above, follow these steps:

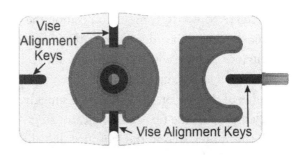

Figure 2-23. Vise alignment keys, *courtesy Kurt Manufacturing*

1. Place the vise on the table and install the t-nuts and studs in position and screw on the flange nuts to lightly touch the mating surface.
2. Tighten the nut on the left side of the vise until snug. Tighten the bolt on the opposite side slightly less snug.
3. Install an empty drill chuck into the spindle.
4. From the MDI mode, key in M19 and then press CYCLE START (this will lock the spindle orientation).
5. Place a dial indicator with a gooseneck connector into the drill chuck (Figure 2-24) and position so the dial is facing forward as shown.

Figure 2-24. Dial indicator with universal gooseneck, *courtesy L.S. Starrett Company*

6. Use HANDLE/JOG mode to position the machine axes in order to bring the indicator tip into contact with the fixed-jaw vertical face and approximately ⅛ inch from the top edge of the jaw on the side that is snuggly clamped. Set the tip of the indicator so it begins to register on the indicator dial and rotate the bezel to zero.

7. Jog the indicator across the fixed-jaw surface and stop near the opposite end of the vise. Take note of the indicator movement and tap the vise (using a soft face or dead blow hammer) in a direction that moves the indicator back to zero.

8. Tighten each of the clamping nuts slightly more.

   Note: With the left nut tighter than the right one, the vise will rotate around this point.

9. Jog the indicator back to the beginning point on the jaw again, and re-set the dial to zero again.

10. Jog back to the opposite side of the vise and tap the vise until your indicator reads zero. Tighten the nuts more.

11. Continue with these steps until the indicator movement stays at zero across the entire indicating surface.

12. Tighten the nuts to manufacturer torque specs.

13. Make one last check across the vise surface to ensure zero indicator movement.

CNC Mill Operation Display

After machine start-up, a common screen is displayed called the Operation: MEM, similar to Figure 2-25. This screen displays data outlined earlier in Figure 2-13. The MEM section displays a portion of the active program. As the program is executed, the program scrolls through the line numbers until the program end (M30) is reached and is then returned to the beginning.

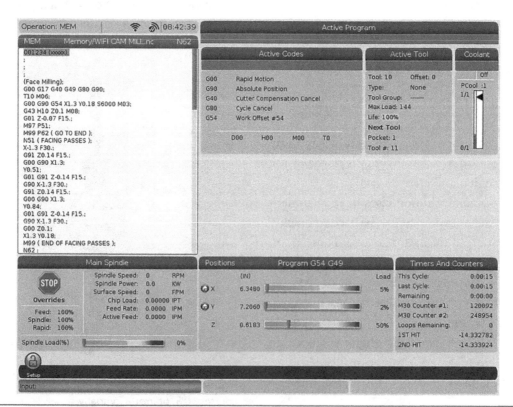

Figure 2-25. CNC mill operation display, *courtesy Haas Automation*

A Program Is Loaded from CNC Memory

The program may be in the List Program directory, but not activated for automatic operation. Follow these steps to activate a program:

1. Press the LIST PROGRAM button to enter the resident memory directory.
2. Use the CURSOR key to navigate to the desired program in the list.
3. With the desired program line highlighted, press the SELECT PROGRAM button.
4. Complete entry of all setup data and work-holding equipment is ready and close machine doors.
5. Press the AUTO cycle key to execute the program.

A Program Is Loaded from USB

Follow these steps to load a file from a USB drive:

1. Press the LIST PROGRAM button to enter the resident memory directory.
2. Use the right cursor to activate the USB tab.
3. Navigate to the desired program from the list using the down arrow.
4. Follow the prompt: F2 to copy selected Files/Programs, Erase Program to delete, Press F1 for command menu and Help for a menu of listings.
5. Press the F2 function key to copy the selected file.
6. Cursor to the Memory tab and press ENTER to set the destination.
   Note: If the Program number already exists in memory, you will be prompted to overwrite it.
7. Press CANCEL to exit.

To access network drives (if available) connected via the ethernet, consult the mill operators manual for directions.

A Program Is Deleted from Memory

1. To delete a program from the controller memory, follow these steps.
2. Press the LIST PROGRAM button.
3. Use the CURSOR key to move to the desired program in the list.
4. With the desired program line highlighted, press the ENTER button to place a check mark in the box.
5. Press the DELETE button. The program will be deleted from memory.

MDI Operations

This mode allows input of small programs via the keypad at the control. Small programs provide an excellent method of executing simple commands like tool changes, controlling the spindle RPM and its rotation direction, etc. To enter the MDI mode of operations, follow these steps:

1. Press the MDI button on the Operator Panel.
2. Type in the CNC codes needed. *Note:* To enter a string of commands (multiple lines of code), press the semicolon button (End-of-Block) on the alpha keypad to separate the lines.
3. Press the ENTER key to accept the entries. More can be added, if needed, by repeating these steps.
4. Press CYCLE START to execute the commands.

For the program number, the control assumes O0000 and the data may be entered. Each block ends with the end-of-block (EOB) character (;) so that individual blocks of information can be kept separately. If you make a typographical error while entering a given block, you can eliminate it by pressing the CANCEL key to cancel the error and then reenter the correct value.

You may execute the MDI program as you do with automatic operation. Please refer to the machine tool manufacturer manual for complete and specific instructions. You can save the MDI program to memory by following these steps:

1. From the EDIT mode, place the cursor on the top line of the program.
2. Press the letter O and enter an unused program number.
3. Press the ALT key. The program will be saved.

You can erase an entire program created in MDI mode by pressing the ERASE PROGRAM button.

To perform an individual MDI operation, use the methods described above.

Example 1

To activate the spindle at 500 RPM in the clockwise direction, key in the command S500 M03.

1. Press ENTER.
2. Press CYCLE START.

Example 2

Use the following directions to position tool number 5 to the active position on the turret (or to install tool 5 into the spindle on a milling machine). Key in the following commands:

1. T5 M6
2. ENTER
3. CYCLE START

Measuring CNC Mill Work Coordinate Offsets

Following is the procedure for setting the Work Offsets (sometimes called Fixture Offsets) for any of the workpiece coordinate systems G54–G59.

Work Offsets can be measured manually by positioning an edge-finding tool to contact with the workpiece zero surface in both X- and Y-axes sequentially. In this procedure, which is called edge-finding, it is nearly always the perpendicular edges (secondary and tertiary datum) of the workpiece that are referenced (Figure 2-26).

Follow these steps for measuring Work Coordinate Offsets:

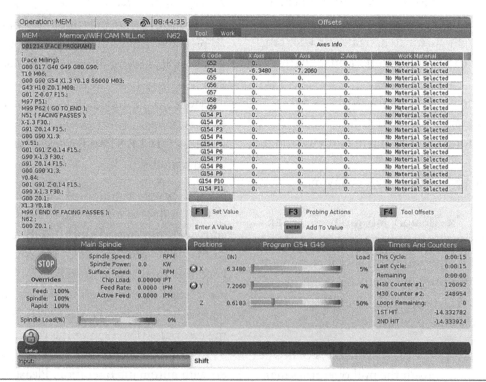

Figure 2-26. CNC mill work offset display, *courtesy Haas Automation*

1. Use the MDI mode to install a drill chuck with an edge-finding device into the spindle (follow the steps in previous Example 2, substituting for the tool number needed).

2. Use the MDI mode to input and RPM of 1000 and start the spindle. (Follow the steps in previous Example 1, substituting for the RPM value to 1000). *Note:* Once you have entered the value and started the spindle, it may be stopped at any time by pressing the RESET key or the STOP button in the Overrides section of the control. To restart the spindle at the same RPM again, merely press the FWD button.

3. Press the HANDLE/JOG button. The offset screen will automatically display and Tool and Work tabs will be present at the top of the page.

4. Position the axes using the HANDLE and the axis increment keys to bring the edge-finding tool into position up to the part surface along the X-axis until the edge-finder tip stops wobbling. Now change to move in small increments until the tip moves slightly (jumps to one side a small amount).

5. Stop the spindle as previously described.

6. Use the arrow keys to navigate to the Work tab (Figure 2-26).

7. Use the arrow keys to navigate to the desired numbered offset (G54–G59).

8. Use the arrow key to move over to the X-axis column and highlight for the relevant work coordinate.

9. Press the PART ZERO SET. The value for the position will be automatically entered and the cursor will advance to the Y-axis column.

10. Use the HANDLE to retract the spindle in the Z-axis to clear the part.

   Note: It is necessary to input the difference between the value input and the edge-finder radius (typically 0.100 inch or 3mm) before automatic operation can be executed.

11. Use the arrow key to return to highlight the column for the X-axis and key in .100 inch and press ENTER. Pressing ENTER adds the incremental value to the offset. This instruction is true when accessing the edge of the part from the left. The value would be -.100 inch if accessing from the edge of the part from the right.

12. Ensure the Y-axis column is highlighted. If it is not, use the arrow key to move to it.

13. Use the HANDLE to jog to a location to access the Y-axis surface that represents the zero coordinate.

14. Jog the Z-axis back to a suitable depth (approximately .100 inch from the top surface).

15. Start the spindle by pressing the FWD button.

16. Repeat the same actions as in step 4 to set the coordinate position for the Y-axis.

17. Repeat the same actions as in step 9.

18. Repeat step 10.

19. Use the arrow key to return to highlight the column for the Y-axis and key in .100 inch and press ENTER to add the incremental value to the offset. This instruction is true when accessing the edge of the part from the front. The value would be -.100 inch if accessing from the edge of the part from the back.

Fixture/Work Offset Adjustments

To change the coordinate values of the work offsets, use the following method.

1. Use the arrow keys to highlight the desired offset.

2. Use the alphanumeric keypad to enter the incremental value for the offset adjustment.

   Note: Always input the value with a decimal.

3. Press the ENTER key.

Note: When the F1 key is used to enter values, the amount entered w ill replace any amount in the register.

When the ENTER key is used, the existing amount in the offset register will be added or subtracted, whichever applies, by the amount entered into it. Once the value is entered here, it is the new Workpiece Zero for the workpiece coordinate system.

Input of Known Work Offset Values

In some cases, the work-holding is clamped to a subplate or other positive locating fixture that allows accurate realignment. In situations such as this, the Work Coordinate values are recorded and will be the same each time the job returns.

1. Position the machine to Machine Zero.
2. Press the OFFSET button.
3. Use the procedure above (steps 6–8, Measuring CNC Mill Work Coordinate Offsets) to find the desired work coordinate display.
4. Key in the known (prerecorded) X value and press the F1 key to input the offset.
5. Repeat step 4 for the known (prerecorded) Y-axis value and press the F1 key to input the offset.

Note: To toggle between the offset tabs, press the F4 key.

Tool Compensation Factors

Important information about the tool must be given to the machine control unit (MCU) to be able to use the tool effectively. In other words, the MCU needs the tool identification number, the tool length offset (TLO), and the specific diameter of each tool. A TLO is a measurement given to the control unit to compensate for the tool length when movements are commanded. The cutter diameter compensation (CDC) offset is used by the control to compensate for the diameter of the tool during commanded movements. The tool number identifies where the tool is located within the storage magazine or turret and often is the order sequence in which it is used. Each is assigned a TLO number. This number correlates with the pocket or turret position number and, in the case of a milling machine, is where the measured offset distance from the cutting tip to the spindle face is stored. For example, Tool No. 1 will have TLO No. 1. Finally, when milling, the diameter of the tool is compensated for and the actual part geometry is programmed in order to facilitate the use of different tool diameters for a specified operation.

In high production shops, tool presetting and tool management systems are used to accurately input the TLO and diameter data values via network connection to the machine tool. This prevents incorrect data from being entered via the keyboard and does not use machine time for measuring. Tools are also often fitted with Radio Frequency Identification (RFID) chips that carry this information to the machine tool.

Tool Length Offsets (TLOs) are referenced in the program with the letter "H". The length values are input into the corresponding offset page data for the tool number, in the columns labeled H(LENGTH) GEOMETRY and are needed to properly position the tool along the Z-axis. When adjustments are needed to compensate for wear, incremental values are input into the WEAR column. Similarly, the CDC values are entered on the offset display register into the column labeled, D(DIAMETER) GEOMETRY and are referenced in the program words beginning with letter "D." These compensations are important for proper radial (X, Y) positioning of the tool. If the values are known, the following sequence can be used to input them into the offset page.

Tool Length Offset Measurement for CNC Mills

Tools length offsets can be measured by manually positioning the tool tip to contact the Workpiece Zero surface (Z-axis). This procedure is called *touching off* and is nearly always the topmost surface of the workpiece. All tools used in the program must have their offsets recorded in the offset register. If there is not a value in the offset register for a programmed tool, the control will not execute for that tool call, an alarm will occur, and the machine will stop.

The following steps are needed for the tool offset measuring procedure:

1. Use the MDI mode to input the desired tool number (described previously in Example 2) and press CYCLE START.

Alternatively, while in the MDI mode, key in the desired tool # and press the ATC FWD or ATC REV button.

2. Press the OFFSET button and select the Tool tab by using the arrow key (Figure 2-27). The tool that is in the spindle will be the active offset.

3. Use the HANDLE to manually position the tool tip to contact the workpiece zero surface (Z-axis). As an aid, a piece of paper may be used as a feeler between the surface and the tool tip.

4. Press the TOOL OFFSET MEASURE key. The Z value for the tool offset will be input to the offset register.

5. Press the NEXT TOOL button to advance to the next tool in the magazine.

6. Repeat the steps above until all tools are measured.

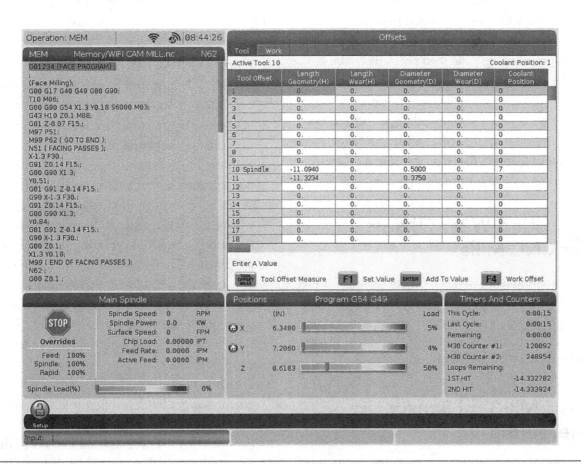

Figure 2-27. CNC mill tool offset display, *courtesy Haas Automation*

Each tool offset needs to have the diameter set to the actual size of the tool. Measure the tool prior to installation and input by following the steps below.

Input of Known Tool Offset Values

In cases where tools have been removed from the magazine (yet not disassembled) and the offset values have been recorded, the tool offsets can be re-entered for use again by following these steps:

1. Press the OFFSET button.
2. Use the cursor keys to position the cursor to the tool offset number to be set.
3. Key in the recorded values for the TLO and CDC and press the F1 key.
4. Continue these steps until all are entered.

In a similar way as just described, Work Offset values can be added or subtracted from an existing offset value. To do this, key in the amount with the appropriate sign and then press the ENTER key.

Diameter compensation values are input as known after measuring their actual size. Depending on the parameter setting for the specific machine used, the value is entered as either tool diameter or radius. Consult the appropriate manufacturer operation manual for exact conditions.

Adjusting Wear Offsets for CNC Mills

For machining centers, WEAR offset is assigned in the direction of the Z-axis for tool length compensation. Variations in the X- and Y-axes are compensated by adjusting the values in the D(DIAMETER) WEAR column.

CNC Mill Tool Sensor Measuring

On most modern machines, a tool sensor is used as opposed to manually measuring each tool length.

When this is the case, all of the programmed tools are manually or automatically positioned to contact the sensor for each tool axis, and the offset values are automatically input into the control. Review the operator manual specific to your machine for exact procedures.

Tool Path Verification of the Program

Tool paths created in computer-aided manufacturing (CAM) are verified by back-plotting and simulation prior to post-processing the NC code as preparation for machining. Additionally, a standard feature of the Haas controller helps verify that the program is ready to use via graphic display of the programmed tool path without any movement of the axes. This visual representation of the programmed tool path provides yet another pre-check for program errors before machining takes place. Follow these steps to access this display:

1. From the MEMORY mode, press the GRAPHICS button. The graphics screen will be displayed. In the lower left-hand corner of the display, prompts are listed that identify function keys and their purpose.

   In the lower-right corner of the display is a rectangular box that simulates the machine table area where the simulation will be displayed and can be used for magnification (Zoom) and centering of the graphics (Focus). The large area above these will be where the actual path will be displayed. Two colors are used to differentiate path types: green for rapid traverse and black for feed moves. Drilling path points are identified with an X.
2. Press the F2 function key, then the HOME button (upper left from cursor keys) to display the entire work envelop.
3. Press CYCLE START and simulation of the programmed tool paths will be displayed on the screen.

4. To adjust the magnification (ZOOM) press the F2 function key. A rectangle will display that can be used to position around the area desired.

5. Use the cursor keys to center the rectangle over the desired focus area and use the PAGE-UP key to increase or PAGE-DOWN key to decrease the size.

6. Press ENTER to accept the changes.

7. Press CYCLE START to run the graphics simulation.

Program Execution in Automatic Cycle Mode

Remember to position the coolant nozzles appropriate to the tools being used to ensure copious flow at the tool before machining starts (either manually or by programmable nozzle position, in the Tool Offset page).

A good practice is to use the previously described SINGLE BLOCK and set the RAPID 5% for the first time each tool is run. After confirmation that the tool length and positioning is correct, turn off SINGLE BLOCK by pressing the button again and set the RAPID OVERRIDE back to 100 percent. The following steps outline the process for executing a program in the automatic cycle mode.

1. Press the MEMORY button. Figure 2-25 shows the Operation: MEM and Active Program display screen that includes important operation information for each content area. When all the steps listed above have been completed, the program is ready to be executed under automatic cycle. A helpful and informative display screen to use during this first cycle is the POSITION, Distance-To-Go display, because you can see programmed moves count down as the tool is positioned.

2. To activate this feature, press the POSITION button to toggle through the five options and select DIST TO GO. After the program

is proven any of the other options may be chosen based on feedback preference.

3. Be sure that the desired program is in the control and active, and that all set-up procedures have been completed. If the program is not, use the steps above to activate by following the directions as stated in the section "A Program is Loaded from CNC Memory."

4. From the MEMORY mode.

5. Press the RESET key on the controller.

6. Set the Position Display as described above.

7. Press the CYCLE START. The automatic cycle will begin and run continuously until complete.

If intervention is necessary, the Feed Hold button can be pressed to stop feed motion or SINGLE BLOCK can be activated. The latter option will complete execution of the current command and stop.

CNC Mill Program Editing Functions

Editing part programs includes inserting, deleting, and altering program words and blocks. There are techniques for number searching, sequence number searching, word searching, and address searching that aid any editing of the program. First, the control needs to be in the proper mode. Details about specific meanings for program codes will be covered in the next section, "Introduction to CNC Mill Programming," of this chapter.

The MDI mode allows editing of the program in the active program display area (item 2, Figure 2-13). This MDI display area is on the left side at all times. Whereas this mode is generally used for non-saved programs, the program can be saved if the need arises (previously described in MDI Operations).

In the EDIT mode, changes can be made to any existing program. When the EDIT button is pressed the right side of the display is activated and program EDITOR appears (Figure 2-28).

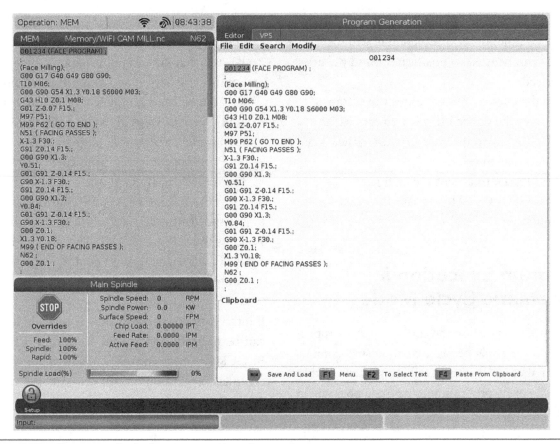

Figure 2-28. Program edit display, *courtesy Haas Automation*

The currently active program is displayed for editing. *Note:* Changes made by editing must be saved to overwrite the original file. To do so, press the MEMORY button to Save and Load the edits. Any changes made by editing the active program will be not be executed until saved.

Setting the Program to the Beginning

By pressing the RESET key while in the EDIT or MEMORY mode, the active program will be returned to the beginning line of the program (program head).

Cursor Scanning

The program may be scanned to an editing location by using the cursor and the page keys. Follow the directions as stated in the Cursor

Keys section. If the program is very large, using the cursor scanning method is not the most efficient method for searching through the program for edit locations. Figure 2-28 displays an example of how the screen might look during the editing process. In this case, the program is displayed with the cursor highlighting the program number.

Sequence Number Searching

If the sequence number in the program that requires editing is known, you can search directly to that program location by following these steps:

1. From the EDIT mode, use the alphanumeric keypad to input the sequence number preceded by the letter address N.
2. Input the desired line sequence number (i.e., N???).

3. Press the CURSOR down arrow.
4. The search direction can be in either direction, down or up, as needed in this and all cases to follow. The cursor will be moved to the identified sequence number.

Word Searching

Much like sequence number searching, you can search to a specific word in the program. For instance, to search to a specific word in the program like 25.0, follow these steps:

1. From the EDIT mode.
2. Key in the desired program word (i.e., 25.0).
3. Press the CURSOR down arrow. The cursor will be moved to the identified program word 25.0.

Address Searching

As with word or sequence number searching, you can search to a specific address in the program. For instance, to search to a specific program address in the program such as M06, follow these directions:

1. From the EDIT mode.
2. Key in the program word M06.
3. Press the CURSOR down arrow.

The cursor will move to the word address M06 or the first instance of the M-address that is found. Pressing the direction arrow again will advance to the next instance, if needed.

Inserting a Program Word

From the EDIT mode, use a searching method to scan the program to the word immediately before the word to be inserted. Follow these steps to insert a program word:

1. Use the methods above to position the cursor to the location desired for insertion.

Note: The insertion will follow the highlighted code.

2. Use the alphanumeric keypad to key the address and the data to be inserted.
3. Press the INSERT button.
4. The new data are inserted.

Edit insert example:

To insert the program word Z.2 on sequence number N4 of the program listed below:

1. Press the EDIT mode button.
2. Key in the word X1.2.
3. Press the CURSOR down arrow.
4. Key in the new word to insert Z.2.
5. Press the INSERT button.

O1234 (edit insert example);

N1 G50 S1000;

N2 T0100;

N3 G96 S600 M03;

N4 G00 <u>X1.2</u>;

The result will be as follows: N4 G00 X1.2 Z.2.

Altering Program Words

From the EDIT mode, use a searching method to scan the program to highlight the word to be altered.

1. Use the alphanumeric keypad to key the new data to be changed.
2. Press the ALTER Edit key.
3. The new data are changed.

Edit alter example:

To change the program word, Z.2, in the edit insert example above to Z.3, follow these steps:

1. Press the EDIT button.
2. Key in the word Z.2.
3. Press the CURSOR down arrow to highlight the Z.2 program word.

4. Key in the new word Z.3.
5. Press ALTER. The result will be as follows:
   N4 G00 X1.2 Z.3.

Deleting a Program Word

From the EDIT mode, use a searching method to scan the program to highlight the word that needs to be deleted. Then press the DELETE Edit key.

To delete the program word, Z.3 from the edit insert example, follow these steps:

1. Press the EDIT button.
2. Key in the program word Z.3.
3. Press the CURSOR down arrow to highlight the Z.3 program word.
4. Press the DELETE button.

Common CNC Mill Operation Scenarios

In this book, it's important to include explanations concerning situations that may arise during actual machining. We will concentrate on the procedures that should be followed when repetition of particular parts of the program for a specific tool is required. We will also review cases when there is a need to use an EMERGENCY STOP button and recovery from this condition. The following program is used for the case studies.

CNC Milling Machine Program

O2345

N1 G40 G80 G90
N2 G54 G00 X0.0 Y1.5 S1520 M03
N3 G43 Z1.0 M08 H01
N4 ... ...
... ...
N29 G91 G28 Z0.0
N30 M01
N31 T02

N32 M06
N33 G90 G54 G00 X.5 Y1.3 S1500 M03
N34 G43 Z1.0 M08 H02
N35 G81 G98 Z-.47 F6.0 R.1
... ...
... ...
N38 G91 G28 Z0.0
N39 M01
N40 T03
N41 M06
N42 G90 G54 G00 X-4.125 Y0.0 S2000 M03
N43 G43 Z1.0 M08 H03
N44   ... ...
      ... ...
N55 G91 G28 Z0.0
N56 T01
N57 M06
N58 G28 X0.0 Y0.0
N59 M30

CNC Milling Machine Scenario 1

Scenario 1 Problem: Execution of the program was interrupted in block N30, and you need to repeat operations performed by tool T01.

Scenario 1 Solution:

1. From the MEMORY mode, press the RESET button.
2. From the MEMORY mode, press CYCLE START.

CNC Milling Machine Scenario 2

Scenario 2 Problem: During the work of tool T02, the tool was damaged. You need to change the tool and repeat all operations performed by this tool.

Scenario 2 Solution:

1. Press the FEED HOLD button.
2. Press RESET to stop spindle rotations and coolant flow.
3. Change to the HANDLE/JOG mode.
4. Use the MPG to move the axes to a clearance point from the part.
5. Press the ZERO RETURN button.
6. Press the letter Z on the alphanumeric keypad.
7. Press the SINGLE button. This will return the Z-Axis to the Home position.
8. Remove and replace the damaged tool and clear any value in the wear offset column.
9. Press the EDIT key.
10. Using the alphanumeric keypad on the control panel and the search methods described earlier, search to block N33.
11. Return to the MEMORY mode and then press CYCLE START.

Note: In both cases, in order to repeat the work of the remaining tools, the OPTIONAL STOP button should be in the OFF condition.

If you need to repeat the work of only one tool, follow the steps listed below:

With the OPTIONAL STOP ON and after machining is completed, for tool T01:

1. From the EDIT mode, press the RESET button.
2. Press the HOME G28 button position on the machine with respect to X, Y, and Z axes.
3. From the MEMORY mode, press CYCLE START.

After machining is completed for tool T02 or T03:

1. From the EDIT mode, press the RESET button.
2. Press the HOME G28 button position the machine with respect to X, Y, and Z-axes.
3. Using alphanumeric keypad on the control panel and the search methods described earlier, search to block N31 for T02 or N40 for T03.
4. Return to the MEMORY Mode and press CYCLE START.

CNC Milling Machine Scenario 3

Scenario 3 Problem: Execution of the whole program is completed, but you need to repeat the operations performed by tool T03.

Scenario 3 Solution:

1. From the EDIT mode.
2. Using alphanumeric keypad on the control panel and the search methods described earlier, search to block N40.
3. Return to the MEMORY mode and press CYCLE START.

Machine Shutdown

At the end of the day or machining shift, the machine should be cleaned as described in Chapter 1. It is a shop best practice to write a brief pass down document to communicate production status and any details that ensure a smooth transition into the next shift.

To shut down the machine:

1. Position the X and Y axes to the center of the travel envelope.
2. Press the Power OFF button.
3. Turn off the air pressure to the machine at the ball valve.

Summary

In this section, we have covered CNC milling machine setup and operation procedures commonly performed to prepare for automatic operation. Please understand that there are a multitude of situations possible during this process. While the items covered here are core to successful use, they may not cover every possibility. For complete details on operation features specific to your machine, consult the manufacturer manual.

CNC Mill Setup and Operation, Study Questions

1. Which button is pressed to set a known offset value into the GEOMETRY column?

   a. ENTER
   b. CYCLE START
   c. F1
   d. F2

2. The counterclockwise direction of rotation is always a negative axis movement when referring to the HANDLE (manual pulse generator). T or F?

3. Which button is used to activate automatic operation of a CNC program?

   a. EMERGENCY STOP
   b. CYCLE STOP
   c. CYCLE START
   d. MEMORY

4. Which display area of the screen lists the CNC program?

   a. POSITION
   b. OFFSET
   c. EDIT
   d. PROGRAM

5. Which operation selection button allows for the execution of a single CNC command?

   a. DRY RUN
   b. SINGLE BLOCK
   c. BLOCK SKIP
   d. OPTIONAL STOP

6. Which mode switch/button enables the operator to make changes to the program?

   a. EDIT
   b. MDI
   c. AUTO
   d. JOG

7. What does the acronym MDI stand for?

   a. Manual Direct Input
   b. Manual Digital Input
   c. Manual Data Input
   d. Memory Driven Information

8. Which display screen is used to enter tool information?

   a. Program
   b. Offset
   c. Work
   d. Edit

9. If the Reset button is pressed during automatic operation, then spindle rotations, feed, and coolant will stop. T or F?

10. During setup, the mode button used to allow for manual movement of the machine axes is:

    a. AUTO
    b. MDI
    c. EDIT
    d. HANDLE/JOG

Introduction to CNC Mill Programming

Objectives

1. Identify and properly use G-codes associated with CNC mill programming.
2. Identify and properly use M-codes associated with CNC mill programming.
3. Input the correct cutting tool feeds and speeds within CNC milling machine programs.
4. Properly use coordinate systems for programming the CNC milling machine.
5. Implement the use of canned cycles.
6. Input and use Cutter Diameter Compensation (G40, G41 and G42) within CNC milling machine programs.
7. Edit existing CNC mill programs with multiple techniques.

In Chapter 2, a brief presentation of Program Structure and Format was given that identified the basic alphabet of CNC mill programming. These components will be included and codes will be introduced and followed to create programs. Individual programming words and codes will be defined and demonstrated in an example program. This section is presented as if the program manuscript is being created manually, line-by-line. This experience will help you understand the programs and how they are made, which will make it easier for you to edit existing programs. In reality, though, the most common method for creating programs is by using CAD/CAM software, which is introduced in Chapter 4, Introduction to Computer Aided Manufacturing.

For a complete and detailed resource for learning CNC Programming refer to my texts *Programming of CNC Machines* (4th ed.) and its companion, *Student Workbook for CNC Programming of CNC Machines* (4th ed.), Ken Evans.

Program Structure for CNC Milling

Program Number

Each program is assigned a number. The capital letter O is reserved for the program number identification and is usually followed by four digits, which specify the actual program number. For example, to create a program with the number 1234, the programmer must input the letter address O, and then the number 1234 (O1234). All programs require this format.

Note: Refer also to Basic CNC Program Structure and Format in Chapter 1.

Program number examples:

O0001 = program number 1

O0014 = program number 14

A common mistake made here is to enter zero (0) instead of the letter O, resulting in an alarm on the control system.

Program Comments

Comments are instructions that help the operator and may be added to the program by using parentheses. The comments or data inside parentheses will not affect the execution of the program in any way. A very common place to add comments is at the program number, to identify a part number, tool change or program stop in order to direct the operator in some way. An EOB character (;) is typically required after the parenthesis if multiple lines of code are to be entered via MDI. No EOB character is needed if the data is entered via an offline text editor or is post processed data from a CAM program.

Program comments examples:

O0001;

(PN587985-B);

N90 M00;

(REMOVE CLAMPS FROM OUTSIDE OF PART);

Block Numbers

The capital letter N is reserved to identify the program line sequence numbers (block numbers), and precedes any other data in a program line. For each line in a program, a block number is normally assigned sequentially. For example, the first block of a program is labeled N10, the second N15, etc. Typically, the program block numbering system is sequenced by an increment other than one. An example of this is sequencing by five, where line one is labeled N5 and line two is labeled N10. The original intent of an increment of other than one is to allow for inserting additional blocks of data between the increments, as needed. These additions can sometimes be advantageous when editing programs. Note that block numbering is not necessary for the program to be executed. The program blocks, even if not numbered, will be executed sequentially during the machine's automatic cycle. Removing block numbers is sometimes helpful when a program is too large to fit into the resident controller memory on older machines—each character of a program takes memory space and large programs can have block numbers into five or more digits.

It is a common practice to place block numbers only at a tool change command. This aids the operator when a restart of machining program at a specific tool change is desired rather than rerunning the entire program. Block numbers enable movement within the program in order to enter offsets, verify data, or search for a block in the program. They are often referenced by the control in case of a programming error that causes an alarm, enabling the programmer to search to the problem directly by block number. Block numbers are not required for the program to work, but they are necessary for restarting the program at a specific place.

Program End (M30)

The difference between M02 and M30 is that M02 refers to the program end, while M30 refers to the program end and a simultaneous return to the program beginning (head). Both commands are found in the final line of the main program only. *Note:* On some controls, M02 behaves the same as M30 for compatibility with older programs.

Tool Function (T-Word)

In the Cutting Tool Selection section of Chapter 1, the proper method for selecting tools was presented. The tools selected will be incorporated into the programming process. Refer again to this section and the *Machinery's Handbook*, and the cutting tool and insert manufacturers' ordering catalogs and various online resources.

The tool function is utilized to prepare and select the appropriate tools from the tool magazine. In order to describe the tool in the program, the address T is followed by one or more digits that refer to the pocket numbers in the tool magazine.

Tool call example:

T05 = tool number 5

Note: Most controllers do not require the use of leading zero in a tool call; thus, T5 has the same meaning as T05.

Tool Changes

A tool change is specified in the program by the miscellaneous function M06. To initiate a tool change, first call for the desired tool number. Then use the miscellaneous function M06 to execute the change.

Tool change example:

N10 T01 (Tool in the Ready position)

N15 M06 (Actual Tool Change)

N20 T02 (Next tool in the Ready position)

N25 . . .

N40 M06 T03

N45 . . .

In block N10 of the example, the requested tool is positioned in the tool magazine to a ready state (waiting position). In block N15, tool T01 is automatically installed into the spindle. In block N20, tool T02 is positioned to a ready state in the tool magazine for the next tool change. In block N25–N40, tool T01 performs programmed work. In block N40, T02 is placed into the spindle, while tool T01 is returned to the tool magazine and tool T03 is positioned to a ready state in the tool magazine for the next tool change. In block N45, tool T03 performs the programmed work until complete.

In cases where Random Access Tool Changers are used, it is not necessary to stage the next tool. When a tool is finished, it is replaced with the next required tool, which is positioned in the same (now empty) pocket. Assigned tool pockets are not required except at initial job setup. The control system keeps track of where each tool is placed in the magazine. These systems attain very fast tool changes because no wait is involved in tool magazine rotation.

Note: On most controllers, it is not necessary to use the leading zero in a miscellaneous function call; thus, M6 has the same meaning as M06. Also note that on machines with umbrella style tool magazines, the tool call (T01) and tool change call (M06) must be stated together.

Umbrella style tool call example:
T01 M06

Feed Function (F-Word)

The F-word is utilized to determine the work feedrates. This program word, which is used to establish feedrate values, precedes a numeric input for the feed amount in inches per minute (in/min), millimeters per minute (mm/min), inches per revolution (in/rev), or millimeters per revolution (mm/rev). The value that is set by this command stays effective until changed by re-entering a new value for the F-word. It is important to always include the decimal point when programming.

Feed function example:
F20. = a feedrate of 20 in/min
F.006 = a feedrate of .006 in/rev

If the function G20 (data in inches) is active, the notation F20. refers to the feed speed of 20 in/min; whereas, with the function G21 (data in millimeters), the notation F20. refers to a feedrate of 20 mm/min.

With rapid traverse G00, the machine traverses at the highest possible feedrate that is specified in control memory (actual rates depend on the design of the machine). In the case of feedrate motion G01, the value of the feedrate must be accurately specified. The machine default setting for feedrate is inches per minute in the United States.

Feed input examples:
F20.0 = 20.0 inches of feed per minute
F500. = 500 millimeters of feed per minute
F2.0 = 2.0 inches of feed per minute
F50. = 50 millimeters of feed per minute
F.02 = twenty thousandths inch of feed per revolution
F0.50 = millimeters of feed per minute
F.002 = two thousandths inch of feed per revolution
F0.050 = millimeters of feed per minute

The proper methods for tool selection and calculation of feedrates were discussed in the Chapter 1 sections, Cutting Tool Selection and Machining Mathematics. Refer again to those sections and the *Machinery's Handbook*, the cutting tool and insert manufacturers' ordering catalogs and the previously mentioned online resources for additional tool and material-specific feed and speed data.

Spindle Speed Function (S-Word)

The letter address S is followed by a specified value in revolutions per minute (r/min).

Spindle speed example:
S2100 (specifies 2100 r/min)

One or more digits following the letter address S are used for the value of the rotational speed. If S0 is input, this command deactivates the spindle rotation and leaves it in a neutral position so that the spindle can be rotated manually, depending on the machine tool. Having this ability is quite useful, especially when using a coaxial indicator for dialing in the X and Y coordinate locations of the workpiece to establish Workpiece Zero.

Preparatory Functions (G-Codes)

G-codes are the preparatory functions that identify the type of activities the machine will execute. A program block may contain one or more G-codes.

The letter address G and specific numerical codes allow communication between the controller and the machine tool. This combination of letters and numerical values is commonly called G-code. In order to perform a specific machining operation, a G-code must be used. There are two types of G-codes: modal and non-modal. Modal commands remain in effect, in multiple blocks, until they are changed by another command from the same group. Non-modal commands are in effect only for the block in which they are stated.

Modal and non-modal examples:

Group 00 (non-modal, one-shot commands)

Group 01 (modal commands)

There are several different groups of G-codes as indicated in column 2 of Chart 2-3. One code from each group may be specified in an individual block. If two codes from the same group are used in the same block, the first will be ignored by the control and the second will be executed. Those G-codes that are active upon startup of the machine are indicated by an asterisk (*) in the chart.

The digits following the letter address G identify the action of the command for that block.

Miscellaneous Functions (M-Codes)

Miscellaneous function or M-codes (Chart 2-4) control the working components and accessories that activate and deactivate coolant flow, spindle rotation, the direction of the spindle rotation and similar activities.

Programming of CNC Mills in Absolute and Incremental Systems

These two coordinate measuring systems—absolute and incremental—are used to determine the values that are input into the programming code for the X, Y and/or Z program words. They can also be used in the same manner for rotary axes A, B and/or C.

Absolute Coordinate Programming (G90) of the CNC Mill

In absolute programming, all coordinate values are relative to a fixed origin of the coordinate system. Axis movement in the positive direction does not require inclusion of the sign whereas negative movements do require signs. This system is by far the most used.

Incremental Coordinate Programming (G91) of the CNC Mill

In incremental systems, every measurement refers to a previously dimensioned position (point-to-point). Incremental dimensions are the distances between two adjacent points.

Code	Group	Function	Code	Group	Function
*G00	01	Rapid Traverse Positioning	*G80	09	Canned Cycle Cancellation
G01	01	Linear Interpolation	G81	09	Drilling Cycle, Spot Drilling
G02	01	Circular and Helical Interpolation CW (clockwise)	G82	09	Drilling Cycle, Counter Boring
G03	01	Circular and Helical Interpolation CCW (counterclockwise)	G83	09	Deep Hole Peck Drilling Cycle
G04	00	Dwell	G84	09	Tapping Cycle
G09	00	Exact Stop	G85	09	Reaming Cycle
G10	00	Programmable Data Setting	G86	09	Boring Cycle
G12	00	Circular Pocket Milling, Clockwise	G87	09	Back Boring Cycle
G13	00	Circular Pocket Milling, Counter Clockwise	G88	09	Boring Cycle
*G17	02	XY Plane Selection	G89	09	Boring Cycle
G18	02	ZX Plane Selection	*G90	03	Absolute Programming
G19	02	YZ Plane Selection	G91	03	Incremental Programming
G20	06	Input in Inches	G92	00	Setting for the Work Coordinate System Shift
G21	06	Input in Millimeters	G93	00	Inverse Time Feed Mode
G28	00	Reference Point Return	*G94	05	Feed per Minute
G29	00	Return From Reference Point	G95	05	Feed per Revolution
G31	00	Feed Until Skip	*G98	10	Canned Cycle Initial Level Return
G35	00	Automatic Tool Diameter Measurement	G99	10	Canned Cycle R=Level Return
G36	00	Automatic Work Offset Measurement	G100	10	Cancel Mirror Image
G37	00	Automatic Tool Length Measurement	G101	10	Enable Mirror Image
*G40	07	Cutter Compensation Cancel			
G41	07	Cutter Compensation, Left			
G42	07	Cutter Compensation, Left			
G43	08	Tool Length Offset Compensation positive (+) direction			
G44	08	Tool Length Offset Compensation negative (-) direction			
G47	00	Text Engraving			
G49	08	Tool Length Offset Compensation Cancel			
*G50	11	Scaling Cancel			
G51	11	Scaling			
G52	00	Local Coordinate System Setting			
G53	00	Non-Modal Machine Coordinate Selection			
*G54-G59	14	Work Coordinate System Selection (G54 default)			
G60	00	Single Direction Positioning			
G61	15	Exact Stop			
*G64	15	G61 Cancel			
G65	00	Macro Subroutine Call Option			
G68	16	Rotation of Coordinate System			
*G69	16	Cancellation of Corrdinate System Rotation			
G70	00	Bolt Hole Circle			
G71	00	Bolt Hole Arc			
G72	00	Bolt Holes Along an Angle			
G73	09	High Speed Peck Drilling Cycle			
G74	09	Reverse Tapping Cycle			
G76	09	Fine Boring Cycle			
G77	09	Back Bore Canned Cycle			

NOTES:

The items marked with an asterisk (*) are active upon startup of the machine or are reinstated when the RESET button has been pressed. Check the specific manufacturer Operation Manuals for your application.

For G00, G01, G90 and G91 the initial code that is active is determined by a parameter setting. These are typically G01 and G90 for startup condition.

G-Codes from groups 00 are one-shot G-Codes.

Multiple G-Codes from different groups can be specified in the same block. If more than one from the same group is specified only the last G-Code listed will be active.

Chart 2-3. CNC Mill Preparatory Functions

M-Code	Function	M-Code	Function
M00	Program Stop	M11	Work Table Rotation Unlocked
M01	Optional Stop	M16	Change of Heavy Tools
M02	Program End Without Rewind	M18	Clamp APC Pallet and Close Door
M03	Spindle ON Clockwise (CW) Rotation	M19	Spindle Orientation
M04	Spindle ON Counterclockwise (CCW) Rotation	M21-M25	Optional User M-Function with M-Fin
M05	Spindle OFF Rotation Stop	M29	Set Output Relay with M-Fin
M06	Tool Change	M30	Program End With Rewind
M07	Mist Coolant ON	M31	Chip Conveyor Forward
M08	Flood Coolant ON	M33	Chip Conveyor Stop
M09	Coolant OFF	M98	Subroutine Call
M10	Work Table Rotation Locked	M99	Return to Main Program From Subroutine

Chart 2-4. CNC mill miscellaneous functions

Work Coordinate Systems (G54 – G59)

Identifying where the workpiece is located in X, Y and Z within the machining envelope is necessary in every program. Functions G54 through G59 are highly effective when multiple coordinate systems are required on one or multiple parts to establish Workpiece Zero coordinates. Workpiece Zero coordinates G54 through G59 are measured from Machine Zero to the Workpiece Zero.

In order to specify multiple Workpiece Zero points for the setup shown in Figure 2-29, use the following functions:

G54: Left vise = first Workpiece Zero

G55: Right vise = second Workpiece Zero

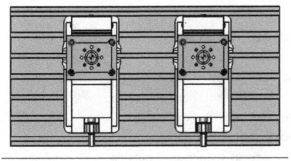

Figure 2-29. Multiple work coordinate systems

When using functions G54 through G59, the coordinates of the program zeros are entered into controller memory on the work offset page for each of the two coordinate systems. Refer to Chapter 2, Measuring CNC Mill Coordinate Offsets, for setting details. For the left vise, highlight offset G54 on the Work Offset page and press the PART ZERO SET button for the X-axis and then the Y-axis. Repeat this process for the right vise by highlighting offset G55 on the work offset page.

Tool Length Compensation (G43, G44, G49)

Workpiece Zero for the X, Y and Z-axes is placed within the machining envelope in relationship to Machine Zero. The Z-axis is typically on the top most surface of the machined part. Workpiece Zero for the Z-axis should be in the same position for all tools. Due to different tool gage lengths, in order to transfer Workpiece Zero along the Z-axis from Machine Zero to the surface of the workpiece, you must apply function G43, a Tool Length Compensation function (tool length offset).

The offset number H?? (H01, H02) is always assigned in the same line with function G43. As a rule, in order to simplify program execution, offset number H should be the same as the tool number for each corresponding tool. The measured value of offset H is entered into the offset registers in the computer memory (example e.g., H01 = –11.1283). The value of the tool length offset for a given tool corresponds to the distance between the tool tip and the surface of the workpiece, as shown in Figure 2-30.

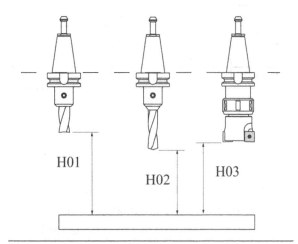

Figure 2-30. Tool length offset (h)

In order to determine the value of offset H for a specific tool, follow the instructions in Chapter 2, Tool Length Offset Measurement for CNC Mills section for the steps required. The value in the position register is the distance that determines the value of offset H for a given tool. This value is registered in the offset table with the number corresponding to the offset number in the program.

Function G43, with the assigned offset number H, must be entered in the program before the tool is used.

By employing the Tool Length Compensation Cancel function (G49) in a safety block at the beginning of the next tool sequence, the offset is canceled before consecutive other tools are used.

G43 = positive tool length offset

Note: the entry in the offset register is negative because the offset represents the distance from the tool tip to the workpiece Z zero.

G44 = negative tool length offset (rarely used)

G49 = cancellation of tool length offset

The tool length offset is called in the program by the letter address H. The tool offset number used is identified by the letter H and two digits.

Tool offset call example:

H01= offset number one

When entering the new offset value to the offset register, the previous offset value is automatically canceled; the machine reads the new value without considering the previous value. To cancel functions G43 and G44, use function G49. This cancellation code (G49) is placed in the safety block as described before.

CNC Mill Program Template

Generally speaking, CNC programs consist of a few characteristic elements that play the following roles:

- Establish the absolute or incremental coordinate system.
- Establish the work offset used.
- Set the spindle r/min and rotation direction.
- Execute the tool length offset given and activate the flow of the coolant.
- Determine the drilling canned cycle.
- Determine the tool's work path to consecutive holes in the pattern to be drilled.
- Positioning of the tool to the consecutive locations.
- Cancel any canned cycles and deactivate the coolant flow.
- Command of the Z, X and Y-axes to return to Machine Zero position.
- Optional Program Stop.
- Tool change, if needed.
- Repeat sequence, if needed.
- Program ending.

Certain sections of the machining program can be repetitious in nature and they are listed, as follows: the program beginning, the safety block, the tool beginning, the tool ending and the program ending.

Note: If you intend to load any of the programs you create into a machine controller for trial and use, you must include a percent (%) sign on a separate line at the beginning and end of the text. This is required for communications purposes.

The Program Beginning

O2406 = program number

Note: 9000 series program numbering is reserved for Macro programs; therefore, avoid using it for your program number.

(Comments) = part number or other identifying information

(Comments) = date or other identifying information

N10 G90 G20 G80 G40 G49

The safety block (See EXPLANATION OF THE SAFETY BLOCK in *Programming of CNC Machines*, Fourth Edition, Part 4, Programming of CNC Machining Centers, for complete details). The necessity for this line of code is diminished because of default codes (see (Figure 2-25) that are active at startup and reset; however, no harm can result from canceling the codes.

The Tool Beginning

(Comments) = Tool identification information

N20 T01 M06

N25 S1000 M03

N30 G54 G00 X0.0 Y0.0

N35 G43 Z1.0 H01

N40 Z.1 M08

Note: The tool number (T01) tool height offset number (H01), the value entered in your

program for spindle r/min (S), and the values entered in your programs for X and Y coordinates listed with G54, will vary dependent on the specific application requirements.

The Tool Ending

N100 G80 Z.1 M09

N105 G53 Z0.0

N110 G53 Y0.0

N115 M01

The Program Ending

N200 G28 X0.0 Y0.0

N205 M30

Note: The coordinate points in line N30 of the above template will be replaced with live data that is relevant to your programming situation.

Preparatory Functions for CNC Milling Machines (G-Codes)

Preparatory functions (listed in Chart 2-3), often called G-codes, are a major part of the programming puzzle. They identify to the controller what type of machining activity is needed. For example, if a hole needs to be drilled, function G81 may be used, or if programming in the absolute coordinate system is required, function G90 is used. These codes, along with other data, control machine motion.

The motion of the axes of a machine may be performed along a straight line, an arc, or a circle.

Codes G00 and G01 allow axes movement along a straight line.

Codes G02 and G03 allow axes movement along a circular path of motion.

Rapid Traverse Positioning (G00)

The rapid traverse function is entered to relocate the tool from position A to position B along a straight line at the fastest possible traverse. The shortest axis movement distance will be accomplished first (sometimes called a dogleg). Therefore, you must always be aware of the workpiece holding the equipment in order to avoid any collision between the tool and the holding equipment.

When possible, work-holding should be included in the CAM model during programming to position the Z-axis to an acceptable clearance plane of 1.0 inch or any amount necessary to move over clamps or obstructions and, thus, reduce the chance of collisions. During CAM tool path simulation, the program will indicate when collisions are imminent. As mentioned before in the first section of this chapter, operators should use Single Block and Rapid Traverse Override during program checking.

Linear Interpolation (G01)

Function G01 is used to move the tool from point A to point B along one or all of the axes simultaneously, along a straight line of motion, and at a given feedrate specified by the F-word.

Linear interpolation code example:
G01 X10.0 Y20.0 F8.0

Circular Interpolation (G02, G03)

Circular interpolation allows tool movements to be programmed to move along the arc of a circle. When applying the circular interpolation, the plane in which the arc is positioned must be determined initially (Figure 2-31). Vertical CNC

Mills default to the X- and Y-axes (G17) upon startup. Programming in any of the other planes require activation of preparatory function G18 or G19. Then, depending on the direction of the machining, select function G02 to make a clockwise movement along the arc and function G03 to make a counterclockwise movement along the arc. In order to describe the movements of the tool along the arc, apply the following two methods:

1. Determine the radius R and the values of the end point coordinates in the given plane.
2. Determine the value of the endpoint of the arc and the values of the incremental distance to the arc center in a given plane.

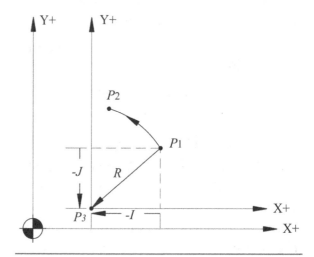

Figure 2-31.   Required data for circular interpolation

P_1 = the start of an arc

P_2 = the end of an arc

P_3 = the center of the arc

R = radius vector

I = the incremental distance to the arc center along the X-axis

J = the incremental distance to the arc center along the Y-axis

The incremental distance is defined as a radius projection onto a given axis. The radius incremental distance is always attached. It begins at the starting point and ends at the center of the circle. It is always directed toward the center of the circle.

- Vector projections onto the X-axis are identified by use of the letter address I.
- Vector projections onto the Y-axis are identified by use of the letter address J.
- Vector projections onto the Z-axis are identified by use of the letter address K.

In the following example, a vector projection is illustrated. The radius is positioned at the origin of the coordinate system, as shown in Figure 2-32.

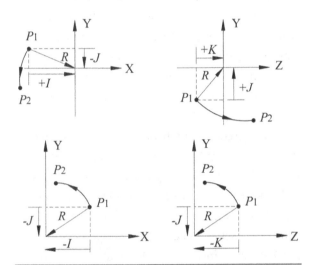

Figure 2-32. Vector projection for circular interpolation

Note: In Figure 2-32, the signs (+, −) of the incremental distances I, J and K depend on the position of the starting point of the arc with respect to the center of the arc—that is, with respect to the coordinate system. If the direction of the vector is consistent with the direction of the assigned axis of the coordinate system, then we apply the positive sign. If not, then we apply the negative sign. Most controllers do not require the use of the positive sign. If no sign is present, the value is considered to be positive.

Note: When circular motion is described with radius function R, no sign is required if the arc is ≤ 180 degrees (the system defaults to positive unless otherwise specified). Assign a negative value to R (−) if the arc is >180 degrees. The

maximum rotation of an arc using R is 359.9 degrees.

A full circle may be accomplished using R by linking two 180-degree arcs. If a full circle of 360 degrees is to be performed, then it is necessary to employ the incremental distances of the arc center points for I, J and K—not radius R—in the program.

For example, do not use I, J or K with R in the same block, because they will be ignored by the control if they are used; the tool will follow the arc with the assigned radius of R. If the value of an entered radius R is zero, an alarm will result. Cutter radius compensation may be used for circular interpolation. However, it must be initiated in a G00 or G01 block preceding the G02/G03 information.

Input in Inches (G20) and Input in Millimeters (G21)

Function G20 or G21 is entered at the beginning of the program in the safety block to establish the measurement system. Either applies to the whole program. Functions G20 and G21 cannot be interchanged during programming. When using functions G20 or G21, the values are the same units of measure for feedrate (F), position of the X, Y and Z-axes and offset values. The default measurement system for most American machines is inches. Therefore, it is not necessary to use the previously mentioned functions unless the metric system is required. The desired system should always be stated within the safety block.

Reference Position Return (G28)

This function is the automatic reference point return through an intermediate point programmed in the X, Y and/or Z-axes. The machine will position at rapid traverse (G00) to the programmed intermediate point coordinate values and then to the reference point of machine zero. This function is commonly used before an

ATC. When you use function G28, you must specify the point through which the tool passes on its way to zero. If the command G28 X0.0 Y0.0 Z0.0 is entered in the incremental mode (G91), the machine will position at rapid traverse to the reference position in all three axes simultaneously. Caution must be exercised so as not to interfere with the work-holding device or part. It is good practice to use G28 with a Z-axis positioning move in a prior block to ensure clearance. Observe the proper use in the CNC Mill Program Template, Tool Ending, above.

Cutter Diameter Compensation (G40, G41, G42)

The use of cutter diameter compensation allows the programmer to use part geometry that exactly matches the engineering drawing/blueprint for programmed coordinates. Without using compensation, the programmer must always know the cutter size and offset the programmed coordinates for the geometry by the amount of the radius. In this scenario, if a different size cutter is used, the part will not be machined correctly. Cutter diameter compensation adds the advantage for using any size cutter as long as the offset amount is input accurately into the offset register. It is also very effectively used for fine-tuning of dimensional results by minor adjustments to the amount in the offset register.

    G40 = Cutter Compensation Cancel

    G41 = Cutter Compensation Left

    G42 = Cutter Compensation Right

Cutter Compensation Cancellation (G40)

Function G40 is used to cancel cutter radius compensation initiated by G41 or G42. It should be programmed after the cut using the compensation completed by moving away from the finished part in a linear (G01) or rapid traverse (G00) move by at least the radius of the tool. Care should be taken here because, if the cancellation is on a line without movement, the cutter will move unpredictably in the opposite direction and may damage the part.

Cutter Radius Compensation Left and Right (G41 and G42)

Functions G41 and G42 offset the programmed tool position to the left (G41) or right (G42), respectively, by the value of the tool radius entered into offset registers and called in the program by the letter address D. For each tool, enter the corresponding offset amount in the Radius Geometry column of the tool data register. In the program, the letter D and the number of the offset (two digits) are input to initiate the compensation call.

The direction the tool is offset, to the left or the right, depends upon which direction the tool is traveling. To accomplish climb cutting with right-hand tools, always use G41; for conventional cutting, use G42. Consider which direction of offset is needed by facing the direction the tool is going to travel, then observe which side of the part the cutter will be—to the left for G41 or to the right for G42 (Figure 2-33).

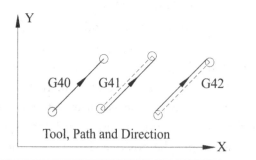

Tool, Path and Direction

Figure 2-33. Cutter diameter compensation

Procedures for Initiating Cutter Compensation

Position the tool in the X- and Y-axes to a point away from the required finished geometry. Then program a linear move that is larger than the radius of the cutter to feed into the part (e.g., G01, G41, or G42 offset direction, X or Y absolute coordinate, and D offset number). To use the cutter compensation properly, there needs to be one full line of movement to position the tool on the proper vector to cut the part. Once this is accomplished, program the part geometry per print.

The tool will not be positioned to the actual programmed point on the geometry. Rather, it will be positioned to a point plus or minus the offset value of the cutter called by D, and the edge of the cutter will be aligned with finished part geometry.

Rules for Cutter Compensation Use

■ When cutter diameter compensation codes are encountered in a program, the control does what is called "look-ahead". To set up the appropriate vector needed to position the offset amount required, the control looks ahead two program lines for each move.

■ Once G41 or G42 commands have been called up, movement must be maintained. By following this rule, over-cutting of the part can be avoided. If two lines of non-movement commands are placed consecutively after cutter diameter compensation is called up, the control will ignore functions G41 or G42; the part will then be cut incorrectly.

■ **Do not** start cutter compensation G41 or G42 when either G02 or G03 is in effect.

■ Before a change from left to right compensation is made, or right to left, you must cancel the first compensation. The second compensation may then be called and, thus, the transition of the tool position vector will not conflict.

■ When machining an inside radius, the radius must be larger than the offset of the tool. Otherwise, the control will stop the program and an alarm will be displayed.

■ The move used to activate cutter diameter compensation must be larger than the radius of the tool used.

■ At the end of the tools work, function G40 must be applied to cancel any previously entered compensation value of offset D.

Canned Cycle Functions

The function of a canned cycle is defined as a set of operations assigned to one block and performed automatically without any possibility of interruption. Usually, it is a set of six operations, as follows:

1. Positioning of the X and Y-axes at rapid traverse.
2. A rapid traverse moves along the Z-axis to an initial clearance level plane (G98).
3. The machining cycle is executed (drill, bore, etc.).
4. A dwell or other operation is executed at the bottom of the hole.
5. A rapid traverse return to the R level plane along the Z-axis (G99).
6. A rapid traverse return to the initial level plane along the Z-axis.

   The block format is as follows:
   N...G...G...X...Y...Z...R...
   Q...P...F...K...L...

   where
   N = the block number

   G = the type of cycle function

   G = initial or R level return G98/G99

   X, Y = the hole position (positioning is carried out by rapid traverse)

   Z = the depth of the hole

   R = the distance between plane R and the surface of the material

Level R refers to the horizontal plane, positioned closely above the material on which the tool tip moves (commonly, .100 in. or .08 mm). The programmed value of R is valid until the new value is entered. It does not have to be included in every block. The tool will return to this level at the end of each hole drilled.

> Q = the depth of cut (drilling) for individual pecks (not used for every drilling cycle)
>
> P = the dwell time for the drill while rotating at the bottom of the hole (not used for every drilling cycle)

The dwell in seconds is for the purpose of complete removal of excess material

> F = feedrate (in/min)
>
> K = the number of repeats
>
> L = the number of holes incrementally spaced

When K is used with G91, L represents the number of holes incrementally spaced by the amount entered as the X or Y position coordinate. (If L does not appear in the block, that means machining of only one hole, L = 1.) On some controls, the letter address K is used in the same manner (Figure 2-34).

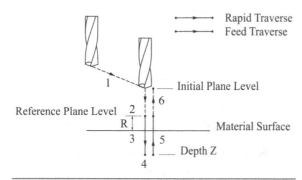

Figure 2-34. Canned cycle drilling functions

1. Positioning with rapid traverse
2. Rapid traverse to level R along Z-axis
3. Feed traverse to Z depth
4. Operations performed on the bottom of the hole
5. Return to level R (G99)
6. Return to plane level (G98)

Canned Drilling Cycle Cancellation (G80)

This command is used to cancel all canned cycles. It should be entered at the end of each canned cycle machining sequence. This code is also entered in the safety block.

Notes for cycles G73 through G89:

- Revolutions of the spindle (right or left), during the cycle, must be entered in the block preceding the canned cycle block.
- Never press the ORIGIN button (zero set) during the canned cycle execution because unpredictable actions may result.
- In order to execute the cycle (the drilling of one hole with the SINGLE BLOCK button ON, after the execution of each block when the machine stops), you must press the CYCLE START button three times to continue.
- If feed hold is pressed during a threading canned cycle operation, the cycle will be completed before stopping.

Canned Cycle, Spot Dwelling (G81)

Block format:

> G81 X . . . Y . . . Z . . . R . . . F . . .

Figure 2-35 explanations:

1. G00 — rapid traverse to R
2. G01 — feed traverse to Z depth
3. G00 — rapid traverse to R (G99) or
4. G00 — rapid traverse to initial level plane (G98)

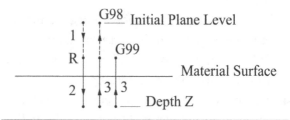

Figure 2-35. Canned cycle spot drilling (G81)

Counter Boring Cycle (G82)

In function G82, the feed is interrupted in the drilling cycle (while the spindle is ON) at the bottom of the hole for time specified by P.

Block format:

G82 X . . . Y . . . Z . . . R . . . P . . . F . . .

Process explanations:

1. G00 — rapid traverse to reference level R
2. G01 — feed traverse to Z depth
3. — interruption of the feed for time duration P, in order to remove the material at the bottom of the hole
4. G00 — rapid traverse to initial plane level for G98
5. G00 — rapid traverse to reference plane level R for G99

Note: On some controls, if no value is assigned for P to function G82, its value will be automatically selected by the control. If you do enter a value for dwell P (e.g., P1000 = 1 second), then the constant value included in the parameters of the machine will be ignored.

Deep Hole Peck Drilling Cycle (G83)

Block format:

G83 X . . . Y . . . Z . . . R . . . Q . . . F . . .

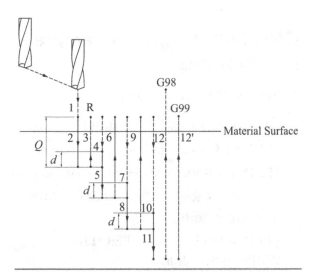

Figure 2-36. Deep hole peck drilling cycle (G83)

Figure 2-36 explanations:

1. G00 — rapid traverse to R level
2. G01 — feed traverse with length Q (peck amount)
3, 6, 9. G00 — rapid return traverse to R
4, 7, 10. G00 — rapid traverse to the depth previously drilled, less the value of *d*
5, 8, 11. G01— feed traverse increased by the value of *d*
12. G00 — rapid traverse to initial plane level for G98
12'. G00 — rapid traverse to reference plane level R for G99

This drilling cycle is used to drill exceptionally deep holes. As the drill reaches the depth identified by Q, the drill then returns at rapid traverse to the R level point, allowing the removal of chips and the delivery of coolant to the bottom of the hole. Entering a given depth of Z into the control enables it to calculate the number of feed traverses necessary for Q. Q can be any incremental amount desired smaller than the total Z-axis travel. The value of Q does not need to have a common factor with the dimension Z. The value of *d* is set by machine parameter.

Tapping Cycle (G84)

Block format:

G84 X . . . Y . . . Z . . . R . . . P . . . F . . .

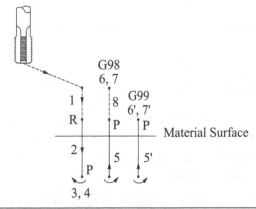

Figure 2-37. Tapping cycle (G84)

Figure 2-37 explanations:

1. G00 — rapid traverse to R level
2. G01 — feed traverse to Z
3. M05 — revolutions stop
4. P. — dwell time at the bottom of the hole
5. M04 — counterclockwise revolutions are ON
6. G01 — feed traverse to R
7. M05 — spindle stop
8. M03 — clockwise revolution is ON
9. G00 — rapid traverse to a level plane for the function G98

Note: The feedrate is calculated as follows:

F = one divided by the # of threads per inch (lead) times the RPM (r/min).

Boring Cycles

Reaming Cycle (G85)

Block format:

G85 X . . . Y . . . Z . . . R . . . F . . . K . . .

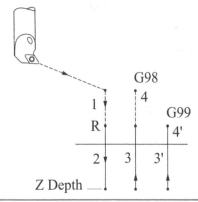

Figure 2-38.   Reaming cycle (G85)

Figure 2-38 explanations:

1. G00 — rapid traverse to R
2. G01 — feed traverse to Z depth
3. G01 — feed traverse to the R level plane and then rapid to the level plane assigned to function G98
3'. G01 — feed traverse to R for function G99
4. and 4'. M03 — the clockwise revolution is ON

Boring Cycle (G86)

Block format:

G86 X . . . Y . . . Z . . . R . . . F . . .

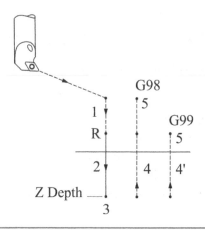

Figure 2-39.   Boring cycle (G86)

Figure 2-39 explanations:

1. G00 — rapid traverse to R
2. G01 — feed traverse to Z depth
3. M05 — spindle revolution is stopped
4. G00 — rapid traverse to the level plane assigned to function G98
5. G00 — rapid traverse to the R level plane for function G99
6. M03 — the clockwise revolution is ON

CNC Mill Program Example Descriptions

%O11719 (CAM Mill Example)

(Using high feed G1 F500. instead of G0.)

(T1 D=2. CR=0. - ZMIN=-0.04 - face mill)

(T2 D=1. CR=0. - ZMIN=-1.04 - flat end mill)

(T3 D=0.5 CR=0. TAPER=90deg - ZMIN=-0.24 - spot drill)

(T4 D=0.5312 CR=0. TAPER=118deg - ZMIN=-1.46 - drill)

(T5 D=0.4219 CR=0. TAPER=118deg - ZMIN=-1.46 - drill)

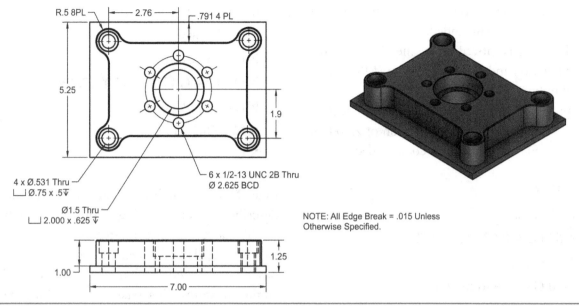

Figure 2-40. CNC mill program example drawing

(T6 D=0.75 CR=0. - ZMIN=-1.29 - flat end mill)

(T7 D=0.5 CR=0. - ZMIN=-1.29 - right hand tap)

(T8 D=0.5 CR=0. TAPER=45deg - ZMIN=-0.12 - chamfer mill)

The segment at the beginning of this program starts with the percentage sign. This is used during program transmission to indicate the beginning of the program code. No other action is performed by this symbol.

Next the program number is listed preceded by the capital letter "O" and then the actual program number.

All of the items in parentheses are program comments listed for the operator to view. These comments contain similar data as the CNC Setup Sheet. Anything within parentheses is not executable.

N10 G90 G94 G17 G20

N15 G53 G0 Z0.

In line N10, the absolute coordinate system (G90), the feed per minute (G94) and the X-Y plane (G17) and the inch measuring system (G20) are activated.

In line N15, the code G53 defines movement at a rapid traverse rate (G0) of the listed axis with respect to the machine coordinate system. The axis, in this case is the Z-axis and the machine coordinate is zero.

(Face1)

N20 T1 M6

N25 S6500 M3

N30 G54 G0 X4.6 Y-1.852

N35 G43 Z0.6 H1

N40 Z0.2 M8

N45 G1 Z-0.04 F65.

In this section of the program, the facing operation is completed. In line N20, tool number one is called (T1) and placed into the spindle (M6).

In line N25, the spindle rotation speed is set to 6500 RPM (S6500) in the clockwise direction (M3).

In line N30, the work coordinate system (G54) is activated so that all machine movements are relative to it. The X and Y-axes are positioned at a rapid traverse rate (G0) to the coordinates listed in line N30.

In line N35, the positive tool length compensation (G43) is activated, and the tool is moved to the Z0.6 position using the value in the height register (H1) is commanded.

The tool is moved along the Z-axis to the Z0.2 position and the flood coolant (M8) is activated in line N40.

In line N45, the linear interpolation code (G1) along with a new Z-axis destination of Z-.04, at feedrate (F) at 65.0 in/min is commanded.

```
N50 X3.5
N55 X-3.5
N60 G2 Y-0.098 I0. J0.877
N65 G1 X3.5
N70 G3 Y1.656 I0. J0.877
N75 G1 X-3.5
N80 X-4.5
N85 M9
```

In lines N50 through N80 removal of .04 thousandths of material from the entire top surface of the stock is performed by the face mill.

At line N85 the flood coolant is turned off (M9).

```
N85 G0 Z0.6 M5
N90 G53 G0 Z0.
N100 M1
```

At line N85, the rapid traverse (G0) is reinitiated, the tool is moved along the Z-axis to the clearance height of 0.6 and the spindle rotation is stopped (M5).

In line N90 code, G53 defines movement at a rapid traverse rate (G0) of the listed axis with respect to the machine coordinate system. The axis, in this case, is the Z-axis and the machine coordinate is zero.

The optional stop code (M1) is activated in line N100. *Note:* The machine will only stop at this code if the Optional Stop button is active on the control refer to the first section of this chapter for details.

```
(2D Contour 1 Rough)
N105 T2 M6
N110 S3820 M3
N115 G54 G0 X4.0047 Y1.7209
N120 G43 Z0.6 H2
N125 Z0.2 M8
N130 G1 Z-1.04 F30.
```

Lines N105 through N130 are similar to those described at the beginning of this program, but in this case, they are related to tool number two. The feeds and speeds and positioning are based on the one-inch diameter end mill. As the comment states, this pass is a roughing pass that leaves material for finishing later with the same tool.

```
N135 X3.9077 Y1.7454 F92.
N140 G3 X3.7863 Y1.6729 I-0.0245 J-0.097
N145 G2 X3.254 Y1.0169 I-0.9734 J0.2459
N150 G1 Y-1.0141
N155 G2 X3.434 Y-2.6251 I-0.4989 J-0.8713
N160 X1.8445 Y-2.3082 I-0.6789 J0.7396
N165 G1 X-1.8443
N170 G2 X-3.431 Y-2.6127 I-0.9057 J0.4332
N175 X-3.254 Y-1.0067 I0.681 J0.7377
N180 G1 Y1.0067
N185 G2 X-3.4432 Y2.6013 I0.504 J0.8683
N190 X-1.8592 Y2.3381 I0.6932 J-0.7263
N195 G1 X1.9006
N200 G2 X3.5142 Y2.6373 I0.9122 J-0.4193
N205 X3.7863 Y1.6729 I-0.7013 J-0.7184
N210 G3 X3.8587 Y1.5514 I0.097 J-0.0245
N215 G1 X3.9557 Y1.5269
N220 G0 Z0.6
```

In lines N175 through N260, tool number two is used to remove the material along the outer contour step of the part. The use of linear interpolation

(G1) and circular interpolation in both clockwise (G2) and counterclockwise directions are used along with arc center coordinate identification using the I and J method.

(2D Contour1 Finish)

N225 G1 X4.0493 Y1.7096 F500.

N230 Z0.2

N235 G1 Z0.0394 F30.

N240 Z-1.04

N245 G41 X4.0871 Y2.2673 D2 F92.

N250 G3 X3.2976 Y1.7963 I-0.1592 J-0.6302

N255 G2 X3.0448 Y1.4759 I-0.4848 J0.1225

N260 G3 X2.75 Y0.9886 I0.2552 J-0.4872

N265 G1 Y-0.9879

N270 G3 X3.0146 Y-1.458 I0.55 J0.

N275 G2 X3.0932 Y-2.2538 I-0.2595 J-0.4274

N280 X2.3077 Y-2.1087 I-0.3381 J0.3684

N285 G3 X1.8156 Y-1.8042 I-0.4921 J-0.2456

N290 G1 X-1.8157

N295 G3 X-2.3051 Y-2.1032 I0. J-0.55

N300 G2 X-3.0891 Y-2.2424 I-0.4449 J0.2282

N305 X-3.0119 Y-1.4491 I0.3391 J0.3674

N310 G3 X-2.75 Y-0.9806 I-0.2881 J0.4685

N315 G1 Y0.9806

N320 G3 X-3.0119 Y1.4491 I-0.55 J0.

N325 G2 X-3.0952 Y2.2367 I0.2619 J0.4259

N330 X-2.3127 Y2.1174 I0.3452 J-0.3617

N335 G3 X-1.8317 Y1.8341 I0.481 J0.2667

N340 G1 X1.8716

N345 G3 X2.3646 Y2.1404 I0. J0.55

N350 G2 X3.1621 Y2.2766 I0.4482 J-0.2216

N355 X3.2976 Y1.7963 I-0.3493 J-0.3578

N360 G3 X3.7686 Y1.0069 I0.6302 J-0.1592

N365 G1 G40 X4.0003 Y1.5157

N370 G0 Z0.6

In lines N225 through N370, tool number two is again used for the finish pass around the outside step contour. Make special note of line N245 where the cutter compensation mode left (G41) is initiated and references the diameter offset value (D2) registered in the tool offset diameter column (refer to the first section of this chapter for directions on setting this value). Then, in line N365 when the work for the tool is completed, the compensation is cancelled with code (G40).

(2D Pocket 1)

N375 G1 X0.2305 Y0.4378 F500.

N380 G0 Z0.6

N385 Z0.2

N390 G1 Z0.1999 F92.

N400 X0.2293 Y0.4385 Z0.183

N405 X0.2254 Y0.4405 Z0.1665

N410 X0.2192 Y0.4437 Z0.151

N415 X0.2105 Y0.4479 Z0.1369

N420 G3 X0.1998 Y0.4528 Z0.1246 I-0.204 J-0.429

N425 X0.1871 Y0.4583 Z0.1145 I-0.1932 J-0.4339

N430 X0.1729 Y0.4638 Z0.1069 I-0.1806 J-0.4393

N435 X0.1576 Y0.4693 Z0.102 I-0.1664 J-0.4449

N440 X0.1414 Y0.4744 Z0.1 I-0.151 J-0.4504

N445 X0.1806 Y0.4609 Z-0.0028 I-0.1349 J-0.4554

N450 X0.2185 Y0.444 Z-0.1055 I-0.1741 J-0.442

N455 X0.2547 Y0.424 Z-0.2083 I-0.2119 J-0.4251

N460 X0.2891 Y0.4008 Z-0.3111 I-0.2482 J-0.405

N465 X0.3213 Y0.3747 Z-0.4139 I-0.2825 J-0.3819

N470 X0.3511 Y0.3459 Z-0.5166 I-0.3147 J-0.3558

N475 X0.3783 Y0.3147 Z-0.6194 I-0.3445 J-0.327

N480 X-0.2785 Y0.3989 Z-0.645 I-0.3717 J-0.2957

N485 X0.2975 Y-0.3691 I0.288 J-0.384

N490 X-0.2785 Y0.3989 I-0.288 J0.384

N495 X-0.2914 Y0.3874 Z-0.6435 I0.06 J-0.08

N500 X-0.3018 Y0.3742 Z-0.639 I0.0729 J-0.0684

N505 X-0.3095 Y0.3604 Z-0.6316 I0.0834 J-0.0552

N510 X-0.3144 Y0.347 Z-0.6216 I0.091 J-0.0414

N515 X-0.3172 Y0.335 Z-0.6093 I0.096 J-0.0281

N520 X-0.3183 Y0.3251 Z-0.595 I0.0987 J-0.0161

N525 G1 X-0.3185 Y0.3177 Z-0.5792

N530 X-0.3183 Y0.3132 Z-0.5624

N535 X-0.3182 Y0.3117 Z-0.545

N540 G0 Z0.6

In lines N375 through N540, the center counter bore (Pocket 1) is machined. Since the tool is the same as used for the outside step contour, no tool, RPM or feedrate change is needed. The tool entry into this cut is done by helical interpolation where all three axes are simultaneously fed around a circular path until full depth is reached. Then the tool is programmed to follow a circular path stepping over radially until the diameter and floor, minus a finish allowance amount is reached.

(2D Contour 2 Finish)

N545 G1 X0.2595 Y-0.0851 F500.

N550 G0 Z0.6

N555 Z0.2

N560 G1 Z0.0394 F30.

N565 Z-0.665

N570 G41 X0.3595 Y-0.6351 D2 F92.

N575 G3 X1.0095 Y0.0149 I0. J0.65

N580 X-0.9905 I-1. J0.

N585 X1.0095 I1. J0.

N590 X0.3595 Y0.6649 I-0.65 J0.

N595 G1 G40 X0.2595 Y0.1149

N600 G0 Z0.6 M5

N605 G53 G0 Z0. M9

N610 M1

In lines N545 through N610, the center counter bore (Contour 2 Finish) is machined. The final finish passes to remove the finish allowances listed above are completed in this section of the program. In line N570, cutter diameter compensation is initiated to the left (G41) for tool number two (D2).

(Spot Drill 1)

N615 T3 M6

N620 S764 M3

N625 G54 G0 X2.8129 Y1.9188

N630 G43 Z0.6 H3

N635 G0 Z0.2 M8

N640 G98 G81 X2.8129 Y1.9188 Z-0.24 R0.145 F1.528

N645 G1 X1.1462 Y0.6712 Z0.2 F500.

N650 G98 G81 X1.1462 Y0.6712 Z-0.24 R0.16 F1.528

N655 X0.0095 Y1.3274

N660 X-1.1271 Y0.6712

N665 Y-0.6413

N670 X0.0095 Y-1.2976

N675 X1.1462 Y-0.6413

N680 G80

N685 G1 X2.7551 Y-1.8854 Z0.2 F500.

N690 G98 G81 X2.7551 Y-1.8854 Z-0.24 R0.145 F1.528

N695 X-2.75 Y-1.875

N700 Y1.875

N705 G80

N785 G0 Z0.6 M5

N790 G53 G0 Z0. M9

N795 M1

The spot drilling operation is performed with tool number three from line numbers N615 through N795. When moves between patterns are required, the canned cycle cancellation code (G80) is implemented (on lines N680 and N705). The drilling cycle (G81) is used in all cases. Refer back to the descriptions of the G81 canned cycle in this chapter.

($^{17}/_{32}$ Drill 2)

N800 T4 M6

N805 S2160 M3

N810 G54 G0 X-2.75 Y1.875

N815 G43 Z0.6 H4

N820 G0 Z0.2 M8

N825 G98 G83 X-2.75 Y1.875 Z-1.46 R0.16 Q0.3 F29.

N830 Y-1.875

N835 X2.7551 Y-1.8854

N840 X2.8129 Y1.9188

N845 G80

N850 G0 Z0.6 M5

N855 G53 G0 Z0. M9

N860 M1

Drilling of the four corner holes is performed from blocks N800 through N860. For these holes, the canned cycle (G83) is used with a peck increment of 0.3 inches. In block N845, the canned cycle is cancelled (G80). Refer back to the descriptions of the G83 canned cycle in this chapter.

($^{27}/_{24}$ Drill 3)

N865 T5 M6

N870 S2720 M3

N875 G54 G0 X-1.1271 Y0.6712

N880 G43 Z0.6 H5

N885 G0 Z0.2 M8

N890 G98 G83 X-1.1271 Y0.6712 Z-1.46 R0.16 Q0.25 F29.

N895 X0.0095 Y1.3274

N900 X1.1462 Y0.6712

N905 Y-0.6413

N910 X0.0095 Y-1.2976

N915 X-1.1271 Y-0.6413

N920 G80

N925 G0 Z0.6 M5

N930 G53 G0 Z0. M9

N935 M1

Drilling of the bolt circle pattern of holes is performed from blocks N865 through N935. Once again, for these holes, the canned cycle (G83) is used, but with a peck increment of 0.25 inches. In block N920, the canned cycle is cancelled (G80). Refer back to the descriptions of the G83 canned cycle in this chapter.

($^3/_4$ Endmill for 2D Pocket 2)

N940 T6 M6

N945 S2440 M3

N950 G54 G0 X-0.0008 Y-0.1959

N955 G43 Z0.6 H6

N960 G0 Z0.2 M8

N965 G1 Z0.1 F55.

N970 G3 X-0.0187 Y-0.1834 Z0.0459 I0.1343 J0.2109

N975 X-0.0354 Y-0.1694 Z-0.0082 I0.1522 J0.1984

N980 X-0.0508 Y-0.154 Z-0.0623 I0.1689 J0.1843

N985 X-0.0649 Y-0.1373 Z-0.1164 I0.1843 J0.1689

N990 X-0.0774 Y-0.1194 Z-0.1705 I0.1983 J0.1522

N995 X-0.0883 Y-0.1005 Z-0.2245 I0.2108 J0.1343

N1000 X-0.0975 Y-0.0808 Z-0.2786 I0.2217 J0.1155

N1005 X-0.105 Y-0.0603 Z-0.3327 I0.231 J0.0957

N1010 X-0.1106 Y-0.0392 Z-0.3868 I0.2384 J0.0752

N1015 X-0.1144 Y-0.0177 Z-0.4409 I0.2441 J0.0541

N1020 X-0.1163 Y0.004 Z-0.495 I0.2479 J0.0327

N1025 Y0.0258 Z-0.5491 I0.2498 J0.0109

N1030 X-0.1144 Y0.0475 Z-0.6032 I0.2498 J-0.0109

N1035 X-0.1106 Y0.069 Z-0.6573 I0.2479 J-0.0326

N1040 X-0.105 Y0.0901 Z-0.7114 I0.2441 J-0.0541

N1045 X-0.0975 Y0.1106 Z-0.7655 I0.2384 J-0.0752

N1050 X-0.0883 Y0.1303 Z-0.8196 I0.231 J-0.0956

N1055 X-0.0774 Y0.1492 Z-0.8736 I0.2218 J-0.1154

N1060 X-0.0649 Y0.1671 Z-0.9277 I0.2109 J-0.1343

N1065 X-0.0509 Y0.1838 Z-0.9818 I0.1984 J-0.1522

N1070 X-0.0354 Y0.1992 Z-1.0359 I0.1843 J-0.1689

N1075 X-0.0187 Y0.2133 Z-1.09 I0.1689 J-0.1843

N1080 X-0.0009 Y0.2258 Z-1.1441 I0.1522 J-0.1983

N1085 X0.018 Y0.2367 Z-1.1982 I0.1343 J-0.2108

N1090 X0.0378 Y0.2459 Z-1.2523 I0.1155 J-0.2217

N1095 X0.3835 Y0.0149 Z-1.29 I0.0957 J-0.231

N1100 X0.347 I-0.0182 J0.

N1105 X0.3845 I0.0188 J0.

N1110 X-0.3655 I-0.375 J0.

N1115 X0.3845 I0.375 J0.

N1120 X0.3095 Y0.0899 I-0.075 J0.

N1125 G0 Z0.6

In blocks N940 through N1125, the center hole is machined with tool six. The tolerance on this hole does not require finishing so it can be done without cutter diameter compensation like that used for the counter bore. Again, helical interpolation is used for the entry into the cut and then a radial step over amount is used to reach the final diameter.

(¾ **End Mill-Drill 4**)

N1130 G1 X2.8129 Y1.9188 F500.

N1135 G0 Z0.6

N1140 Z0.2

N1145 G98 G82 X2.8129 Y1.9188 Z-0.54 R0.145 P2000 F30.

N1150 X2.7551 Y-1.8854

N1155 X-2.75 Y-1.875

N1160 Y1.875

N1165 G80

N1170 G0 Z0.6 M5

N1175 G53 G0 Z0. M9

N1180 M1

The same tool is used to counterbore the four corner holes to depth so a tool change is not needed for this operation. In line N1145, the canned drilling cycle (G82) is used. This cycle allows for a 2-second dwell to be introduced once the tool reaches the bottom of the hole. Refer back to the descriptions of the G82 canned cycle in this chapter.

(½-13 Tap)
N1185 T7 M6
N1190 S500 M3
N1195 G54 G0 X-1.1271 Y-0.6413
N1200 G43 Z0.6 H7
N1205 G0 Z0.2 M8
N1210 G98 G84 X-1.1271 Y-0.6413 Z-1.29 R0.16 F38.4616
N1215 X0.0095 Y-1.2976
N1220 X1.1462 Y-0.6413
N1225 Y0.6712
N1230 X0.0095 Y1.3274
N1235 X-1.1271 Y0.6712
N1240 G80
N1245 G0 Z0.6 M5
N1250 G53 G0 Z0. M9
N1255 M1

In lines N1185 through N1255, the bolt circle holes are tapped with the ½-13 thread using the canned cycle for tapping (G84) given on line N1210. *Note:* The calculation of the feedrate for tapping is critical. Refer back to the descriptions of the G84 canned cycle in this chapter.

(2D Chamfer 2)
N1260 T8 M6
N1265 S2650 M3
N1270 G54 G0 X0.9345 Y0.0149
N1275 G43 Z0.6 H8
N1280 G0 Z0.2 M8
N1285 G1 Z0.08 F18.

N1290 Z-0.12
N1295 X0.9845
N1300 G3 X-0.9655 I-0.975 J0. F36.
N1305 X0.9845 I0.975 J0.
N1310 G1 X0.9345 F18.
N1315 G0 Z0.2
N1320 G1 X3.3553 Y1.7795 F500.
N1325 Z0.08 F18.
N1330 Z-0.12
N1335 X3.3068 Y1.7919
N1340 G2 X3.0358 Y1.4601 I-0.494 J0.1269 F36.
N1345 G3 X2.76 Y1.0194 I0.2142 J-0.4407
N1350 G1 Y-1.0165
N1355 G3 X3.0075 Y-1.4422 I0.49 J0.
N1360 G2 X2.292 Y-2.099 I-0.2524 J-0.4432
N1365 G3 X1.847 Y-1.8142 I-0.4449 J-0.2052
N1370 G1 X-1.8468
N1375 G3 X-2.2894 Y-2.0939 I0. J-0.49
N1380 G2 X-3.005 Y-1.4333 I-0.4606 J0.2189
N1385 G3 X-2.76 Y-1.009 I-0.245 J0.4244
N1390 G1 Y1.009
N1395 G3 X-3.005 Y1.4333 I-0.49 J0.
N1400 G2 X-2.2969 Y2.1092 I0.255 J0.4417
N1405 G3 X-1.8616 Y1.8441 I0.4353 J0.225
N1410 G1 X1.9032
N1415 G3 X2.3489 Y2.1306 I0. J0.49
N1420 G2 X3.3068 Y1.7919 I0.4639 J-0.2118
N1425 G1 X3.3553 Y1.7795 F18.
N1430 G0 Z0.2
N1435 G1 X3.0551 Y-1.8854 F500.
N1440 Z0.08 F18.
N1445 Z-0.12
N1450 X3.1051
N1455 G3 X2.4051 I-0.35 J0. F36.

N1460 X3.1051 I0.35 J0.

N1465 G1 X3.0551 F18.

N1470 G0 Z0.2

N1475 G1 X-2.45 Y-1.875 F500.

N1480 Z0.08 F18.

N1485 Z-0.12

N1490 X-2.4

N1495 G3 X-3.1 I-0.35 J0. F36.

N1500 X-2.4 I0.35 J0.

N1510 G1 X-2.45 F18.

N1515 G0 Z0.2

N1520 G1 X-2.4501 Y1.8824 F500.

N1525 Z0.08 F18.

N1530 Z-0.12

N1535 X-2.4001 Y1.8836

N1540 G3 X-3.0999 Y1.8664 I-0.3499 J-0.0086 F36.

N1545 X-2.4001 Y1.8836 I0.3499 J0.0086

N1550 G1 X-2.4501 Y1.8824 F18.

N1555 G0 Z0.6 M5

N1560 G53 G0 Z0. M9

N1565 G53 G0 Y0.

N1570 M30

%

In lines N1260 through N1570, the chamfering tool is used to deburr all of the top intersecting edges. On line N1565, the machine is positioned so the table is forward to allow for easier part access. In line N1570, the miscellaneous code (M30) is given to return the program to its beginning. Lastly, the percentage sign is given to signify the end of the program file data for transmission purposes.

Summary

In this section, the basic structure and common content of CNC programming have been presented for CNC milling. In the Introduction to CNC Lathe Programming section of Chapter 3, basic programming for CNC lathes will be also be presented with specific differences. The data that makes up the programs will undoubtedly be processed from a CAM program but the understanding gained about the individual pieces that make up the code may need to be edited at some point at the machine. The methods presented here for editing are similar no matter which controller is used. In Chapter 4, computer aided manufacturing will be presented, including post-processing of the code, some of which you just learned.

Introduction to CNC Mill Programming, Study Questions

1. What other program word is necessary when programming G01 linear interpolation?

   a. S
   b. F
   c. T
   d. H

2. Which code activates the tool length offset?

   a. G54
   b. G40
   c. G41
   d. G43

3. When programming an arc, which letters identify the arc center location?

   a. X, Y and Z
   b. A, B and C
   c. I, J and K
   d. Q and P

4. Cutter diameter compensation, G41 and G42 offset the cutter to the left or the right. Which command is used for climb milling?

   a. G40
   b. G41
   c. G42
   d. G43

5. When programming an arc, an additional method exists that does not use I, J and K. Which program word is used?

   a. A
   b. B
   c. C
   d. R

6. A block of codes at the beginning of the program are used to cancel modal commands and are called the "safety block." They are:

   a. G90 G54 G00
   b. G20 G90 G00
   c. G40 G80 G49
   d. G91 G28 G00

7. When using the canned drilling cycle G83, which letter identifies the peck amount?

   a. Z
   b. P
   c. K
   d. Q

8. What character is inserted at the End of a Block when the program is loaded into the controller?

   a. ;
   b. /
   c. (
   d.)

9. The letter address H is used to indicate a tool length offset register number. Which preparatory function is used in conjunction with it?

   a. G54
   b. G43
   c. G42
   d. G41

10. When using canned drilling cycles, which of the following codes are used to return the drill to the initial plane?

    a. G99
    b. G98
    c. G90
    d. G92

11. When using canned drilling cycles, which of the following codes are used to return the drill to the reference plane?

a. G99
b. G98
c. G90
d. G92

12. When using canned drilling cycles, what other letter address is necessary to identify the reference plane position?

a. P
b. Q
c. R
d. S

13. When using cutter diameter compensation in a program, what letter address is used to identify the location of the value of the offset?

a. A
b. R
c. H
d. D

14. Which G-code is used to cancel Cutter Diameter Compensation?

a. G40
b. G41
c. G42
d. G43

15. If a linear move is programmed G01 X1.5 Y1.5, what is the angle of the resulting cut?

a. 30 degrees
b. 180 degrees
c. 45 degrees
d. 90 degrees

3

Setup, Operation and Programming of CNC Lathes

CNC Lathe Setup and Operation

Objectives

1. Identify and use common CNC lathe operator panel functions.
2. Identify and use common machine control panel functions for CNC lathes.
3. Identify work-holding components and related accessories.
4. To perform common setup and operation functions at the CNC lathe machine control.
5. Use the controls to input setup data, including tool and work offsets.
6. Use the EDIT mode to edit programs.
7. Identify and solve some common problem situations during setup and operation of CNC lathes.

The CNC lathe is used to machine cylindrical shaped parts to precision dimensions.

Chapter 2, CNC Mill Setup and Operation, Operator Control Pendant Features, provides function descriptions that are identical for the CNC lathe. Those functions that differ for the CNC lathe will be presented here.

The control panel used for a lathe would be essentially identical to that for a mill, except for the axis keys for X and Z only. You should always consult the applicable manufacturer manual for detailed descriptions that match your needs.

Another item found on a CNC lathe (not shown here) is the *chuck foot pedal* which is used for hydraulic clamping and unclamping of the work part. Parts can be clamped by the outside or inside diameter by switching the chucking direction from external to internal by parameter setting (refer to the machine Operator Manual for directions).

Machine Safety for CNC Lathes

Lathe Machine Guarding

CNC lathes, just like CNC mills, are fully guarded and include sliding doors which are fitted with safety interlocks and windows. It cannot be overemphasized that you should never operate any machine tool without proper guards in place and never alter any door safety features!

Just as with CNC mills, there should be a power disconnect mounted on a wall near the back of the machine to remove power from the machine and enable Lockout/Tagout (LOTO) during maintenance activities. This switch must be on prior to machine startup.

CNC Lathe Operator Control Panel

In Chapter 2, the CNC mill operation control panel (operator pendant) was shown and

described. The same information holds true for the CNC lathe with some differences. In this section, items that have identical function as the CNC mill will not be addressed again; instead, only buttons, keys and functions specific to the lathe will be presented. If necessary, refer to Chapter 2, to refresh your memory.

CNC Lathe Operator Control Pendant Features

Figure 3-1 shows the CNC lathe pendant. Notice the nearly identical layout to the CNC mill. This similarity makes learning the second configuration that much easier. The CNC lathe information presented here is limited to the 2-axes of X and Z only.

The following operator control items are the same for the CNC mill: Power On and Power Off, Emergency Stop, Handle/Jog (MPG), Cycle Start, Feed Hold, Universal Serial Bus, USB port; Memory Lock; Setup Mode; Second Home; Servo Auto Door Override; Work Light; and Beacon Light.

CNC Lathe Operator Control Pendant, Keyboard Descriptions

Functional descriptions for most of the operator panel buttons are the same for the mill and lathe; however, specific differences are noted in each section.

The following sections describe the Operator Control buttons that are unique to the CNC lathe (Figure 3-2): Function Keys-1, Cursor Keys-2, Display Keys-3, Mode Keys-4, Numeric Keys-5, Alpha Keys-6, Jog Keys-7 and Override Keys-8.

Function Keys

RECOVER.   This button is used to display instruction necessary to recover from issues with tapping.

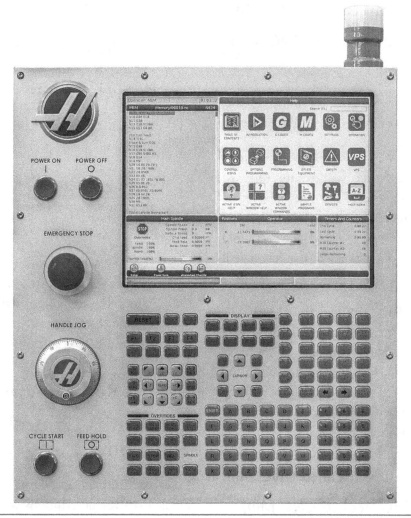

Figure 3-1. CNC lathe operator pendant, *courtesy Haas CNC Automation*

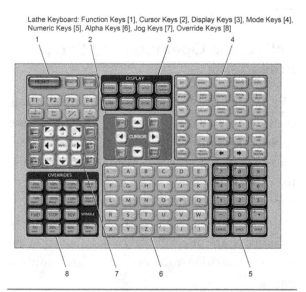

Lathe Keyboard: Function Keys [1], Cursor Keys [2], Display Keys [3], Mode Keys [4], Numeric Keys [5], Alpha Keys [6], Jog Keys [7], Override Keys [8]

Figure 3-2. CNC lathe operator pendant, *courtesy Haas CNC Automation*

X DIAMETER MEASURE. When this button is pressed, you will be prompted to enter the actual diameter and then the current position of the X-axis is recorded into the X Geometry offset column of the Tool Offset.

X/Z. This button is used to toggle between the X and Z-axes while in the HANDLE/JOG modes during part setup.

Z FACE MEASURE. During the process of tool measurement, this button is used to record the current position of the Z-axis for the active tool into the Z Geometry column of the Tool Offset display. This button is also used during measuring of the work offset.

Cursor Keys (Identical to CNC Mill)

Display Keys (Identical to CNC Mill)

Mode Keys

All the mode keys are the same for mill and lathe, with the exception of two of the MDI buttons:

TURRET FWD. When this button is pressed, the turret will rotate forward (number sequence) to the next tool position. The control must be in the MDI mode. If a specific tool number is entered and the button is pressed, the turret will advance to that tool.

TURRET REV. When this button is pressed, the turret will rotate in the reverse direction (number sequence) to the next tool position. The control must be in the MDI mode. Again, if a specific tool number is entered and the button is pressed, the turret will advance to that tool.

Numeric Keys (Identical to CNC Mill)

Alpha Keys (Identical to CNC Mill)

Jog Keys

The axis jog buttons function in the same fashion as described in the CNC mill section, except for the following buttons:

TS ←. When this button is pressed, the Tail Stock will advance towards the spindle until released.

TS RAPID. Pressing this button while simultaneously pressing either TS ← or TS → will move the Tail Stock at a rapid rate based on the direction chosen until released.

TS →. When this button is pressed the Tail Stock will retract away from the spindle until released.

–C. This button allows the user to jog the C-axis chuck rotation (if equipped) manually, in the negative direction at a specific feedrate as determined by the HANDLE/JOG increment chosen.

RAPID. Pressing this button while simultaneously pressing either the X or Z-axes buttons in either positive or negative direction will cause movement at a rapid rate based on the direction chosen until released.

+C. This button allows the user to jog the C-axis chuck rotation (if equipped) manually, in the positive direction at a specific feedrate as determined by the HANDLE/JOG increment chosen.

Override Keys (Identical to CNC Mill)

CNC Lathe Control Panel Functions

Refer to Chapter 2, Control Panel Functions, items 1–13 for descriptions of each area of the display shown in Figure 2-13.

CNC Lathe Display

CNC Lathe Operation and Setup Procedures

The following explanations are for operations considered routine for users of CNC machine tools and are given in their sequence of use. Please note that these procedures are specific to the Haas "New Generation Controller" depicted in Figure 3-3. The procedures for another type of control may be similar. Be sure to consult the manufacturer manuals specific to your machine tool operation and control panel.

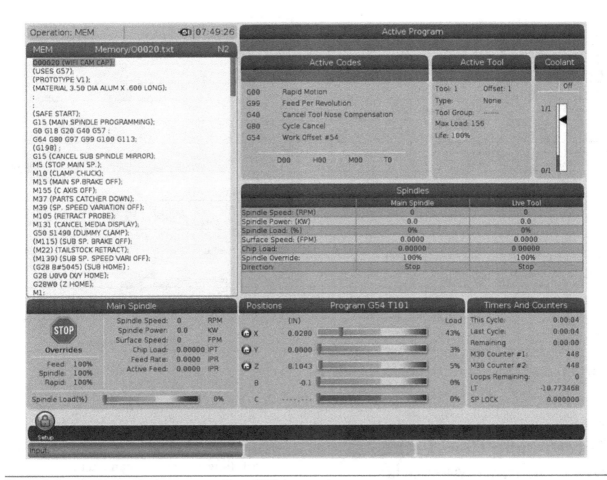

Figure 3-3. CNC lathe display, *courtesy Haas CNC Automation*

Process Planning Documents for CNC Turning

CNC Lathe Process Planning Operation Sheet

In the same fashion as covered in Chapter 2, the Process Planning Operation Sheet for the CNC lathe (Chart 3-1) lays out the order of operations for the machinist to follow. Chart 3-2a, b and c represent an example of a setup sheet.

CNC Lathe Quality Control Check Sheet

Refer to the example given in Chapter 1, Figure 1-284.

Applying the Coordinate Systems for Turning

Both absolute and incremental coordinate systems are used when programming of CNC lathes. Two-axis lathes are limited to X (diameter controlling) and Z-axes (parallel with the spindle axis). When milling, the incremental coordinate system is activated with the code G91. Unlike milling, CNC lathes instead use the letters U and W to initiate an incremental movement for X and Z, respectively. A good example of application are external grooving geometries. Please refer to my text, *Programming of CNC Machines* (4th ed.), for complete details for using this coordinate method.

Date: Today			Prepared By: You	
Part Name: CNC Lathe Project			Part Number: 1234	
Order Quantity: 100			Sheet __ of __	
Material: 4340 Alloy Steel				
Raw Stock Size: 3.0625" x 2.50" Diameter Bar Stock				
Operation Number	Machine Used	Operation Description		Time Standard
1	Saw	Cut the Bar Stock to 3.0625" lengths.		3 minutes each
2	CNC Lathe	Machine complete to dimensions.		5 minutes each
3	Bench	Deburr as Needed.		5 minutes each
4	Quality Control	Final Quality Control Inspection.		10 min. each

Chart 3-1. CNC Lathe Process Planning Operation Sheet

Determining CNC Lathe Work-Holding Methods

Tool and work-holding methods used on manual lathes are similar to those used on CNC lathes. Just as with milling, the material to be machined has a direct effect on what tools will be used, the type of coolant necessary, and the selection of proper speeds and feeds for the metal-cutting operation.

Cylindrically shaped workpiece geometry is by far the norm when using lathes and the work-holding method that will be used is most often a 3-jaw universal chuck (Figure 3-4). The clamping method is important for CNC work because of the high performance expected. Again, it must hold the workpiece securely, be rigid, and minimize the possibility of any flex or movement of the part.

Note: It cannot be overstated that safety is extremely important when using the CNC lathe. Improper clamping can cause the part to come out of the chuck while machining is in progress

causing the potential for loss of life! It is important to follow these and manufacturer recommended best practices when clamping for CNC lathes and always err on the side of caution.

The 3-jaw universal chuck is the most commonly used work-holding device for CNC lathes. They are clamped and unclamped by hydraulic actuation of a threaded draw tube. The clamping pressure is adjusted by a knob located on the machine spindle assembly that includes a gauge to aid accurate pressure setting. This setting is a critical component of the system because too much clamping pressure can crush thin-walled parts and too little pressure could allow slippage or even part flyout. Check with the chuck manufacturer for recommended pressure settings and adhere to them strictly.

Installation and Removal of Work-Holding Devices

CNC lathe chucks are heavier than 50 pounds in most cases, so proper lifting techniques and/

Chart 3-2a

Setup

WCS: #0

STOCK:
DX: 2.5in
DY: 2.5in
DZ: 3in

PART:
DX: 2.5in
DY: 2.5in
DZ: 3in

STOCK LOWER IN WCS #0:
X: -1.25in
Y: -1.25in
Z: -3in

STOCK UPPER IN WCS #0:
X: 1.25in
Y: 1.25in
Z: 0in

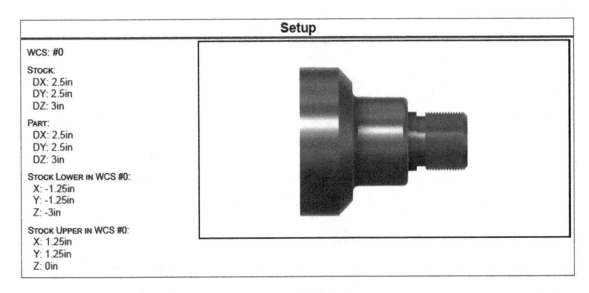

Total

NUMBER OF OPERATIONS: 6
NUMBER OF TOOLS: 4
TOOLS: T1 T4 T5 T11
MAXIMUM Z: 0.197in
MINIMUM Z: -3.2in
MAXIMUM FEEDRATE: 40in/min
MAXIMUM SPINDLE SPEED: 1000rpm
CUTTING DISTANCE: 36.559in
RAPID DISTANCE: 54.995in
ESTIMATED CYCLE TIME: 11m:52s

Chart 3-2b

Tools

T1 D1
TYPE: general turning
INSERT: ISO C 80deg
INSCRIBED CIRCLE: 0.5in
NOSE RADIUS: 0.031in
CROSS SECTION: G
TOLERANCE: M
RELIEF: N 0deg
COMPENSATION: Tip tangent
DESCRIPTION: RH Turning/Facing Tool: MCLNR 12-4C
VENDOR: Tormach
PRODUCT: <u>33130</u>

MINIMUM Z: -2.362in
MAXIMUM FEED: 25in/min
MAXIMUM SPINDLE SPEED: 500rpm
CUTTING DISTANCE: 22.612in
RAPID DISTANCE: 24.679in
ESTIMATED CYCLE TIME: 5m:49s (49%)

HOLDER: ISO L Right

Chart 3-2a, 2b and 2c.   Example Setup Sheet

T4 D4
TYPE: general turning
INSERT: ISO V 35deg
INSCRIBED CIRCLE: 0.375in
NOSE RADIUS: 0.031in
CROSS SECTION: T
TOLERANCE: M
RELIEF: N 0deg
COMPENSATION: Tip tangent
DESCRIPTION: RH Turning/Facing Tool: MVJNR 12-3C
VENDOR: Tormach
PRODUCT: 33133

MINIMUM Z: -2.379in
MAXIMUM FEED: 7.12339in/min
MAXIMUM SPINDLE SPEED: 500rpm
CUTTING DISTANCE: 3.096in
RAPID DISTANCE: 4.165in
ESTIMATED CYCLE TIME: 43s (6%)

HOLDER: ISO J Right

T5 D5
TYPE: thread turning
INSERT: ISO triple
COMPENSATION: Tip tangent
DESCRIPTION: OD Threading Tool: CER0750K16
VENDOR: Tormach
PRODUCT: 33144

MINIMUM Z: -0.87in
MAXIMUM FEED: 40in/min
MAXIMUM SPINDLE SPEED: 1000rpm
CUTTING DISTANCE: 5.085in
RAPID DISTANCE: 17.072in
ESTIMATED CYCLE TIME: 8s (1.1%)

HOLDER: Straight Right

T11 D11
TYPE: groove turning
INSERT: Square
WIDTH: 0.16in
NOSE RADIUS: 0.004in
COMPENSATION: Tip tangent
DESCRIPTION: RH Grooving Tool: SGTHR 19-4
VENDOR: Tormach
PRODUCT: 33347

MINIMUM Z: -3.2in
MAXIMUM FEED: 10in/min
MAXIMUM SPINDLE SPEED: 1000rpm
CUTTING DISTANCE: 5.766in
RAPID DISTANCE: 9.079in
ESTIMATED CYCLE TIME: 3m:57s (33.3%)

HOLDER: External Right

Chart 3-2c

Operations

Operation 1/6
DESCRIPTION: Face1
STRATEGY: Turning Face
WCS: #0
TOLERANCE: 0in

MAXIMUM Z: 0.197in
MINIMUM Z: -0.04in
SURFACE SPEED: 350ft/min
FEEDRATE PER REV: 0.005in
CUTTING DISTANCE: 1.585in
RAPID DISTANCE: 2.585in
ESTIMATED CYCLE TIME: 20s (2.8%)
COOLANT: Flood

T1 D1
TYPE: general turning
INSERT: ISO C 80deg
INSCRIBED CIRCLE: 0.5in
NOSE RADIUS: 0.031in
CROSS SECTION: G
TOLERANCE: M
RELIEF: N 0deg
COMPENSATION: Tip tangent
DESCRIPTION: RH Turning/Facing Tool: MCLNR 12-4C
VENDOR: Tormach
PRODUCT: 33130

Operation 2/6
DESCRIPTION: Profile2
STRATEGY: Turning Profile
WCS: #0
TOLERANCE: 0in
STOCK TO LEAVE: 0.004in
MAXIMUM STEPDOWN: 0.08in

MAXIMUM Z: 0.197in
MINIMUM Z: -2.362in
SURFACE SPEED: 350ft/min
FEEDRATE PER REV: 0.005in
CUTTING DISTANCE: 21.027in
RAPID DISTANCE: 22.093in
ESTIMATED CYCLE TIME: 5m:28s (46.2%)
COOLANT: Flood

T1 D1
TYPE: general turning
INSERT: ISO C 80deg
INSCRIBED CIRCLE: 0.5in
NOSE RADIUS: 0.031in
CROSS SECTION: G
TOLERANCE: M
RELIEF: N 0deg
COMPENSATION: Tip tangent
DESCRIPTION: RH Turning/Facing Tool: MCLNR 12-4C
VENDOR: Tormach
PRODUCT: 33130

Chart 3-2a, 2b and 2c. Example Setup Sheet

Operation 3/6		T4 D4
DESCRIPTION: Profile3	MAXIMUM Z: 0.197in	TYPE: general turning
STRATEGY: Turning Profile	MINIMUM Z: -2.379in	INSERT: ISO V 35deg
WCS: #0	SURFACE SPEED: 350ft/min	INSCRIBED CIRCLE: 0.375in
TOLERANCE: 0in	FEEDRATE PER REV: 0.005in	NOSE RADIUS: 0.031in
STOCK TO LEAVE: 0in	CUTTING DISTANCE: 3.096in	CROSS SECTION: T
MAXIMUM STEPOVER: 0.04in	RAPID DISTANCE: 4.165in	TOLERANCE: M
COMPENSATION: wear (center)	ESTIMATED CYCLE TIME: 43s (6%)	RELIEF: N 0deg
SAFE TOOL DIAMETER: < 0in	COOLANT: Flood	COMPENSATION: Tip tangent
		DESCRIPTION: RH Turning/Facing Tool: MVJNR 12-3C
		VENDOR: Tormach
		PRODUCT: 33133

Operation 4/6		T11 D11
DESCRIPTION: Single Groove1	MAXIMUM Z: 0.197in	TYPE: groove turning
STRATEGY: Turning Groove	MINIMUM Z: -0.946in	INSERT: Square
WCS: #0	SURFACE SPEED: 350ft/min	WIDTH: 0.16in
TOLERANCE: 0in	FEEDRATE PER REV: 0.002in	NOSE RADIUS: 0.004in
STOCK TO LEAVE: 0in	CUTTING DISTANCE: 2.458in	COMPENSATION: Tip tangent
	RAPID DISTANCE: 2.286in	DESCRIPTION: RH Grooving Tool: SGTHR 19-4
	ESTIMATED CYCLE TIME: 1m:54s (16.1%)	VENDOR: Tormach
	COOLANT: Flood	PRODUCT: 33347

Operation 5/6		T5 D5
DESCRIPTION: Thread1	MAXIMUM Z: 0.197in	TYPE: thread turning
STRATEGY: Turning Thread	MINIMUM Z: -0.87in	INSERT: ISO triple
WCS: #0	MAXIMUM SPINDLE SPEED: 1000rpm	COMPENSATION: Tip tangent
TOLERANCE: 0in	FEEDRATE PER REV: 0.04in	DESCRIPTION: OD Threading Tool: CER0750K16
	CUTTING DISTANCE: 5.085in	VENDOR: Tormach
	RAPID DISTANCE: 17.072in	PRODUCT: 33144
	ESTIMATED CYCLE TIME: 8s (1.1%)	
	COOLANT: Flood	

Operation 6/6		T11 D11
DESCRIPTION: Part1	MAXIMUM Z: 0.197in	TYPE: groove turning
STRATEGY: Turning Part	MINIMUM Z: -3.2in	INSERT: Square
WCS: #0	SURFACE SPEED: 350ft/min	WIDTH: 0.16in
TOLERANCE: 0in	FEEDRATE PER REV: 0.002in	NOSE RADIUS: 0.004in
	CUTTING DISTANCE: 3.308in	COMPENSATION: Tip tangent
	RAPID DISTANCE: 6.794in	DESCRIPTION: RH Grooving Tool: SGTHR 19-4
	ESTIMATED CYCLE TIME: 2m:3s (17.3%)	VENDOR: Tormach
	COOLANT: Flood	PRODUCT: 33347

Chart 3-2a, 2b and 2c. Example Setup Sheet

Figure 3-4. 3-Jaw universal chuck

or mechanisms should be used to avoid injury. Most chucks are fitted with a threaded hole on the main body for installation of a lifting ring to be used with a hook and hoist. Don't forget to remove the lifting ring after you are finished using it.

Before installation of any work-holding systems, all of the mating surfaces should be cleaned and inspected for burrs. Lift the chuck into position and install the clamping screws. Tighten the screws to manufacturer recommended torqued settings using a torque wrench.

Adjusting Jaws on 3-Jaw Chucks

The top or soft jaws are mounted to the chuck using socket head cap screws inserted into the sliding t-nut in the adjustment slide. The accurate and equal positioning of each jaw is critical and controlled by the serration engagement position. Never allow the chuck jaw clamping nuts to extend beyond the outer diameter of the chuck.

When adjusting for proper gripping of the work, it is important to keep the position of the master jaw within the appropriate stroke range. Setting the grip position in the center of the stroke is the most stable for the mechanism, and ensures the best precision. *Note:* If gripping is set near either end of the stroke, the work may not be gripped sufficiently and creates a dangerous situation that may allow the work to come loose during cutting.

Check that the reference mark on the side of the number one master-jaw clamping is within the range of the entire stroke, as shown in Figure 3-5. Be aware of clamping diameter limitations, the chuck jaws should never extend too far beyond the outer diameter of the chuck. Check the chuck manufacturer recommendations for proper clamping pressure and set accordingly.

Using Soft Jaws on 3-Jaw Chucks

CNC lathe chucks are usually supplied with a set of top-jaws that are hardened steel jaws that are used only for gripping of raw material for single end operations.

All the same rules apply as for the previous section. It is also common to use soft jaws that are machined considering the exact dimensional shape and are less likely to damage the surface finish of pre-existing machined surfaces. Cutting soft jaws also make it possible to relocate each part to the exact depth against a machined shoulder (Figure 3-6). Accuracy of less than .001 inch total indicator runout (TIR) can be attained using soft jaws. Machining soft jaws to match the desired clamping diameter requires understanding of boring techniques on the CNC lathe. Cutting soft jaws can be done manually, but programming the cuts is most suitable and may require a setup or journey machinist to execute the task. Jaws can be cut to match for outside holding or inside holding situations. To cut the jaws, the chuck must be in a clamped condition on a piece of material or with use of a special fixture.

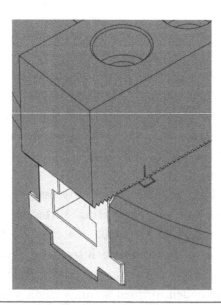

Figure 3-5.   3-jaw stroke adjustment

Figure 3-6.   Soft jaws cut

3-Jaw Chucks or Collets

So far, we have discussed the use of the Universal 3-Jaw Chuck, some of the reasons for choosing this method are summarized next.

Justification for Selecting a 3-Jaw Chuck

- The part diameter to be machined is greater than collet capacity.
- Small diameter parts are rarely manufactured in your facility.
- Investment in a collet chucking system (including sets of collets) is not justified for the situation.
- The parts being machined are not well-suited for collet use, such as uneven surfaces present in castings.
- Parts being machined are unique in shape and require use of special chuck jaws or a 4-jaw independent chuck.
- A collet chucking system is unavailable for the machine in use.

Justification for Selecting a Collet

- Collets offer the greater surface contact area and, thus, the best and most uniform gripping force.
- The highest precision can be accomplished more readily due to the compact nature of the collet being closer to the spindle nose, thus minimizing clamping system runout.
- Collets are available in incremental sizes covering a wide range of diameter capacities.
- Collets are precision ground and hardened which extends their useful life.
- Clamping pressure is not affected by centrifugal force like 3-jaw chucks, which allows for higher spindle speeds.
- Collet systems are more compact and offer more clearance for tooling and less chance of collisions.

Collet Advantages

Collet Concentricity

One of the major advantages gained with a collet is concentricity. The accuracy of this can be measured by inserting a gauge pin or a ground dowel pin in the collet, clamping the chuck and then placing a dial indicator at several positions along the part length that is extended out from the collet (Figure 3-7). The Total Indicator Runout (TIR) should be less than .001 inch.

Figure 3-7.   5C collet concentricity

The Collet Work-Holding System

ER style collets for tool holding were described in Chapter 1. Collet systems for the CNC lathe are used to clamp the rotating workpiece and come in different sizes like 5C, 16C and 3J. Like their CNC milling tool holding counterparts come in specific precision ground incremented sizes. A representation of a 5C style collet is shown in Figure 3-8.

The 5C, 16C or 3J collet chuck is mounted directly into the spindle nose and then the collet is screwed into the drawbar-nut (Figure 3-9). The part is inserted in the collet and then the CNC

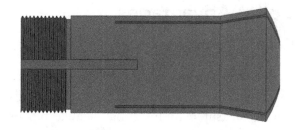

Figure 3-8. 10-degree straight slot collet

lathe foot pedal is pressed to activate the hydraulic draw-bar, which draws the collet into the taper of the collet-chuck. This actuation is repeated to check part holding pressure and is adjusted by further engaging the screw threads into the draw-bar nut until adequate part clamping is accomplished. Actual pressure settings are adjusted to the appropriate gauge PSI level based on manufacturer recommendations and the part being machined.

Figure 3-9. 5C collet chuck

Locking Slot Alignment

In Figure 3-8, a slot is facing upwards on the screw end of the collet. This slot is used to lock the collet in position once correct clamping actuation adjustment is completed. This slot must align with the set screw hole shown in Figure 3-9. Remove the set screws completely

in order to visually align the slot. Then reinsert the dog-point set screw and tighten it just enough to prevent further rotation of the collet from happening. *Do **not*** overtighten. Once the dog-point set screw is in place, you can insert the second set screw to lock both into place and tighten it.

Just as we discussed in Chapter 2, for ER-style collets there are limits to how far a collet can clamp either an undersized or oversized part. To attain proper clamping and to avoid damage to the collet, avoid trying to clamp parts more than 1/64 smaller than the collet size. For oversize parts, it may only be .003 inch larger than the collet size.

Finally, collets can be purchased in shapes other than round, such as hexagonal or square.

Physical Part Stops

When using collets during production runs, a means of precisely locating each part exactly the same is needed. At the back of the collet, the external threads are where the draw-bar nut engages to the collet and internal threads allow for an adjustable part stop to be attached (Figure 3-10).

Figure 3-10. 5C adjustable collet stop

Parts Catchers

A parts catcher is an accessory that is a mechanically activated bucket apparatus that is programmable to position up next to the part during the final part-off section of the program. This allows the finished part to fall into the bucket and then be placed into a catch tray that is accessible without opening the machine doors to retrieve the parts. The parts catcher has a manually adjustable position for use with 3-jaw or collet systems. Care must be taken when setting this position to avoid interference with the chuck jaws.

Part Positioning in CNC Lathe Work-Holding

Regardless of the work-holding method chosen, positioning the part for clamping is absolutely critical to performance and safety.

■ Extend enough material to allow for the desired geometry to be machined.
■ Never allow more than 2.5 times the diameter of the work to be extended out of the chuck/collet without using a tailstock for support.
■ Always clamp on at least one times the diameter or as much material as possible.
■ Avoid taking heavy facing cuts (along the X-axis).
■ Never extend long parts from the back of the chuck without using support bushings in the spindle tube. Use the jaw clamping length to estimate an acceptable extension of no more than 2.5 times their length.

Mounting Tools in the Turret

There are several different turret styles used with CNC lathes and some have multiple turrets. In this text, the focus will be limited to what is called the Bolt on Turret (BOT), as shown in Figure 3-11. CNC lathe turrets commonly have either 8 or 12 positions. Every other location around the perimeter has holders that receive internal operation tools like drills and boring bars while the remaining holders are used for external operations like facing, turning and grooving. They must be tightened to

manufacturer specifications and once the holders are mounted to the turret, they usually are not removed.

Figure 3-11. BOT style tool turret with tool holders, *courtesy Haas CNC Automation, Command Tooling Systems and Sandvik Coromant*

For the internal holders, bushings are added to downsize and fit to specific tool diameter shanks and clamp into the holders bolted to the turret.

Most external toolholders use a wedge-type locking system to clamp the shank of the tool. Depending on the size of the machine, the shank size varies. A one-inch square shank is a common size.

In Chapter 1, Cutting Tool Selection for Turning, external inserted turning tools are generally described, including the following: an 80-degree diamond rough turning and facing tool; a 55-degree diamond finish profiling tool; a grooving tool; a 60-degree threading tool; and a part-off blade. Refer to catalogs and online data provided by tool manufacturers like Sandvik Cormant for a complete menu of tooling for specific applications (Figures 3-12, 3-13 and 3-14).

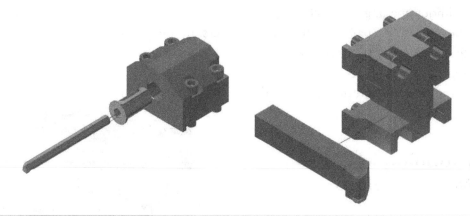

Figure 3-12.   CNC lathe tool assemblies, *courtesy Command Tooling Systems and Sandvik Coromant*

Routine CNC Lathe Procedures

Machine Startup for the CNC Lathe

Complete the following steps to start the machine:

1. Confirm that the power disconnect to the machine is ON.

2. On the machine electrical enclosure, turn on the main power switch.

3. At the machine control pendant, press the Power ON button.

   As the machine computer starts, a sequence of warnings will display on the screen. Heed these warnings at all times. When the boot-up finishes, the display will be active and an alarm

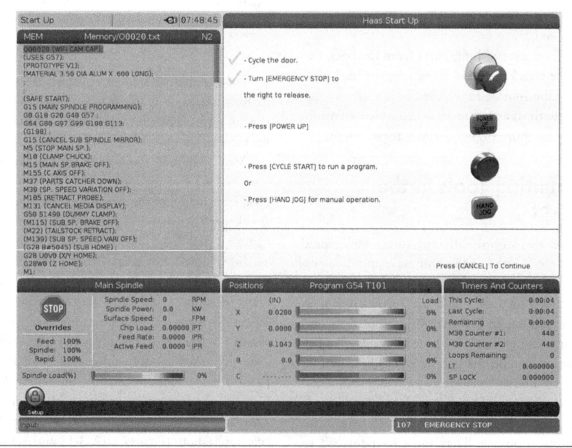

Figure 3-13.   CNC lathe first startup display, *courtesy Haas CNC Automation*

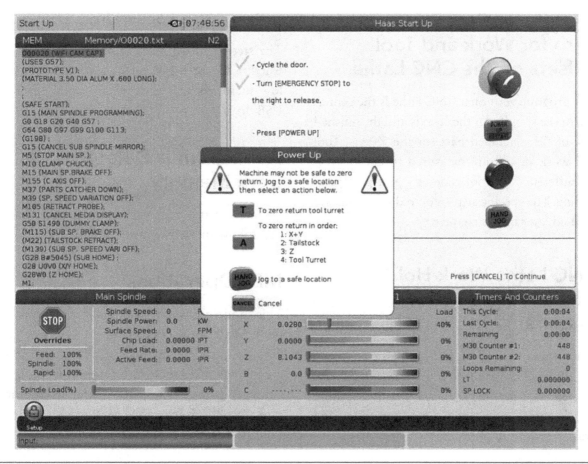

Figure 3-14. CNC lathe second startup display, *courtesy Haas CNC Automation*

will be displayed (highlighted red) in the lower right-hand corner (alarm status field) that reads "102 Servos Turned Off". Press the CANCEL button at any time during the startup instruction screens to dismiss the information.

4. Check the E-Stop button to see if it has been pressed and release it by turning it clockwise.

5. Press the RESET button to clear the Alarm status.

Note: Always be sure to carefully examine the work envelope prior to performing the Power-Up procedure for potential of impact with accessories such as: the tail stock, the tool post or turret, tool sensor or other accessories.

6. With the machine doors closed, press the POWER UP button.

The machine will begin moving the X-axis to its maximum positive location and then the

Z-axis will move to its maximum positive locations which is Machine Home. At this point, the machine is ready for use.

Once power-up is complete and the MEMORY button has been pressed, the display areas will read: 1. Operation: MEM - Program; 2. The active Program will be displayed; 3. Program Tool Offsets; 4. Active Codes; 5. Active Tool; 6. Coolant; and 10. The Position readout display will be on: machine and all axes will be zero (similar to Figure 3-3).

Machine Warm Up for the CNC Lathe

Refer to Chapter 2, Machine Warm Up, for instructions. For the CNC lathe, the Spindle Warm Up program is a different number (09220) than the mill.

Identifying Part Program Zero for Work and Tool Offsets on the CNC Lathe

Part program zero on a CNC lathe is the centerline of the spindle for the X-axis and the finished face of the machined part for the Z-axis. Tool and work offsets are measured to compensate for differences in tool geometry and part positioning. The specific steps required to do this are covered later in this chapter.

CNC Lathe Work-Holding Device Alignment Requirements

When using the CNC lathe, there are just a few things that would affect alignment. Be sure to clean all mating surfaces when installing the clamping device and use proper clamping practices.

Operations Performed at the CNC Control

The following explanations are for operations considered routine for users of CNC machine tools and are given in their sequence of use. Please note that these procedures are specific to the type of controller depicted here (Haas New Generation Control). The procedures for another type of control may be similar. Be sure to consult the manufacturer manuals specific to your machine tool operation and control panel.

A Program Is Loaded from CNC Memory

The steps to load and activate a CNC program are listed in Chapter 2. Refer back to these and follow these steps to activate a program.

A Program Is Loaded from USB

The steps to load a file from a USB drive are listed in Chapter 2. Refer back to these and follow these steps to activate a program from a USB drive.

A Program Is Deleted from Memory

To delete a program from the controller memory, follow the steps given in Chapter 2.

MDI Operations

This mode allows input of small programs via the keypad at the control. Small programs provide an excellent method of executing simple commands like tool turret position changes, controlling the spindle RPM and its rotation direction, etc. To enter the MDI mode of operations, follow these steps:

1. Press the MDI button on the operator panel.
2. Type in the CNC codes needed. *Note:* To enter a string of commands (multiple lines of code), press the semicolon button (End-of-Block) on the alpha keypad to separate the lines.
3. Press the ENTER key to accept the entries. More can be added, if needed, by repeating these steps.
4. Press CYCLE START to execute the commands.

For the program number, the control assumes O0000 and the data may be entered. Each block ends with the end-of-block (EOB) character (;) so that individual blocks of information can be kept separately.

If you make a typographical error while entering a given block, you can eliminate it by pressing the CANCEL key to cancel the error and then reenter the correct value.

You may execute the MDI program as you do with automatic operation. Please refer to the machine tool manufacturer manual for complete and specific instructions.

You can save the MDI program to memory by following these steps:

1. From the EDIT mode, place the cursor on the top line of the program.
2. Press the letter O and enter an unused program number.
3. Press the ALT key. The program will be saved.

You can erase an entire program created in MDI mode by pressing the ERASE PROGRAM button. To perform an individual MDI operation, use the methods described above.

CNC Lathe MDI Example 1
1. To activate the spindle at 500 RPM in the clockwise direction, key in the following command:
2. S500 M03
3. ENTER
4. CYCLE START

CNC Lathe MDI Example 2
To position tool number 5 to the active position on the turret (or to rotate the tool turret to position 5 on a CNC lathe), key in the following commands:

1. T0500
2. ENTER
3. CYCLE START

Measuring Tool Offsets for the CNC Lathe

On CNC lathes, the tool offsets are measured in two directions: X and Z. These values represent the difference between the relationship of Machine Zero for the tool turret and the actual position of a tool tip used as the programmed tool point. The values are input into the X GEOMETRY and Z GEOMETRY columns for each tool (Figure 3-15). Follow these steps to input the measured tool offset value.

First, measure the X-Axis Geometry Offset by doing the following:

1. Clamp the raw material workpiece into the chuck with the appropriate extension.
2. Press the OFFSET button. Ensure the Tool tab is active.
3. Press the HANDLE/JOG button.
4. Press the NEXT TOOL button to rotate the turret one station at a time, until the desired tool is in the cutting position.

   Alternatively, use MDI to input tool number by following the instructions in the *CNC Lathe MDI Example 2* above.
5. Close the machine doors. Key in 500 and press the FWD button to activate the spindle.
6. Manually position the cutting tool to carefully make a cut along the Z-axis to create a diameter on the workpiece.
7. Without moving the X-axis, press the STOP button to stop the spindle and then move the tool away from the part in the Z-axis direction.
8. Open the machine doors and measure the diameter just cut on the workpiece.
9. Press the X DIAMETER MEASURE button to record the X-axis value into the offset.
10. Key in the measured value for the diameter into the prompt area and press ENTER. The offset in the X Geometry column will be updated for the tool just measured.

Next, measure the Z-Axis Geometry Offset by doing the following:

11. Close the machine doors and press the FWD button to restart the spindle at 500 RPM.

12. Manually position the cutting tool and make a small cut on the face of the workpiece.

13. Without moving the Z-axis, press the STOP button to stop the spindle and move the tool away from the part in the X-axis direction.

14. Press the Z FACE MEASURE button to record the Z-axis value into the Z Geometry column for the offset.

Apply steps 1–14 for all of the remaining tools used in the program. The offset values are automatically calculated and set.

Adjusting CNC Lathe Tool Offsets

The values measured into the X and Z Geometry columns can be adjusted by positioning the cursor on the desired axis, then keying in the signed incremental value of the adjustment and press the ENTER key. A more appropriate place to make fine tuning adjustments is in the Wear column for the desired axis.

Tool Tip Orientation

In order for the cutting tool to be compensated properly when using Tool Nose Radius Compensation (TNRC) is used in the CNC program (G41 and G42), the tool tip orientation must be defined in the Tip Direction column of the offset display. The number ranges from zero to 9 and depends on the cutting direction of the tool. In Figure 3-16, the example on the right has the tool tip superimposed over the nose radius. For example, an OD turning and facing tool would use number two (2), and an ID boring bar would use number three (3). Use number 0 or 9 for programming the center of the tool nose.

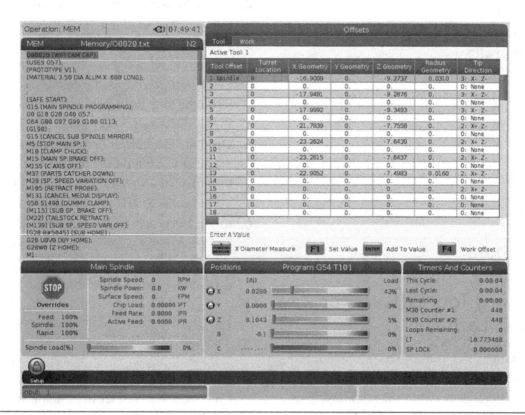

Figure 3-15. CNC lathe tool offset display, *courtesy Haas CNC Automation*

When straight facing or turning cuts (parallel to either the X or Z-axes) are made, they do not require the use of TNRC, but in the case of tapered or circular contouring cuts and radii, TNRC is essential.

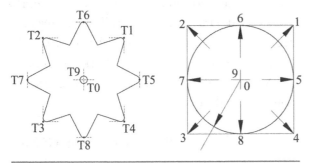

Figure 3-16.   CNC lathe tool tip orientation

Adjusting for Tool Nose Radius Compensation

The actual value of Tool Nose Radius is input into the Radius Geometry column on the OFFSET display on the Tool tab. The correct entry is required here based on the machining insert chosen for the operation (found on the box label of Sandvik inserts). Input of an incorrect value here will have an effect on the finished part when tapers and radii are turned. Refer back to Chapter 1, Cutting Tool Selection, for insert identification and selection.

Adjusting Wear Offsets for CNC Lathes

Wear offsets are used to correct the dimensions of the workpiece that change because of cutting tool wear. For a CNC lathe, the X direction offset corresponds to the diameter. For example, if the X wear offset for a tool is .001 inch, an incremental change of minus .001 inch refers to a decrease of the diameter by .001 inch and an incremental change of plus .001 inch refers to an increase of the diameter by .001 inch.

To adjust the WEAR offsets, follow these steps:

1. Press the OFFSET button.
2. Use the cursor to highlight the Tool tab.
3. Move the cursor to the right until the X Geometry Wear column is visible (or Z Geometry Wear, if applicable).
4. Input the desired adjustment and press ENTER.

Examples of Adjusting Wear Offsets

For the following examples, the operator should have the OFFSET display, Tool tab and Geometry Wear column active and the cursor should be positioned to the tool and axis requiring adjustment.

Wear Offset Example
After machining the workpiece with a diameter dimension of 1.000 inch and the measured value exceeds the tolerance (for example, 1.003 inches), enter the adjustment value offset with a negative sign, –.003 inch in the wear offset by following these steps:

1. Position the cursor on the Geometry Wear column for the X-axis.
2. Key in –.003.
3. Press ENTER.

Then, after machining several more pieces, the diameter increases due to tool wear. If the measured diameter is 1.005 inches, enter the wear offset adjustment as follows:

1. Position the cursor on the Geometry Wear column for the X-axis.
2. Key in –.002. (remember that the values are added incrementally).
3. Press ENTER.

A similar approach is applicable in the direction of the Z-axis.

For the part in question, there is a requirement of 1.500-inch shoulder dimension. If the actual measured length is 1.492 inches, then the value of the offset entered needs to be −.008 inch.

1. Position the cursor on the Geometry Wear column for the Z-axis.
2. Key in −.008.
3. Press ENTER.

After running several parts, the part is checked again resulting in a new measured length of 1.494 inches then the adjustment value of the offset of .002.

1. Position the cursor on the Geometry Wear column for the Z-axis.
2. Key in .002.
3. Press ENTER.

CNC Lathe Tool Sensor Measuring

On a Haas CNC lathe that is equipped with the optional Automatic Tool Presetting function, tool measurement can be greatly simplified. The sensor incorporates a touch pad that the tool tip is brought into contact with instead of machining the diameter and face of the material. It can also be programmed to automatically be positioned to contact the sensor for each axis and the offset values are automatically input to the control.

The operator must still manually enter the Tool Nose Radius compensation values in the Radius Geometry column of the OFFSET/GEOMETRY register and Tool Tip Orientation the Tip Direction column. Review the operator manual specific to your machine for exact procedures.

Measuring the Work Coordinate Offset for the CNC Lathe

It is necessary to establish a relationship between the machine coordinate system and the workpiece coordinate system. The following steps are necessary to input the measured values for the workpiece zero to the control's Work Offset page.

Measure the Z-axis work coordinate by doing the following:

1. Identify the coordinate system G54–G59 to be used.
2. Use the HANDLE/JOG MODE to manually position a (measured) cutting tool and make a light cut on the face of the workpiece. This finished surface will represent the Z-axis zero point.
3. Without moving the Z-axis, stop the spindle and move the tool away from the part in the X-axis direction.
4. Identify the distance along the Z-axis from cut surface to the desired zero point, if needed.
5. Press the OFFSET button to display the OFFSETS, WORK tab is displayed (Figure 3-17).
6. Use the cursor to position to the desired workpiece offset to be set (i.e., G54).
7. Use the cursor to position to the Z-axis column for measurement.
8. Press the Z FACE MEASURE button.

The value of the work coordinate for the Z-axis will be input.

Adjusting CNC Lathe Work Offsets

Occasionally it becomes necessary to adjust (shift) the measured offset value by some amount. Most commonly, this is done for the

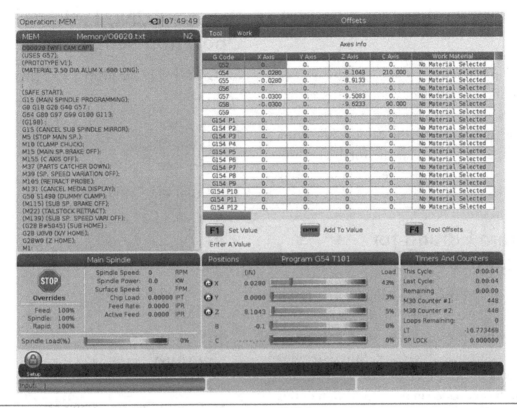

Figure 3-17. CNC lathe work offset display, *courtesy Haas CNC Automation*

Z-axis to ensure full cleanup cut on the face of the part. Please note that any adjustment made to the Work Offset affects the entire program and all of its tools. Follow these steps to shift the work coordinate for the Z-axis:

1. Press the OFFSET button and ensure the Work tab is highlighted.
2. Position the cursor in the Z-Axis column.
3. Key in the desired shift amount (i.e., –.030).
4. Press ENTER.

Input of Known Work Offset Values for CNC Lathes

In cases where the offset value is known for the Work Coordinate, it can be directly input as follows:

1. Press the OFFSET button and ensure the Work tab is highlighted.

2. Position the cursor in the Z-Axis column.
3. Key in the numerical value for the offset (i.e., –8.1043).
4. Press the F1 button.

CNC Lathe Tool Path Verification of the Program

In Chapter 2, Tool Path Verification of the Program was presented for the CNC mill. The same exact steps are required to perform a graphic simulation of the tool path on the CNC lathe.

CNC Lathe Program Execution in the Automatic Cycle Mode

In Chapter 2, Program Execution in Automatic Cycle Mode was presented for the CNC mill. The same exact steps are required to perform execution of the Automatic Cycle Mode for the CNC lathe. Adjustment of the coolant nozzles are done manually on each tool holder, for each tool, unless the tools used have the through coolant option. Also, as one final check of the programmed path, the work offset can be shifted in the positive direction to a distance greater than the total length of the part to dry-run the cycle, without cutting the part.

Program Editing for CNC Lathes

In Chapter 2, program editing functions for the CNC mill were covered in detail. Functions covered work in exactly the same fashion on the lathe controller. There are, however, some specific codes that are unique to the lathe and these will be addressed in the next section of this chapter. Another difference is the axes used for this text only. The X (U) and Z (W) axes will be addressed (Figure 3-18).

Common CNC Lathe Operation Scenarios

CNC Lathe Machining Program Example

O1234

N1 G50 S2000

N2 T0100

N3 G96 S400 M03

N4 G00 X1.25 Z.2 T0101 M08

N5 ... ...

N6 ... ...

... ...

N17 M01

N18 G50 S1000

N19 T0200

N20 G96 S200 M03

N21 G00 X.75 Z.1 T0202 M08

N22 ... ...

... ...

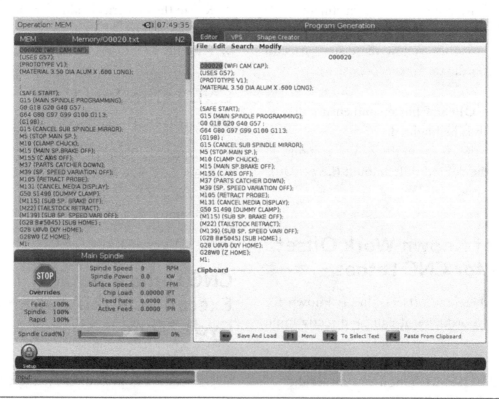

Figure 3-18. CNC lathe editor display, *courtesy Haas CNC Automation*

N39 M01

N40 G50 S2500

N41 T0300

N42 G96 S600 M03

N43 G00 X2.2 Z.05 T0303 M08

N44 ... ...

... ...

   N45 M30

Using the program from this example, let us review procedures you should follow if you need to repeat a part of the program for tools T01, T02, and T03.

CNC Lathe Scenario 1

Scenario 1 Problem: Execution of the program was interrupted in block N17 and you need to repeat from the beginning all operations performed by tool T01.

Scenario 1 Solution:

1. From the MEMORY or EDIT mode, press the RESET button.
   This will cause a cancellation of CNC commands under control and return the program to the beginning.
2. From the MEMORY mode, press CYCLE START.

CNC Lathe Scenario 2

Scenario 2 Problem: Execution of the program was interrupted in block N39, and you need to execute a program for tool T02.

Scenario 2 Solution:

1. From the EDIT mode, press the RESET button.

2. Using the alphanumeric keypad on the control panel and the search methods described earlier, search to block N18.
3. Change to the MEMORY mode and press CYCLE START.

Note: In both cases, if you intend to execute the program to the end without interruption, the OPTIONAL STOP button must be turned off. However, if you intend to execute only part of the program corresponding to work of tool T01 or T02, the procedure is as follows:

1. Press the OPTIONAL STOP button to the ON condition. After the work is completed by the desired tool.
2. Press the RESET key while in the MEMORY or EDIT mode.

The machine is ready for automatic cycle once again from the program beginning. The program will stop after reading an M01 code.

CNC Lathe Scenario 3

Scenario 3 Problem: Execution of the whole program is completed but you need to repeat operations performed by tool T03.

Scenario 3 Solution: From the EDIT mode:

1. Press the RESET button (if the program is completed and at its head, the RESET is not required).
2. Using alphanumeric keypad on the control panel and the search methods described earlier, search to block N40.
3. From the MEMEORY mode, press CYCLE START.

CNC Lathe Scenario 4

Scenario 4 Problem: Execution of the program is interrupted by the use of the EMERGENCY STOP button.

Scenario 4 Solution: In this case, follow the same procedure as mentioned above. However, in this case, you must HOME the machine to reset the machine coordinate system with respect to the X- and Z-axes by pressing the POWER UP button.

CNC Lathe Machine Shutdown

The procedure for shutting down the CNC lathe is the same as with the CNC mill described in Chapter 2, except the operator would press the HOME G28 button to position the axes at the machine zero in the X and Z axes. The remaining steps are the same as for the CNC mill.

Summary

CNC lathe machining accounts for a large percentage of all machining done in manufacturing. A wide range of diameters with very precise dimensions and fine surface finishes can be machined. In this chapter, common setup methods, work-holding and tool types were introduced and routine procedures and adjustment activities were presented. Keep in mind, the intent here is to begin learning. There are many more things to learn and many more CNC lathe configurations than the scope of this text will allow. Always continue to research best practices as you go forward, and **always** think safety first.

CNC Lathe Setup and Operation, Study Questions

1. CNC lathes consist of which axes?

   a. A, B and C

   b. X, Y and Z

   c. U, V and W

   d. X and Z

2. On a CNC lathe, which axis controls the diameter?

   a. X

   b. Y

   c. Z

   d. U

3. The use of the incremental coordinate system on CNC lathes involves the use of:

   a. G91 code

   b. G54 code

   c. U and W coordinate values

   d. X and Z coordinate values

4. The most commonly used work-holding device on a CNC lathe is the

   a. universal 3-jaw chuck

   b. 6-jaw chuck

   c. 4-jaw chuck

   d. vise

5. How is the stroke range adjustment measured when using a 3-jaw universal chuck?

   a. With a 6" steel rule

   b. By checking alignment of the reference marks on the side of the number one master-jaw

   c. By counting serration engagement position agreement for each jaw.

   d. Both b and c

6. Why would soft jaws be used on 3-jaw chucks?

   a. Because they can be machined to exact dimensional shape.

   b. They are less likely to mar preexisting surface finishes.

   c. They allow quick accurate relocation of successive parts.

   d. All of the above.

7. Collet systems are used when?

   a. Part diameter is too small for the 3-jaw to chuck

   b. When high precision of .001 inch and low TIR is required

   c. Both a and b

   d. None of the above.

8. Which of these is not important to part positioning in the work-holding device for CNC lathes?

   a. *Never* extend long parts from the back of the chuck.

   b. *Never* allow more than 2.5 times the diameter of the work to extend out of the chuck without using a tailstock for support.

   c. Avoid taking heavy facing cuts (along the X-axis).

   d. None of the above.

9. What will be affected if an improper Tool Nose Radius value is entered into the OFFSET, Radius Geometry?

    a. The finished diameter.
    b. The finished length.
    c. Finished taper and radius cut values.
    d. All of the above.

10. When using a collet on the CNC lathe, how can positive repetition of the depth location for the part be ensured?

    a. By measuring each part.
    b. By using a Parts Catcher.
    c. By using an adjustable collet stop.
    d. All of the above.

Introduction to CNC Lathe Programming

Objectives

1. Identify and properly use G-codes associated with CNC lathe programming.
2. Identify and properly use M-codes associated with CNC lathe programming.
3. Recognize the correct codes needed to control cutting tool speeds and feeds within CNC lathe programs.
4. Use coordinate systems for programming CNC lathes.
5. Implement the use of common multiple repetitive cycles.
6. Properly input tool nose radius compensation (G40, G41 and G42) for CNC lathe programs.

In this section, we will focus our attention on programming the two-axis CNC lathes. These machines are the base configuration for CNC turning; a solid foundation is laid for further learning about more advanced machinery by gaining full understanding of the techniques presented here. Again, just as in Chapter 2, Introduction to CNC Mill Programming, the information is presented here as if the program manuscript is being created manually, line-by-line. This will help you understand the program contents and how they work together and make it easier for you to edit existing programs. As stated earlier in Chapter 2, the most common method for creating programs is by using CAD/CAM software, which is introduced in Chapter 4, Introduction to Computer Aided Manufacturing. For a complete and detailed resource for learning CNC Programming, refer to my texts: *Programming of CNC Machines* (4th ed.), and its companion, *Student Workbook for CNC Programming of CNC Machines* (4th ed.).

We begin by introducing the program codes in the language commonly called "G-Code."

Program Structure for CNC Lathes

Many of the elements for programming for CNC lathes are the same as described earlier in Chapter 2. Only the items that differ for CNC lathes will be presented here.

Tool Function

Tool pocket location in turrets on CNC lathes are assigned numbers, 1–12. The number coding system for each location on the turret is fixed and cannot be erased even if the machine is turned off. In the process of programming, tool function is commanded by the four digits that follow the letter address T. The first two digits are related to the tool pocket number and the

remaining two digits signify the tools Geometry and Wear Offset numbers, as illustrated below. For example, T0101 signifies tool pocket number one and geometry offset number one.

Tool numbers may vary from 1 to the maximum number of pockets in the tool turret, for example, from 1 to 12, etc. If tool offset number 01 is assigned to tool pocket 01, this rules out the possibility of assigning the same tool offset number to another tool. This common convention simplifies operating procedures for the operator.

It should also be noted that when the controller encounters the T01 in the program code, the turret is rotated to that position and when the full code, T0101 is encountered in the program, the tool is positioned according to the existing geometry offset and the listed program coordinates.

A program call of T0100 in a program cancels the geometry offset for the active tool (T0200, T0300, etc.). This code is commonly placed at the end of said tool's use, prior to the next tool call.

Application of Tool Wear Offsets

Tool wear compensation is a procedure aimed at correcting dimensional variations along the X or Z- axes caused by tool wear or deflection.

Tool X and Z geometry offset numbers are values that relate to the tool tip to Workpiece Zero and Machine Home. These geometry offsets should not be used to make adjustments for wear; only wear offsets are used for this purpose (described in Chapter 2).

Small correctional values are input to a specific tool wear offset number for X or Z to compensate for any cutting variations. For example, to decrease a diametrical measurement of the part, one should input a negative value into the wear offset column for the X-axis (i.e., –.001 reduces a diameter by .001 inch). In a similar fashion, to increase a shoulder depth measurement along the Z-axis, again, a negative value should be input into the wear offset column for the Z-axis of the tool.

Feed Function

The feed function determines the axes movement feedrate for cutting tools in the machining process. Feed is programmed using the letter address "F," followed by up to four digits in the metric system and five digits in the inch system. These digits represent certain values of feed. The following examples are two methods of designating feedrate:

1. Inches per revolutions (in/rev), or millimeters per revolution (mm/rev), of the spindle (G99). In this case, in order to obtain feed with a certain assigned value of speed, the spindle, as well as the workpiece must be rotating.

CNC Lathe IPR Feedrate Examples
F1.1205   in/rev

F0.05   in/rev

F0.001   in/rev

When the spindle speed is changed after the constant surface speed per revolution (G96) has been called, the feedrate will change for a certain period of time. Therefore, feed is directly coupled with spindle speed.

The following notes apply to the feed function, G99 . . . F.

a. The values entered into the program for feed remain active until replaced by another feedrate, or cancelled by the G00 rapid traverse call.
b. The input value of speed is equivalent to the actual speed if the feedrate override on the operation panel is set to 100 percent. See Chapter 2 for a detailed description of feedrate override.

2. Feedrate per time is measured (programmed) in inches per minute (in/min), or millimeters per minute (mm/min) (G98). If the feed function in the program contains feedrate per time period, then any change in the spindle speed has no effect on the feedrate because the feedrate and the spindle speed are not coupled.

CNC Lathe IPM Feedrate Examples
F121.15   in/min

F1.05   in/min

F0.5   in/min

Generally, all CNC lathes are set to a default of feed per revolution of the spindle at machine start-up. In order to establish feedrate per minute (in/min or mm/min), the G98 function must be used. This function remains effective (modal) until cancelled by function G99, or until the machine is turned off.

The following notes apply to feed function G98:

a. The values entered into the program for feed remain active until replaced by another feedrate, or cancelled by the G00 rapid traverse call.
b. The input value of speed is equivalent to the actual speed if the feedrate override on the control panel is set to 100 percent.
c. Feed functions containing feed in inches per minute are not applicable to threading cycles.

Spindle Function

Spindle function, for the CNC lathe is similar to the CNC mill and is commanded by the letter address S, followed by a number up to four digits, as shown below. The spindle rotation direction clockwise (M03), or counterclockwise (M04) is typically in the same block.
S50, S150, S3000

Some machines are equipped with two gear ranges for speeds: low or high. Depending on the commanded value of the rotational speed, the machine automatically adjusts to the appropriate range, as seen in the following rotational speeds. In practice, most manufacturers overlap the low and high range, for example:

30–1200 (rev/min) low range (M41)

80–4000 (rev/min) high range (M42)

For CNC lathe, two functions are applicable to the control of spindle speed. These are:

G96:   Constant surface speed control

G97:   Constant surface speed control cancellation (sometimes referred to as constant spindle speed)

Both functions appear together with function S, for example:

G97 S500 = 500 r/min

G96 S400 = 400 Surface Speed (*V*) in ft/min (or m/min)

Constant spindle speed (G97) is applied in the case of threading cycles and in the machining of a workpiece, with the diameter remaining relatively constant. It is also used for all operations on the center line, like drilling, etc. If the situation calls for several changes of spindle speed in a given program, new values for the S function are assigned.

G97 Spindle Speed Example

G97S1000 is active for diameter one and sets the constant spindle speed.

S800 changes the r/min for diameter two.

S300 changes the r/min for diameter three.

Constant Surface Speed Control (G96)

To further examine this concept, study the diagram of peripheral speed distribution in Figure 3-19. The situation appears during a facing cut, using function G97 S1000. The following formulas calculate peripheral surface speed for each diameter of 1.00, 2.00, and 3.00 inches. The result is not desirable because of a decrease in surface speed as the diameter gets smaller.

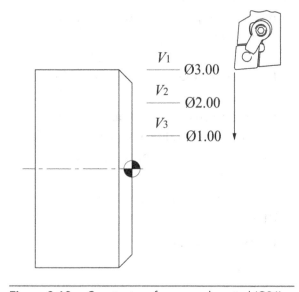

Figure 3-19.   Constant surface speed control (G96)

$$V_1 = \frac{\pi \times D \times n}{12} = \frac{3.14 \times 3.0 \times 1000}{12} = 785 \text{ (SFPM)}$$

$$V_2 = \frac{\pi \times D \times n}{12} = \frac{3.14 \times 2.0 \times 1000}{12} = 523 \text{ (SFPM)}$$

$$V_3 = \frac{\pi \times D \times n}{12} = \frac{3.14 \times 1.0 \times 1000}{12} = 261 \text{ (SFPM)}$$

where:

n = RPM or r/min

D = diameter

π = 3.14

V = Cutting speed (FPM)

The diagram shows that surface speed decreases if G97 is used—as a diameter decreases and reaches zero at the center line of the part. Is this phenomenon of any advantage to us in the process of facing? Before answering this question, look at the advice of cutting tool manufacturers who recommend specific cutting speeds for different types of machined materials. In the case of function G97, such a condition will be fulfilled only with respect to one diameter. As mentioned previously, Constant Surface Speed control (G96) is one of the factors that can be included in programs for CNC lathes.

If the surface speed must remain constant, then the spindle speed has to increase with a decreasing diameter. Spindle speed for each consecutive diameter is calculated by the control, according to the formula: $n(12 \times V)/(\pi \times D)$. Thus, a closer look at this formula leads to the conclusion that, theoretically, as the diameter decreases to zero, the spindle speed increases to infinity. In reality, the spindle speed range is limited by the maximum r/min capacity of the machine.

In practical terms then, function G96 is very useful during facing and also in all cutting that involves a change in the diameter of the workpiece. As the diameter of the machined workpiece decreases, spindle speed increases; conversely, as the diameter increases, the spindle speed decreases.

Maximum Spindle Speed Setting (G50) or G50 S

When the letter address S is given within a program block and is preceded with function G50, it refers to the maximum spindle speed setting that can be applied in the current operation for a given tool. As previously mentioned, the spindle speed is calculated according to technological metal-cutting conditions for a given tool, or for any particular material. In some cases,

the work-holding arrangement may require special equipment, which is mounted onto the conventional holding equipment. Such work-holding equipment creates conditions that do not permit utilization of the full range of spindle speeds, especially maximum spindle speed for a given machine. Because of this fact, a maximum spindle speed for a particular operation can be assigned by using the function G50. Therefore, if, in the machining process, metal-cutting conditions arise that require a higher spindle speed, an increase of the spindle speed will not take place. Using function G50 is sometimes called "clamping the spindle speed" at a safe maximum r/min.

Note: If G50 is not used in conjunction with the spindle speed command (such as G50 S1250) when G96 is commanded, then the machine will increase the r/min as the diameter decreases, up to the machine's maximum capability. This condition could result in an excessive r/min and damage could occur.

Preparatory Functions (G-Codes)

Preparatory functions are programmed with the letter address G, normally followed by two digits, to establish the mode of operation in which the tool moves. In Chart 3-3 and throughout this text, the G-codes listed and explained, after Miscellaneous functions that refer to the most commonly used Haas CNC system. There are some variations in the use of the other controls, but most of the codes are identical. When system programming, consult the programming manual for the specific control prior to selecting the type. Some of the most commonly used codes are presented here.

More detailed descriptions and application examples are given in Chapter 3 of my book *Programming of CNC Machines* (4th ed.).

Code	Group	Function	Code	Group	Function
G00*	01	Rapid Traverse Positioning	G97*	13	Constant Surface Speed Control Cancellation
G01	01	Linear Interpolation	G98	10	Feed per Minute
G02	01	Circular Interpolation CW (clockwise)	G99*	10	Feed per Revolution
G03	01	Circular Interpolation CCW (counterclockwise)	G100	00	Disable Mirror Image
G04	00	Dwell	G101	00	Enable Mirror Image
G09	00	Exact Stop	G103	00	Limit Block Lookahead
G10	00	Set Offsets	G105	09	Servo Bar Command
G14	17	Secondary Spindle Swap			**NOTES:**
G15	17	Secondary Spindle Swap Cancel			
G17	02	X Y Plane			*1. In the table, G-Codes marked with an asterick (*) are active upon startup of the machine.*
G18*	02	X Z Plane			
G19	02	Y Z Plane			*2. At machine startup or after pressing reset, the inch (G20) or metric (G21) measuring system last activated remains in effect.*
G20	06	Input in Inches			
G21	06	Input in Millimeters			
G28	00	Machine Zero Point Return			*3. G-Codes of group 00 represent "one shot" G-Codes, and they are effective only to the designated blocks.*
G29	00	Return From Reference Point			
G31	00	Skip Function			
G32	01	Thread Cutting			*4. Modal G-Codes remain in effect until they are replaced by another command from the same group.*
G40*	07	Tool Nose Radius Compensation Cancel			
G41	07	Tool Nose Radius Compensation, Left			*5. If modal G-Codes from the same group are specified in the same block, the last one listed is in effect.*
G42	07	Tool Nose Radius Compensation, Right			
G50	00	Spindle Speed Clamp			*6.Modal G-Codes of different groups can be specified in the same block.*
G52	00	Local Coordinate System Setting			
G53	00	Machine Coordinate System Selection			
G54-59	12	Work Coordinate System One Selection			*7. If a G-Code from group 01 is specified within a canned drilling cycle block, the cycle will be cancelled just as if a G80 canned cycle cancellation code were called.*
G61	15	Exact Stop Modal			
G64*	15	Exact Stop Cancel G61			
G65	00	Macro Subprogram Call Option			*More detailed descriptions and application examples are given in Programming of CNC Machines, 4th Edition, in the section "Overview of Preparatory Fucntions for CNC Turning Centers". And of course, always refer to the operation and programming manuals that are provided with the machine tool in use.*
G70	00	Finishing Cycle			
G71	00	Stock Removal in Turning			
G72	00	Stock Removal in Facing			
G73	00	Path Pattern Repeating Cycle			
G74	00	Peck Drilling Cycle/Face Grooving			
G75	00	O.D. and I.D Groove Cutting Cycle			
G76	00	Multiple Pass Thread Cutting Cycle			
G80*	09	Canned Drilling Cycle Cancellation			
G81	09	Drill Canned Cycle			
G82	09	Spot Drill Canned Cycle			
G83	09	Normal Peck Drilling Canned Cycle			
G84	09	Face Tapping Cycle			
G85	09	Boring Canned Cycle			
G86	09	Bore and Stop Canned Cycle			
G89	09	Bore and Dwell Canned Cycle			
G90	01	Outer/Inner Diameter Turning Cycle			
G92	01	Thread Cutting Cycle			
G94	01	End Facing Cycle			
G95	09	Live Tooling Rigid Tap (Face)			
G96	13	Constant Surface Speed Control ON			

Chart 3-3.   Preparatory Functions "G-Codes" for CNC Lathes

Miscellaneous Functions (M-Codes)

Miscellaneous functions are used to command various operations. Two commonly used M-codes—M03 and M08—are used for starting spindle rotation in the clockwise direction and activating coolant flow, respectively. The code consists of the letter M typically followed by two digits. Normally, one block will contain only one M-code function; however, up to three M-codes may be in a block depending upon parameter settings. Most of the common M-codes are listed in Chart 3-4, Miscellaneous Functions (M-codes) Specific to CNC Lathes. Many machine tool builders assign other codes for specific purposes relative to their equipment. Always consult the manufacturer manuals specific to the machine in use for pertinent M-codes.

The following descriptions are given for many of the most commonly used M-codes in the chart. Also, observe their use within the example program. Please consult the appropriate operator and programming manual for machine-specific information about M-codes, for your application.

Program Stop (M00). When the M00 code is encountered in the program, spindle r/min, feeds, and coolant flow stop. This function interrupts the automatic work cycle, in order to allow for the possibility of the following:

1. In-process inspection and gauging
2. Visual inspection of tool wear and other components
3. Removal of chips
4. Interruption of the cycle, in order to relocate the workpiece when the workpiece is being machined from both sides during one operation

After use of this code, the program block following must include either M03 or M04 as well as M08 to reactivate these functions. Functions G96, G97, S, and F are NOT canceled by M00.

M-Code	Function	M-Code	Function
M00	Program Stop	M19	Spindle Orient
M01	Optional Stop	M21	Tailstock Direction Forward
M02	Program End Without Rewind	M22	Tailstock Direction Reverse
M03	Spindle ON Forward (FWD)	M23	Thread Finishing With Chamfering
M04	Spindle ON Reverse (REV)	M24	Thread Finishing With Right Angle
M05	Spindle Rotation Stop	M30	Program End With Rewind
M08	Flood Coolant ON	M36	Parts Catcher On (Optional)
M09	Coolant OFF	M37	Parts Catcher Off (Optional)
M10	Chuck Clamp	M38	Spindle Speed Variation On
M11	Chuck Unclamp	M39	Spindle Speed Variation Off
M12	Auto Jet Air Blast ON (Option)	M41	Spindle Low Gear Range (Optional)
M13	Auto Jet Air Blast OFF (Option)	M42	Spindle High Gear Range (Optional)
M14	Main Spindle Brake On (Optional C-Axis)	M98	Subroutine Call
M15	Main Spindle Brake Off (Optional C-Axis)	M99	Return to Main Program From Subroutine
M17	Rotation of Tool Turret Forward	M104	Probe Arm Extend (Optional)
M18	Rotation of Tool Turret Reverse	M105	Probe Arm Retract (Optional)

Chart 3-4. Miscellaneous Functions M-Codes Specific to CNC Lathes

Optional Program Stop (M01). This function is nearly the same as M00 with one significant difference: it is applied only by activating the OPTIONAL STOP button to switch the ON condition (e.g., to stop the machine so that measurements can be taken, or to remove chips at the discretion of the machine operator). If the OPTIONAL STOP button is not active, the machine will ignore this command even if it is present in the program.

Program End (M02). Function M02 cancels the automatic work cycle, interrupts revolutions, stops feedrate and coolant flow, and cancels the control system of the CNC. Repetition of a programmed operation is not possible without RESET of the control. To repeat the program, the M30 function is used. The M02 command was used primarily on NC tape machines.

Spindle On (Clockwise) (M03). This function signals the machine to activate the spindle with clockwise revolutions at a value indicated by the S function. M03 is cancelled by M04, M05, M00, M01, M02 and M30.

Spindle On (Counterclockwise) (M04). To activate or cancel this function, follow the same instructions as for M03.

Spindle Off (M05). This function is cancelled by M03 and M04.

Flood Coolant On (M08). Function M08 activates the flood coolant flow.

Coolant Off (M09). Function M09 deactivates the coolant flow.

Chuck Close and Open (M10 and M11). M10 automatically closes the spindle chuck jaws and M11 automatically opens the spindle chuck jaws. Settings are used to toggle whether clamping is internal or external. M10 and M11 are used in certain cases when there is a special part puller or gripper used for the insertion and removal of the workpiece from the spindle chuck. Such devices are used in automated operations where mass production is the primary focus.

Rotation of the Tool Turret Forward and Reverse (M17 and M18). Miscellaneous function M17 rotates the tool turret in the normal (FWD) direction whereas M18 rotates the tool turret in the opposite direction (REV). Function M17 is valid at machine start-up. Function M18 implies a change in direction of the rotation opposite to the one set previously. These commands may be useful when special tooling is mounted in the tool turret or chuck where clearance issues must be considered during turret rotation.

Tailstock Direction Forward and Reverse (M21 and M22). Programmable shifting of the tailstock as a whole is an optional function on some types of machines, especially if the extended length of the tailstock spindle is not sufficient to perform the operation and/or the lathe is large and has a long Z-axis stroke. These codes are a factory option only.

Thread Finishing (M23 and M24). Miscellaneous function M23 should be applied to the G92 Thread Cutting cycle when the threading tool usually exits at a 45-degree angle. M24 is necessary when the ending thread is followed by a greater diameter or a recess groove. M23 is the default state after machine power is turned on. These functions are used in conjunction with the G92 Tread Cutting Cycle at a thread end for either a 90-degree or a 45-degree retract, respectively.

End of Program (M30). This function returns the program to its beginning.

Work Coordinate System Setting (G54–G59). After measurement of all cutting tools has been completed, as outlined in Chapter 3, the work offset is then measured. The operator or setup person merely identifies and measures the finished face of the part along the Z-axis in order to establish the work coordinate system and input the position on the WORK offset display for the Z Geometry register of the program identified offset. Consult the appropriate operation manual for your machine to find specific details for setting methods.

CNC Lathe Program Template

Certain sections of the machining program can be repetitious in nature and they are listed here as follows: the program beginning, the tool beginning, the tool ending and, the program ending. *Note:* If you intend to load any of the programs you create into a machine controller for trial and use, you must include a percent (%) sign on a separate line at the beginning and end of the text. This is required for communications purposes.

The program block structure for each of these sections is:

The Program Beginning.

O7306 = program number
(Comments) = part number or other identifying information
(Comments) = date or other identifying information

Note: 9000 series program numbering is reserved for Macro programs; therefore, avoid using it for your program number.
Note: The question marks (T????) in the template sections below represent variable data that are specific to the situation and need to be replaced with live data that is relevant to your programming situation.
N10 G90 G80 G40
N15 G00 G28 U0.0 W0.0
N20 G97 S???? M3
N25 F.???

The Tool Beginning.

(Comments) = Tool identification information
N95 G28 U0.0 W0.0
N100 T????
N105 G50 S???
N110 G96 S??? M03
N115 G00 G54 X???? Z???? M08

The Tool Ending.

N200 G00 G40 X???? Z.1 M09
N205 G28 U0.0 W0.0 T??00
N210 M01

The Program Ending.

N300 G28 U0.0 W0.0 M09
N305 M30

Preparatory Functions for CNC Lathes (G-Codes)

Preparatory functions (listed in Chart 3-3), G-Codes, are a major part of the programming puzzle. They identify to the controller what type of machining activity is needed. For example, if a facing operation needs to be completed, function G72 may be used, or if the part profile needs to be turned, function G73 may be used.

The scope of this text is not intended to cover every code in Chart 3-3 in detail. A few of the most commonly used codes are listed below.

Rapid Traverse Positioning (G00). G00 is *Rapid Traverse Positioning*; it is used for position changes without machining. Function G00 can be combined with the M, S or T functions. It is a modal G-code and will remain in effect until replaced by another command from the same group.

CAUTION: If G00 rapid positioning is programmed in the direction of both axes, note that the tool will not advance to a specified point following the shortest possible path. The tool path is determined by the speed of rapid traverse, with respect to each axis. In most cases, the speed for the X-axis is much greater than that for the Z-axis. The axis that must travel the least distance will be reached first.

Work-holding devices (chuck jaws) quite often extend beyond the work-holding equipment (chuck). Therefore, careful consideration should be given to the position of work-holding devices (including tailstock centers and quill extension position), so that rapid traverse paths of the tool do not interfere with them and cause a crash that may damage the machine, clamping device or part.

Linear Interpolation (G01). Straight line feed moves, linear interpolation, is programmed by using the G01 function; it may be applied simultaneously for both axes. The G01 function commands the movement of the tool from a given position to the position with the assigned coordinates having feedrate specified by the F-Word. G01 is a modal command that stays in effect until replaced by another command from the same group. The block format for linear interpolation is given as follows:

CNC Lathe Linear Interpolation Program Block
     G01 X(U)... Z(W)... F ...

The interpolator in the control system calculates various speeds for the motion axis so that the resulting speed is equivalent to the programmed feedrate. Remember, U and W represent incremental movement for the X or Z-axes, respectively.

Circular Interpolation (G02 and G03). Circular interpolation allows programmed tool movements along an arc. In order to define the circular interpolation function, the following four conditions must be met:

1. Selection of the direction of interpolation:
   G02 = Clockwise
   G03 = Counterclockwise
2. Position coordinates of the starting point of the arc.
3. Position coordinates of the ending or point of the arc. The ending point coordinates may be omitted if they correspond to the coordinates of the starting point (half circle, circle).
4. The dimension corresponding to the distance between the center of the tool nose radius and the center of the arc from the starting point of each axis must be given.

The incremental distance for the X-axis is defined by the value of letter address I, and the Z-axis is defined by the value of the letter address K. Values for I and K may be omitted from the program if they are equal to zero. The block format for circular interpolation is given as follows:

CNC Lathe Circular Interpolation Program Block
  G02 X(U)...Z(W)...I...K...F...
  G03 X(U)...Z(W)...I...K...F...

U/2 = Incremental distance from the arc starting point to the ending point along the X-axis
 W = Incremental distance to the arc end point along the Z-axis
X/2 = Absolute coordinate for the ending point of the arc along the X-axis
 Z = Absolute coordinate for the ending point of the arc along the Z-axis

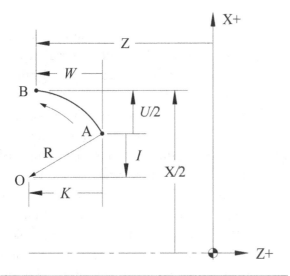

Figure 3-20. Arc center identification for circular interpolation

Figure 3-20 graphically identifies all of the components necessary for programming arcs and their descriptions.

  A = Arc starting point
  B = Arc ending point
  I = Incremental distance from start point (A) to the arc center (O) along the X-axis
  K = Incremental distance from start point (A) to the arc center (O) along the Z-axis
  O = Arc center
  R = Arc radius (a negative signed value will produce a concave arc)

In order to establish signs for I and K, consider the following directions: Imagine a line is drawn from the arc starting point to the arc center point with a direction vector toward the center of the arc. Next, project this vector onto the axes of the coordinate system with the origin at the arc start point. If the resulting projections of the vector are oriented in the same direction as the corresponding axis of the absolute coordinate system, the plus sign (+) is applicable. Otherwise, the minus sign (−) is applicable. If no sign is given in the coordinate entry, the control assumes the sign is positive.

 The CNC control includes an additional capability to use R in place of I and K. R is the distance from the center of the tool radius to the center of the following arc. If an arc is smaller than, or equal to 180 degrees, then R assumes a positive sign; if it is greater than 180 degrees, then R assumes a negative sign. The block format for circular interpolation using R is given as follows:

CNC Lathe Circular Interpolation using R, Program Block
  G02 X(U)...Z(W)...R...F...
  G03 X(U)...Z(W)...R...F...

Circular interpolation can be performed these two ways: the first using I and K, the second using R. However, the application of address R is less difficult.

Where: *R = Radius of the arc*

Note: Tool nose radius compensation must be used (G41 or G42 covered later) in the program to make the exact radius of the arc.

If I and K are applied, then machine control is fed with precise information about the position of the center of the radius of the performed arc. In this case, the coordinates of the arc ending point must correspond to a position on the programmed circle. However, if the given values of the coordinates are incorrect, the tool will not respond by following an arc.

If the second method using R is employed, the tool will follow an arc equal to the value given.

Dwell (G04).   Dwell is initiated by use of function G04, and the length of time for the dwell is specified by P, X, or U, (depending on the control type) as follows: G04 P . . . (in milliseconds)

Dwell Example
    G04 P2500.

In the examples above, the dwell time values are the equivalent of 2.5 seconds. Also note, when using P to address the amount of time for dwell, a decimal point must be used. The value of time is measured in milliseconds (ms), 1000 ms = 1 second. Function G04 is a "one shot" command; it is active only in the block in which it is called. The dwell is activated at the end

of the feed move and should be the only contents of the block. To be sure of the exact method used, study the manufacturer programming manual specific to the equipment. A common use for dwell is in the process of machining internal or external grooves.

Dwell Program Example
    G01 X2.0 F.008
    G04 P1500.
    G00 X2.5

In the process of making the groove, you must remove a layer of material with a thickness corresponding to the depth of the cut and equivalent to the feed per revolution. In order to avoid an egg-shaped workpiece, a certain amount of time must be allowed for the cut to be completed for the full circumference as the tool tip reaches the diameter indicated in the program. If function G00 follows function G01, the resulting shape of the workpiece will be that of an egg because the tool is removed from the groove before all of the material can be cut. In the example above, the programmed dwell is 1.5 seconds, which allows a sufficient pause necessary to clean up the bottom of the groove before retracting from it.

Return to Machine Zero Point (G28).   One of the ways to command a tool to return to the machine zero point after completion of a cutting path is through the application of function G28. The block format for G28 function is given as follows:
    G28 X(U). . . Z(W). . .

In this block, the values in X(U) and Z(W) are the coordinates for an intermediate point through which the tool will simultaneously pass on its way to machine zero. The tool will position at a rapid traverse rate in nonlinear form. Therefore, the last programmed move before calling G28 should move off of the part, so that the tool is clear of the part. In the next line, when using G28, tool offsets should be cancelled in a prior block as described earlier.

Return to Machine Zero Examples:
    G28 X0. Z0.

This command moves the tool from its current position to (through) the Work Offset position and then, to the machine zero point.
    G28 U0. W0.

This command moves the tool from its current position directly to the machine zero point because of implementing incremental move of zero in both axes first. This is the preferred method.

Application of Tool Nose Radius Compensation (TNRC) (G40, G41 and G42). Even if a value is entered into the Radius Geometry column of the tool offset tab the tool nose radius will not be compensated for when turning tapers or radii unless codes G41 or G42 are used. In the cases of straight facing along the X-axis or straight turning along the Z-axis, the tool contact point is the same as the measured point on the tool nose; therefore, no compensation is required. However, tool nose compensation is required for chamfers, tapers and arcs. Without using compensation, the aforementioned would be undercut or overcut by a small amount that varies, depending on the radius of the tool nose.

Using TNRC adds the ability to control the dimensional quality of geometry features when using indexable inserts, or any tool with a nose radius. It also aids in the programming process by making it necessary only to program the workpiece profile radii and tapers, without shifting the tool path to compensate for the tool nose radius. A dimensionally accurate workpiece will result when a program is properly written using functions G41 or G42, setting the values correctly in the Radius Geometry column of the offset register of the control. Using TNRC also accounts for values entered into the X or Z Geometry Wear column for the affected tool. See Chapter 3, Adjusting Wear Offsets for CNC Lathes, for wear offset adjustment details.

The following explanations are given for the critical information needed for using functions G40, G41 and G42.

Tool Nose Radius Compensation Cancel (G40). Earlier in this chapter, the CNC lathe we learned that the program template includes the G40 command in line N10, the first line actual program code and at the Tool Ending section in line N200. These two strategic locations ensure that cancellation of any active compensation is complete. Note that when the machine is first started, the G40 command is activated by default.

In order to end the use of G41 or G42, TNRC may be cancelled by using function G40. To program the cancellation of tool nose radius compensation, the command is generally input on a move that is in a departing vector from the machined profile. This move may be either G00 or G01, and cancellation may be initiated with the G28 command, where the compensation will be cancelled upon reaching the intermediate point.

Selecting Which Code to Use—G41 or G42. Selection of either G41 or G42 is based on the side of the part profile that the tool needs to be on in order to create the desired results. Think of the part profile as the center line of a highway. Then, based on the direction of travel, decide which side of the road to drive on. For the left side, G41 is selected and, for the right side, G42 is used. This procedure may be applied whether the cut is internal or external.

Initiating Tool Nose Radius Compensation with G41 or G42. To initiate the use of tool nose radius compensation, the G41 or G42 should occur on a G00 rapid positioning move that is at least .100 inch, or 2 mm away from the part profile. This move need only be in one axis direction, but it can include both.

Tool Nose Radius Compensation Left (G41). For facing and turning inside diameter/bore profiles, G41 is used.

The block format is G00 G41 X(U)... Z(W)...

Tool Nose Radius Compensation Right (G42). For turning of the outside diameter profiles uses G42. The block format is G00 G42 X(U)... Z(W)...

Notes on Using G41 and G42. If the values for the Tool Nose Radius (radius geometry) and the Tip Direction are omitted in the offset register, the desired results will not be obtained (zero radius is assumed in this case).

Work Coordinate Systems (G54–G59). Using one of the G54–G59 codes identifies the workpiece zero for the finished face of the part along the Z-axis. All of the programmed tool path is referenced to this origin. The X-axis value is rarely changed from zero, because the center line of the spindle axis is the origin.

Refer to the "Measuring the Work Coordinate Offset for the CNC Lathe" section of this chapter and Figure 3-15 for details on setting the work offset. Note: Figure 3-15 includes Y-axis and C-axis columns that would not be present on a two-axis CNC lathe.

Multiple Repetitive Cycles

Finishing Cycle (G70). The Finishing Cycle is used to make finishing cut paths for removal of the remaining stock allowance with the previously applied cycles G71, G72, and G73 (descriptions to follow), in a single pass, along the defined profile. Stock allowances left for finishing (U, W) may be removed by the same tool used in rough cutting. However, it is a common practice to use a different tool for the finishing pass.

The block format for function G70 is G70 P ... Q ... F ... S ...

where

P = number of the first block of the finished profile (as shown in Figure 3-22 as position *b*)

Q = number of the last block of the finished profile (as shown in Figure 3-22 as position *c*)

F = feedrate, effective for blocks P through Q

S = spindle speed, effective for blocks P through Q

Notice from the block diagram, that it is only necessary to enter position coordinates of the first block (b) through last block (c) of the previously used roughing cycle which defines the finished profile. This will cause an automatic return to the earlier part of the program for the coordinates needed for the completion of the process removing allowances U and W.

Note: The feed values used in blocks for function G71 are ignored for the finishing cycle, but those included in the with function G70 are valid.

Stock Removal Turning Cycle (G71). Function G71 is the stock removal cycle for turning that removes metal along the direction of the Z-axis. In a case where there is a lot of material to be removed, such as bar stock, this cycle provides an easy method for programming.

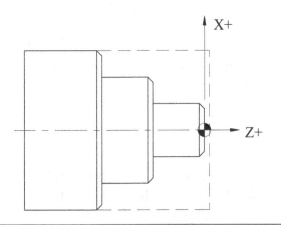

Figure 3-21. Stock removal cycle (G71)

In Figure 3-21, the dotted lines refer to the initial shape of the workpiece, whereas the solid line refers to the final product. Programming of individual blocks to perform all the individual cuts for roughing is cumbersome. By using function G71, programming of the final shape of the

workpiece is defined and fewer lines of code are required. Material is removed automatically in each pass.

For Fanuc controls there are two types of program format for stock removal using function G71: single block and double block. The CNC control model used determines which type will be needed. Both of these are covered in detail in my text *Programming of CNC Machines* (4th ed.). The single block method for Haas controls will be presented here. Consult the manufacturer programming manual specific to the machine you're using to determine the required method.

Function G71. There is only one program block required for function G71, when using this method. Much more freedom is allowed in regards to programmable shapes. In this case, it is not necessary to program a steadily increasing or decreasing pattern in both axes, it is only required along the Z-axis, and up to ten concave figures are allowed.

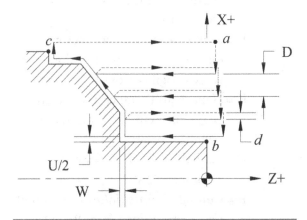

Figure 3-22. Stock removal turning cycle (G71) diagram

In Figure 3-22, the letter *d* represents the amount of X-axis retract programmed for clearance amount set by parameter

(Haas Setting 73). Solid lines represent feed moves and the dashed lines represent rapid traverse moves.

> a = the starting clearance point for the X- and Z-axes for the given cycle.
> b = the sequence number of the first programmed point for the finish profile, which corresponds with P number of the second G71 block.
> c = the sequence number of the last programmed point for the finish profile.

1. Block format is G71 P...Q...U...W...I...K...D...F...S... *where*

> P = the sequence number of the first block in the program, which defines the finish part profile.
> Q = the sequence number of the last block of the program, which defines the finish part profile.
> U = the stock allowance to be left for a finishing pass in the X-axis direction (diameter sign is + or –).
> W = the stock allowance to be left for finishing in the Z-axis direction (sign is + or –).
> I = (Optional) Radial distance and direction of the final cut along the X-axis (sign is + or –).
> K = (Optional) Distance and direction of the final cut along the Z-axis (sign is + or –).
> D = Radial depth of cut for each roughing pass.
> F = Cutting feedrate (in/rev or mm/rev) for blocks defined from P to Q
> S = Spindle speed (ft/min or m/min) for blocks defined from P to Q (Optional: RPM (S) is usually defined prior to starting this cycle)

2. Two axes of movement may be programmed in the first block

identified by the parameter P if the TYPE II method is used (consult the Haas CNC programming manuals).

3. The signs attached to symbols U and W may have negative or positive values, depending on the orientation of the coordinate system and the direction in which the allowance is assumed (Figure 3-23).

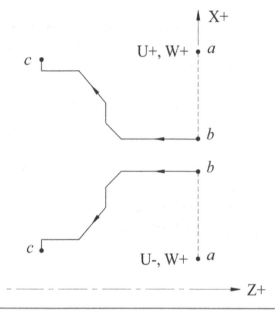

Figure 3-23. Stock removal cycle (G71) sign designation

Notes:

1. Changes in the feed between blocks P and Q will be ignored in G71. Only feed F, indicated by function G71, is valid.
2. The first tool path movement of the programmed cycle from point a to point b cannot include any displacement in the direction of the Z-axis.
3. The tool path between point b and point c must be a steadily increasing or steadily decreasing pattern in both axes.
4. Both linear and circular interpolation are allowed.

Figure 3-23 illustrates that if the allowance for finishing is located on the positive side (in the direction of the X and Z axes) with respect to the programmed contour, no sign is used; if on the negative side, the negative sign (–) is used.

Stock Removal Facing Cycle (G72). The properties for function G72 are similar to G71 described above (Figure 3-22). The only difference is in the change of the cutting direction to facing as detailed in Figure 3-24 below. The finish allowance is intended to be removed by finishing cycle G70 described above. The block format is G72 P... Q... U...W... D...F...S...

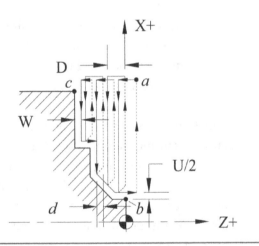

Figure 3-24. Stock removal facing cycle (G72) diagram

The parameters for this function have the same meaning as those described for function G71.
Notes:

1. The first block of the programmed cycle should not include any displacement in direction of the X-axis.
2. The remaining notes for this function are identical to those for function G71.

3. The principle defining the choice between a positive or negative sign for U and W is identical to function G71.

Pattern Repeating Function (G73). Function G73 permits the repeated cutting of a fixed pattern, with displacement of the axes position by an amount determined by the total material to be removed, divided by the number of passes desired. This cycle is well suited for previously formed castings, forgings, or rough machined materials. This machining method assumes that an equal amount of material is to be removed from all surfaces. It can still be used if the amounts are not equal. However, caution should be applied concerning excessive depths of cut, and there may also be occasions of air cutting. DO NOT use this cycle for solid bar stock operations. The finish allowance is intended to be removed by finishing cycle G70 described above.

The block diagram for function G73 is G73 P... Q... U... W... I... K... D...F...S...
where

P = the sequence number of the first block of the finished profile (given in Figure 3-25 above as position *b*).

Q = the sequence number of the last block of the finished profile (given in Figure 3-25 as position *c*).

U = finish stock allowance in the direction of the X-axis (sign + or –), referred to stock left on the diameter.

W = finish stock allowance in the direction of the Z-axis (sign + or –).

I = total stock removal amount in the direction of the X-axis (radial value with sign + or –).

K = total stock removal amount in the direction of the Z-axis (sign + or –).

D = the number of rough cutting passes.

F = feedrate, effective for blocks P through Q.

S = spindle speed, effective for blocks P through Q.

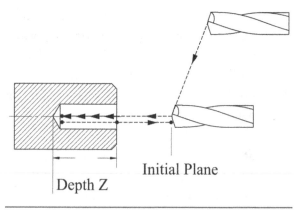

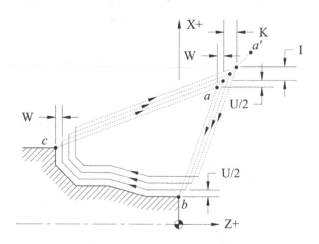

Figure 3-25. Pattern repeating cycle (G73) diagram

Figure 3-26. Peck drilling cycle (G74) diagram

Point *a* on the drawing in Figure 3-25 is the starting point. In executing this cycle, the tool travels from point *a'* to point *a*. Axis displacement amounts are defined by the values of I and K in Figure 3-25. Finish stock allowances are defined with U and W. The cycle begins equivalent cutting passes with the number of passes determined by D. At the end of the cycle, the tool automatically returns to point *a*. Points *b* to *c* define the finished profile to be machined by function G70, as described earlier in this section. Address I and K are measured on the machined workpiece. The principle defining choice of signs is similar to that for U and W.

Peck Drilling Cycle (G74). The most common use of this function is for drilling deep holes that require an interruption in the feed in order to break long stringy chips. A block diagram of this function, as well as the movements of the tool, is illustrated in Figure 3-26.

The block diagram for function G74 is
G74 Z . . . K . . . F . . .
where

Z = final Z cut depth in the absolute system.

K = depth of cut per peck in Z direction (no sign) (retract distance determined by Setting 22).

F = cutting feedrate.

Function G74 may be applied to face groove cutting (along the Z-axis) that exhibit hard breaking chips as well. The amount of the clearance (indicated by *d* in Figure 3-27) is set by a setting parameter. The amount of the return (indicated by *e* in Figure 3-27) is also set by a setting parameter. The block diagram for face grooving function G74 is G74 X . . . Z . . . U . . . W . . . I . . . K . . . D . . . F . . .
where

X = diameter value of the workpiece at the minimum absolute location diameter of the groove.

Z = final Z cut depth in the absolute system.

U = the incremental distance between the groove starting and end points along the X-axis.

W = the Z-axis incremental cut peck depth.

I = the incremental depth of cut per side in X direction (radius value, no sign).

K = the incremental depth of cut in the Z direction (no sign).

D = the retract amount when returning to the clearance plane.

F = cutting feedrate.

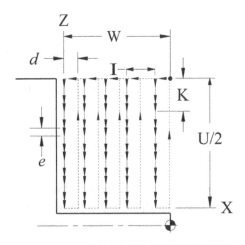

Figure 3-28. Groove cutting cycle (G75) diagram

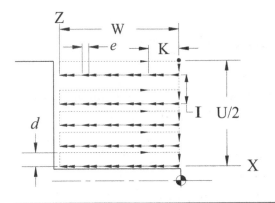

Figure 3-27. End face grooving cycle (G74) diagram

Groove Cutting Cycle (G75). Function G75, in its form, is very similar to function G74. Its difference is seen in the direction of the tool movement, which is orthogonal to that in function G74. Function G75 is used for cutting grooves that require an interrupted cut along the X-axis (Figure 3-28). The block diagram for function G75 is G75 X . . . Z . . . U . . . W . . . I . . . K . . . D . . . F . . .

All notations assumed for this function are defined exactly as in function G74.

Note: At the end of the cycle, the tool returns to the starting point in both functions G74 and G75.

Multiple Thread Cutting Cycle (G76). The G76 multiple thread cutting cycle is used for most external and internal threading applications. All of the information needed to complete the desired thread is input in one block rather than multiple blocks. By inputting the appropriate data for a particular type of thread in the program blocks, the number of cutting passes is automatically calculated by the control. Because of the limited number of blocks required, this method is very easy to program and edit. Figure 3-29 represents the basic functions of the cycle.

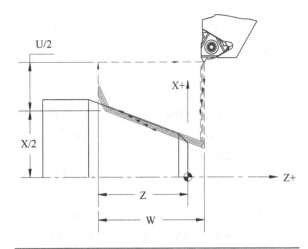

Figure 3-29. Multiple thread cutting cycle (G76) diagram

The block diagram for function G76 is G76 X . . . Z . . . U . . . W . . . K . . . I . . . D . . . P . . . A . . . F . . .

where

X = the finish diameter of thread (last diameter cut).

U = the total incremental distance from the major diameter to the finish diameter.

Z = the full length of the cut thread (end of thread position in absolute).

W = the total incremental distance from the start to the end of the thread.

K = the radial thread height (always positive).

I = the difference of thread radius (+ or −) from start to finish (for tapered threading).

D = the depth of first threading cut (always positive).

A = the thread angle (matches the included angle of the threading insert and is always positive). No decimal is needed.

F = the cutting feedrate is the lead of thread or the distance traveled in the Z-axis per one revolution. To attain this value, take one divided by the number of threads per inch.

Notes:

1. Changing the spindle speed or feedrate via override, while within the threading cycle, is not effective.
2. A dry-run condition is applicable and effective.
3. The use of constant rotational programming G97 S . . . . is required!
4. The depth of a first cut D is approximately .003 to .018, depending on machining conditions.
5. With a small value of first cut, the number of passes increases and, inversely, with a greater value of the first cut, the number of passes decreases.
6. The selection of factor D depends on the type of thread, the material, and the cutting tool tip material. Consult tooling manufacturer recommendations for cutting data. Depending on the value of the first cut specified by D, a certain number of passes will be obtained, as defined by the parameters that are set for the machine.

Example Program

The following explanations are given for the specific program sections of the sample program part in Figure 3-30 and Chart 3-5:

CNC lathe Example Program O0020

O0020

(CNC lathe Programming Example)

(Date: Today, By: You)

(Tool #2, Rough Turning Tool)

N10 G50 S6000

N20 T0200

N30 G96 S615 M3

N40 G00 G54 X2.6 Z.1 T0202 M08

N50 G1 Z0.0 F.0125

N60 X−.02

N70 G00 Z.1

N80 X.2.6

N90 G71 P100 Q200 U.015 W.005

N100 G00 X.90

N110 G01 Z0.0 F.0265

N120 X1.0 Z−.05

N130 Z−1.0

N140 X1.3

N150 G03 X1.5 Z−1.1 R.1

N160 G01 Z−1.9

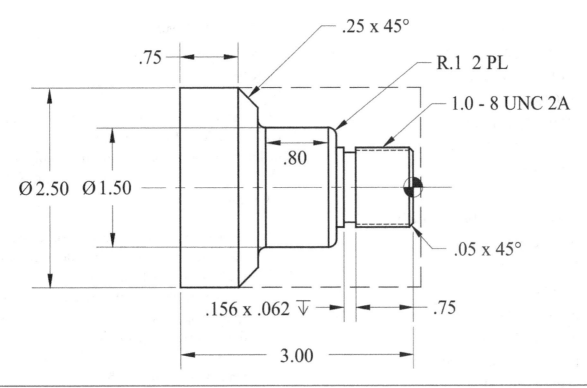

Figure 3-30.   Example program part

Date:	Today	Prepared By:	You
Part Name:	CNC Lathe Example Program	Part Number:	1234
Machine:	CNC Lathe	Program Number:	20

Workpiece Zero: X = Centerline Y = NA Z = Finished Face

Setup Description:

Part Material is 4340 Alloy Steel
Clamp the part in a 3-jaw chuck with soft-jaws, 2.35" minimum extended out of the chuck

Tool #	Tool Description	Offset #	Comments
T02	Rough Turning Tool .031 TNR	T0202	SFPM = 525-705 Feed = .008-.017
T04	Finish Turning Tool .015 TNR	T0404	SFPM = 525-705 Feed = .008-.017
T06	O.D. Grooving Tool	T0606	.118 Wide .005 TNRC
T08	O.D. Threading Tool	T0808	SFPM = 525-705 Feed = .008-.017

Chart 3-5.   CNC Lathe Example Setup Sheet

N170 G2 X1.7 Z–2.0 R.1

N180 G01 X2.0 Z–2.0

N190 X2.5 Z–2.25

N200 X2.6

N210 T0200 M09

N220 G28 G40 U0.0 W0.0

N230 M01

(Tool #4, Finish Turning Tool)

N240 G50 S6000

N250 T0400

N260 G96 S615 M3

N270 G00 G54 G42 X.90 Z.1 T0404 M08

N280 G1 Z0.0 F.022

N290 G70 P100 Q200

N300 T0400 M09

N310 G28 G40 U0.0 W0.0

N320 M01

(Tool #6, OD Grooving Tool .118 Wide)

N330 T0600

N340 G96 S615 M3

N350 G00 G54 X1.6 Z–.906 T0606 M08

N360 X1.1

N370 G75 X.875 I.06

N380 G00 X2.5

N390 T0600 M09

N400 G28 G40 U0.0 W0.0

N420 M01

(Tool #7, OD Threading Tool)

N430 T0800

N440 G97 S1074 M3

N450 G00 G54 X1.1 Z.1 T0808 M08

N460 G76 X.8492 Z–.8 K.0707 D0150

A60 F.125

N470 T0800 M09

N480 G28 G40 U0.0 W0.0

N490M30

The first item listed is the program number: O0020. Followed by comments in parentheses, including information for the first tool to be used in the program. Lines N10 through N460 are sequence or block numbers.

N10 G50 S6000. Block N10 contains the program words G50 S6000. Because G50 is specified simultaneously in the block with the S6000, it sets a maximum spindle speed of 6000 r/min applicable during machining for tool No. 2, called by T0200 in the following block. This maximum value is based on machine specification maximum and or specific clamping limitations.

N20 T0200. Tool function T0200 commands tool number two to the ready position. If the turret is in another position, it will rotate tool the number two into the ready position. The number 00 cancels any wear offset so that no offset compensation is used for this tool, at this time.

N30 G96 S615 M3. Block N30 contains function G96 commands that the machining process will take place at a constant cutting speed of 615 ft/min and the spindle speed will be adjusted automatically, based on the diameter of the workpiece up to a maximum of 6000 r/min. Miscellaneous code M03 refers to the direction of spindle rotation and activation of the spindle, which in this case is clockwise. The SFPM is based on the average of part and tool material recommendations from the *Machinery's Handbook.*

N40 G00 G54 G42 X2.6 Z.1 T0202 M08. Block N40 activates the rapid traverse (G00) tool movement to the X and Z coordinates listed in relation to the G54 work offset and T0202 tool offset in the controller. Typically, this first move positions the tool at a safe distance near the work diameter and the part face. The miscellaneous function M08 activates the flood coolant.

N50 G1 Z0.0 F.0125. The information contained in this block N50 activates execution of a linear feed move of the tool to the position specified by the coordinates Z0.0. The feedrate is based on the average of part and tool material recommendations from the *Machinery's Handbook.*

N60 X-.02. In block N60, the cutting tool is fed to the X-.02 location (a little past center line) to face the front of the part.

N70 G00 Z.1. Block N70 positions the cutting tool to a clearance position of .100" from the finished part face at rapid traverse rate (G00).

N80 X.2.6. Block N80 positions the cutting tool to the X2.6 position (beginning location for the next cycle) at rapid traverse.

N90 G71 P100 Q200 U.015 W.005 D.08. Block N90 activates the multiple repetitive cycle (G71) to remove the material profile defined by blocks N100 (P100 and Q200) – N200. The finish allowance is defined in this block with U.015 for the X and W.005 for the Z. The depth of cut per pass is set to by the address D.08. Lastly, the feedrate is adjusted at block N110 within the profile description.

N100 G00 X.90
N110 G01 Z0.0 F.010
N120 X1.0 Z-.05
N130 Z-1.0
N140 X1.3
N150 G03 X1.5 Z-1.1 R.1
N160 G01 Z-1.9
N170 G2 X1.7 Z-2.0 R.1
N180 G01 X2.0 Z-2.0
N190 X2.5 Z-2.25
N200 X2.6

N210 T0200 M09

In block N215 the tool offset is cancelled and the flood coolant is stopped.

N220 G28 G40 U0.0 W0.0. Block N220 activates function G28, Return to Machine Zero Point and cancels TNRC with G40. The incremental coordinates, U0.0 and W0.0 cause the machine to ignore the intermediate point as the tool returns to machine zero.

N230 M01. In block N230, the miscellaneous command M01, Optional Stop is activated. The machine will stop and await the operator to push cycle start again before proceeding, if the Optional Stop button is activated.

(Tool #4, Finish Turning Tool)
N240 G50 S6000
N250 T0400
N260 G96 S615 M3

The comment identifies the next tool information. Blocks N240–N260 set the maximum RPM, cancel tool offset data for tool four and turn on the spindle rotation at a constant SFM of 615 in the clockwise direction.

N270 G00 G54 G42 X.90 Z.1 T0404 M08. Block N270 activates the rapid traverse mode (G00) and the TNRC left mode (G42) while positioning the axes to X.90 and Z.1, with relation to the work offset (G54) and tool offset data called by T0404. Flood coolant is activated by the M08 command.

N280 G1 Z0.0 F.017. Block N280 initiates the linear feed mode (G01) and commands movement to Z0.0 at a feedrate of .017 IPR.

N290 G70 P100 Q200. In block N290, the Multiple Repetitive finishing cycle is called and executed, based on the data given in blocks N100–N200, removing only the remaining material of .015 inch on diameters and .005 inch on shoulders.

N300 T0400 M09
N310 G28 G40 U0.0 W0.0
N320 M01

Blocks N300–N320 are tool ending blocks for tool #4.

(Tool #6, OD Grooving Tool .118 inch Wide)
N330 T0600
N340 G96 S615 M3

The comment identifies the next tool information. Blocks N330–N340 set the maximum RPM, cancel tool offset data for tool six and turn on the spindle rotation at a constant SFM of 615, in the clockwise direction.

N350 G00 G54 X1.6 Z–.906 T0606 M08. Block N350 activates the rapid traverse mode (G00) while positioning the axes to X1.6 and Z–.906, with relation

to the work offset (G54) and tool offset data called by T0606. Flood coolant is activated by the M08 command.

N360 X1.1. Block N360 further positions the grooving tool to within .100" of the diameter where the groove will be cut.

N370 G75 X.875 I.06 F.008. In block N370 the single grooving cycle command is activated (G75) and the finished diameter of .875 is given with a pecking amount of .06" (I.06) at a feedrate of .008 IPR (F.008).

N380 G00 X2.5. Block N380 positions the grooving tool to a clearance location along the X-axis.

N390 T0600 M09
N400 G28 G40 U0.0 W0.0
N420 M01

Blocks N390–N420 are tool ending blocks for tool #6.

(Tool #7, OD Threading Tool)
N430 T0800

N440 G97 S2005 M3. The comment in parentheses, identifies the next tool information. Blocks N430–N440 cancel tool offset data for tool eight, cancel (G97) Constant Surface Speed (G96) mode and set the spindle to 2005 RPM, in the clockwise direction.

N450 G00 G54 X1.1 Z.1 T0808 M08. In block N450, tool #8 is positioned at rapid traverse to the X1.1 and Z.1 coordinates, in relation to the work and tool offsets given. The flood coolant is also activated.

N460 G76 X.875 Z–.8 K.0625 D.015 A60 F.125. In block N460, Thread Cutting

Cycle (G76) is activated. The X and Z coordinates are given for the final points for each. The radial thread height is given (K.0625) and the first pass amount is listed (D.015). The thread angle is given (A60) and the feedrate is listed (F.125).

N470 T0800 M09
N480 G28 G40 U0.0 W0.0

Blocks N470–N480 are tool ending blocks for tool #8.

N490M30. Miscellaneous function M30 in block N490 ends the program and resets the program to its beginning point, or head.

Summary

In this section, you learned about the basic structure of CNC programs for lathes. Emphasis here has been on the most commonly used programming functions. There are many more capabilities available beyond the scope of this introductory chapter. Please consult operator and programming manuals that are supplied with the machine tool you are using. For more detailed programming descriptions, refer to my text *Programming of CNC Machines* (4th ed.). In the next chapter, an introduction will be given to CAD/CAM. This subject will probably be how your designs and programs are processed on-the-job.

Introduction to CNC Lathe Programming, Study Questions

1. In programming, tool function is commanded by the four digits that follow the letter address T (example T0404). What do these two sets of numbers refer to?
   a. Work Offset and Tool Offset
   b. Tool Number and Tool Offset
   c. Tool Number and Work Offset
   d. Tool number 404

2. Of the following choices, which is the best method for compensating for dimensional inaccuracies caused by tool deflection or wear?
   a. Geometry Offsets
   b. Wear Offsets
   c. Tool Length Offsets
   d. Absolute Position Register

3. When the rough turning cycle G71 is used, which letters identify the amount of stock to leave for the finish pass, X-axis and Z-axis respectively?
   a. U and V
   b. X and Z
   c. P and Q
   d. U and W

4. The default (at machine start-up) feedrate on lathes is typically measured in?
   a. Cutting speed
   b. Inches per revolution in/rev
   c. Inches per minute in/min
   d. Constant surface speed

5. The advantage of using G96 Constant Cutting Speed in turning is that as the diameter changes (position of the tool changes in relation to the center line), the r/min increases or decreases to accomplish the programmed cutting speed.

   T or F

6. The preparatory function G50 relates to:
   a. maximum spindle r/min setting.
   b. constant spindle r/min.
   c. work offset and tool offsets.
   d. constant surface speed.

7. When incremental programming is required in turning diameters and facing, which letters identify the axis movements respectively?
   a. I and J
   b. U and W
   c. I and K
   d. U and V

8. What two letters identify the incremental distance from the starting point to the arcs center in G02/G03 CNC lathe programming?
   a. I and J
   b. I and K
   c. J and K
   d. X and R

9. When using the Multiple Repetitive Cycle, Rough Cutting Cycle, G71, U and W represent the stock allowance for finishing. What cycle is required to remove the stock allowance?
   a. G90
   b. G73
   c. G76
   d. G70

10. What is the major reason for selection of the Pattern Repeating Cycle, G73, as opposed to the Rough Cutting Cycle, G71?

    a. It is required to make the finish allowance cut on X and Z.

    b. This cycle is well suited where an equal amount of material is to be removed from all surfaces.

    c. G73 is limited to orthogonal cuts while G71 can cut radius and chamfers.

    d. G71 is limited to rough cutting only while G73 is required to remove stock allowances.

11. When programming arc and angular cuts using tool nose radius compensation G41 and G42, which is used for facing and turning?

    a. G41 facing, G42 turning.

    b. G42 facing, G41 turning.

    c. Neither. The tool nose radius amount must be calculated and programmed to compensate.

    d. G41 for facing if contours or angles are involved, G42 for turning if contours or angles are involved.

4 | Introduction to Computer-Aided Manufacturing

Objectives

1. Define the capabilities of CAD/CAM.
2. Navigate and utilize the CAD/CAM user interface.
3. Create tool paths for CNC milling machines and CNC lathes.
4. Verify tool paths by simulating and make changes, as needed.
5. Create setup documentation from CAM.
6. Post-process G-codes for CNC milling and CNC lathes.

Computer-aided design and computer-aided manufacturing (CAD/CAM) utilize computers to design part and assembly models for manufacture. CAD is limited in nature to the generation of engineering drawings, whereas CAD/CAM combines both design and manufacturing capabilities. With CAM, the design model from CAD is used to identify specific cutting tool data and create cutting tool paths for CNC machining. Selection of model geometry identifies what the features are and how they are going to be machined in order to develop cutting tool paths and machine code (the part program) for CNC machine tools. One of the major benefits of CAD/CAM is the time saved because it is much more efficient than manually writing CNC code line-by-line. Designers can create the models and engineering drawings needed and share them electronically with the manufacturing department. Model files are saved in native format or exported to a common file format, such as the Initial Graphics Exchange Specification (IGES), Standard for the Exchange of Product model data (STEP) and Stereo Lithographic (STL), to name a few.

Most CAM systems support direct import of solid model files for tool path creation (drag-and-drop functionality), allowing the manufacturing engineer/CNC programmer to create the tool path and assign cutting tool information relative to the desired results very quickly. When using CAD/CAM, the drawing may be created from scratch or imported from a CAD program using one of the file formats mentioned earlier. It is not necessary to have the drawing dimensioned for the CAM operation, but the full scale of the part is required. To make best use of CAD/CAM, it is important to fully understand the machining processes presented in the previous chapters. Remember: the overall objective of CAD/CAM is to generate a tool path for a CNC machine in the form of a CNC program. It is imperative to have full understanding of the rectangular and polar coordinate systems, cutting tool selection, speeds and feeds. Nearly all CAD/CAM programs will automatically develop speed and feeds data based on the tool and work material selection saved in the software database; however, adjustments are commonly made to fine tune and improve results. This database of information can and should be updated to match the requirements for your shop for the best results.

In this chapter, Autodesk® Fusion 360 is featured as the CAD/CAM software used for the programming examples. This chapter is intended only as an introduction to CAD/CAM. To cover the full extent of its capabilities would require an entire text, if not volumes.

Note: Autodesk® Fusion 360 is available for use in Trial, Student and Startup licensing forms. The software is Cloud-Based so internet access is required. Files are saved in the cloud but can also be exported to My Computer and drive locations via the Export command.

Autodesk® Fusion 360 Startup

After first downloading the software and setting up a new account, double-click the Autodesk® Fusion 360 icon and log in. The software will start and the user interface (UI) will display (Figure 4-1).

CAD/CAM User Interface

If a CAD model is available, use the file drop-down, and Open to load the file. Hover over Open in the drop-down to reveal the flyout menu of the extensive list of file types that can be loaded. Once files have been saved to the cloud in project folders, the Data Panel (left-most from the tab) can be used to open files.

Brief descriptions for each area of the interface (from top left to right) are below:

> **Untitled tab.** The untitled tab will be displayed until the current file is saved or another file is loaded, then the tab will inherit the name of the file. Left-clicking the X on the tab closes the file and the plus sign starts a new file.

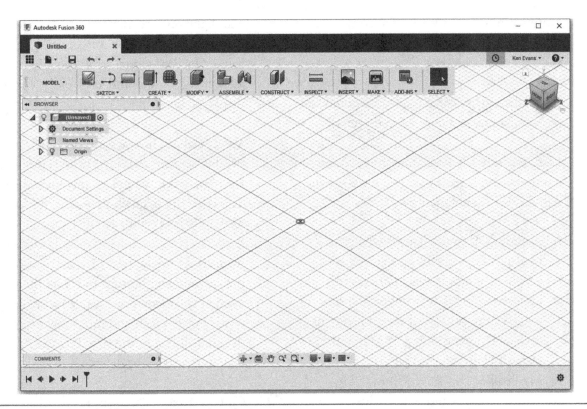

Figure 4-1. Autodesk® Fusion 360 user interface

Data Panel. This icon gives access to file management functions. The data panel icon is located below the "Untitled or File Name" tab (nine square dots). Left-click to expand and display saved projects. The first time you start Fusion 360, Demo Project and My First Project will be displayed at the top section and Libraries: Assets and Samples are below that. In the upper right, there is a button to start a New Project. This is a good place to start by setting up a project folder for your work. Pressing the Data button will list all of the saved files in the active project folder and pressing the People button allows you to invite others to have access to the files in order to work collaboratively. There are also buttons on the right to upload files or make a new folder within the active project and the Settings Gear icon to adjust the way files are listed. As you work and save files, a thumbnail of the versioned file will be displayed here in the cloud. Any of these saved files can be opened by double-clicking

them from the Data Panel. In the upper-left corner next to the project name, there is a back arrow used to leave the Data Details panel. Lastly, to close the Data panel, click on the X in the upper-right corner. Any time you hover over an item with the mouse, a hint will be displayed to help you remember its function.

File. The file drop-down gives access to: start a New Design or a New Drawing; Open a file; Recover Documents, or, Upload a file. In the middle section of the dialog are: Save, Save as and Export. *Export is the method you use to save you file to your local drive.* The next separator includes: 3D Print; Capture Image and Share. Capture image allows you to save a screen shot of what is displayed for setup or other communication needs. In the lower section, the View settings can be adjusted (note the shortcut keystrokes are listed).

Notifications. If any, they'll be displayed to the left of the Job Status icon (none shown in Figure 4-1).

Job Status. This icon is used to see progress on Simulations (not covered in this text).

User Account. The account holder (your) name will be listed here. The drop-down will display the following:

- **Autodesk Account.** This is where your profile, security settings linked accounts and preferences are saved.
- **Preferences.** This very important dialog is where control of the UI behavior such as Drawing size, part Material, Default Units, etc. are set.
- **My Profile.** This screen shows your Projects, Assets and Activity history.
- **Sign Out.** This is where you sign out. Be sure to save your work first.

Help. By pressing the Help icon (?), the Show Learning Panel, Learning and Help, Quick Setup, Community, Support and Diagnostics, What's New, and About functions are accessed. There are volumes of informative help a mouse-click away.

Browser. Document Settings = Units (can be changed here on the fly).

- Named Views = Standard: TOP, FRONT, RIGHT and HOME.
- Origin = Screen origin, X, Y and X-axis origins and the XY, XZ and YZ planes.

The light bulb to the left can be pressed to turn off or back on any of these at any time. Once a model and tool path are added, the model and tool path tree are displayed here.

View Cube. In the upper-right corner of the screen and just below the User Account (your name) is the View Cube. The cube offers quick access to the views: TOP, FRONT and RIGHT, as default. Left-click on each to display the model from that orientation. Experiment with the surrounding buttons to rotate the view from any orientation you choose. Hold down the left-click and move the mouse to dynamically rotate the view. Press the Home icon to return to the default Isometric view.

Canvas Area. The Canvas Area is where the model is displayed. At the bottom center of the canvas is a group of icons that offer access to: Orbit types, Look At, Pan, Zoom functions, Display Settings, Grids and Snaps and Viewports. All of these have drop-down functions that allow the user to adjust viewing preferences.

Toolbar Ribbon. As you position the mouse over the drop-down of the Model button on the Toolbar Ribbon (Figure 4-2), access to the Change Workspace is made available. Presentation of tool path creation for an existing model in Manufacture will be covered in this text.

Manufacture. Once a model file is created or an existing file is loaded, click

Figure 4-2. Toolbar Ribbon

on the Change Workspace drop-down icon (leftmost) from the Toolbar Ribbon and select the Manufacture icon (Figure 4-3) to generate a tool path and post-process the G-code. Steps to make a simple tool path for the created model are given below.

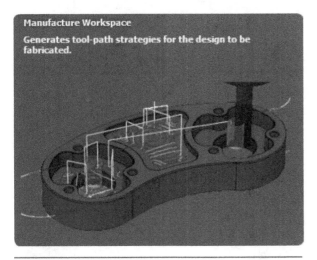

Manufacture Workspace

Generates tool-path strategies for the design to be fabricated.

Figure 4-3. Manufacture flyout

Manufacture Setup for an Existing Model

Follow these steps to complete set up for manufacturing.

1. Open the model file in Autodesk® Fusion 360.
2. Use the drop-down for User Account, select Preferences, Material, then change the Physical Material Category to Metal and the Physical Material Name to Aluminum 6061. The Appearance Category should also be changed to Metal/Aluminum. These settings ensure proper calculation of feeds and speeds based on the cutting tool selected.
3. Change the Workspace to Manufacture.
4. In the Browser tree, under CAM Root, check the Units to confirm that inch is selected and change if necessary.
5. Press the Setup button (intuitively, the next icon from the left on the toolbar). This step automatically generates some stock data

based on the part geometry boundaries and default settings in the software, and the Setup dialog is displayed.

Change the default settings to match what is shown in Figures 4-4, 4-5, and 4-6. Each of the dialog items should be changed to suit specific needs.

Operation Type is, in fact, milling for this example.

The following sub-steps will be implemented for the Work Coordinate System (WCS) of our manufacture setup.

a. **Orientation.** Click the drop-down and change to Select Z-axis/plane and X-axis, then pick the topmost face of the model. If the X-axis Gnomon is pointing toward the right, this is correct. If it is not, check the flip X-axis checkbox.

b. **Origin.** Click the drop-down and change to Model box point. This will change the location of the Gnomon to the finished surface of the model. What this means is that the part zero (G54) is set to the center of the part for the X- and Y-axes and, at the finished surface, in the Z-axis.

c. **Stock tab.** Change the default settings to match the values shown in Figure 4-7.

d. **Mode.** There are seven options to choose from. Use: Relative size box.

e. **Stock Offset Mode.** From the three options choose: Add stock to sides and top-bottom.

f. **Post-Process tab.** In Figure 4-8, the Program Name/Number is changed to what is required within the numbering system at your shop. Program Comment can be a simple description or part number.

g. **Machine WCS.** These settings define the Work Coordinate System numbering, G54–G59. To output G54, zero is the setting.

6. Finally, press OK to accept all of the Setup 1 information entered.

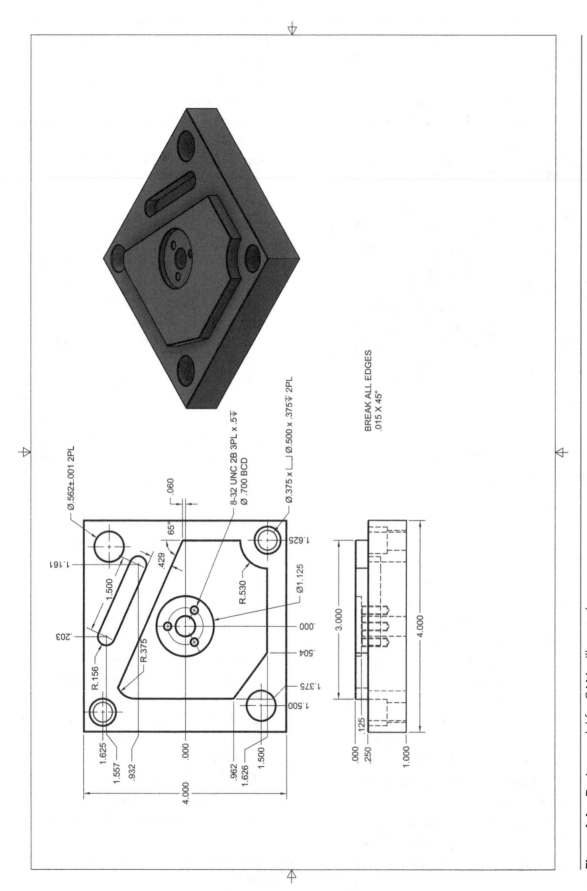

BREAK ALL EDGES
.015 X 45°

Figure 4-4. Design model for CAM mill example

298

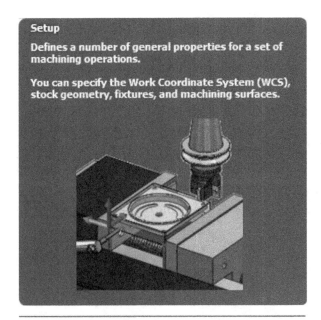

Figure 4-5. Setup flyout

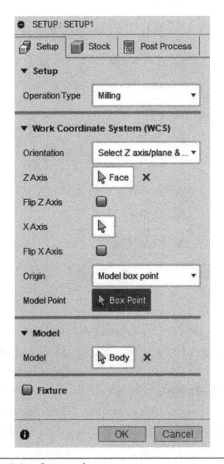

Figure 4-6. Setup tab

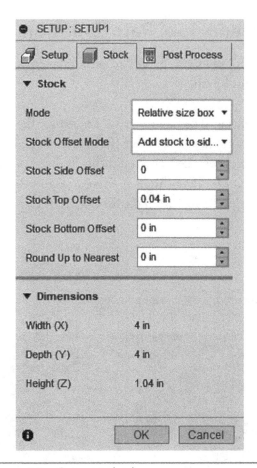

Figure 4-7. Setup stock tab

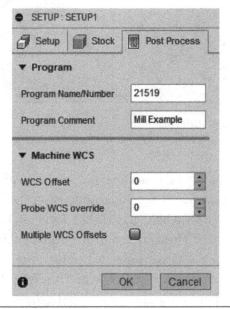

Figure 4-8. Post-process stock tab

Milling Tool Path Creation

The following steps will be used to create a milling tool path.

1. **Facing.** The first tool path needed for the example part is Facing the part to the finished thickness. Left-click on the drop-down on the 2D icon in the Toolbar Ribbon (Figure 4-9). Select "Face" from the list (Figure 4-11).

on Samples from the Libraries section of the Select Tool dialog. All of the tools in the Samples Library are listed by name and cutting diameter in the center of the dialog. Next, press the Type button to set a filter to show only Face Milling cutters (Figure 4-13).

Use the scroll bar to move to the bottom of the list and select the 2-inch face mill (2" Face Mill) from the list and then press

Figure 4-9. Manufacture Toolbar Ribbon

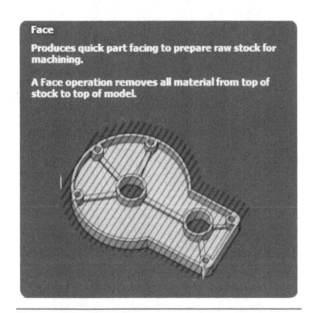

Figure 4-10. Face flyout

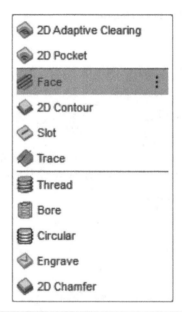

Figure 4-11. 2D Face

Note: As you move the mouse over Face (and each of the other options) in the list, a descriptive help graphic like Figure 4-10 will appear. This is true also for fields within the dialogs. These provide helpful information about function of each item. The FACE dialog will display a 2D face dialog (Figure 4-12).

2. **Tool tab.** The tabs at the top of the dialog will be adjusted sequentially until all are completed. First, the Tool type must be chosen. Press the Select button and left-click

the OK button (Figure 4-14). A translucent image of the selected tool will be displayed in the Canvas area above the model at the X-Y origin. Also, once the tool is selected, the speeds and feeds are adjusted to match the tool and material chosen in preferences earlier. Make sure Flood Coolant is selected and observe the other options for later.

3. **Geometry tab.** For the Face, the facing stock contours are automatically chosen but can be changed by selecting the desired

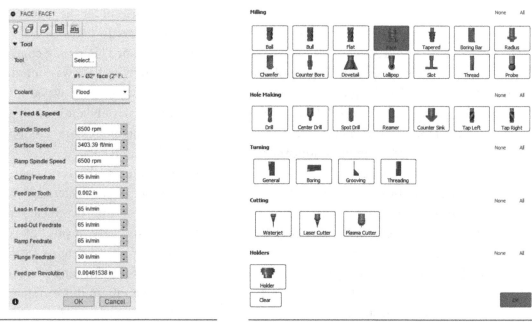

Figure 4-12.   2D Face dialog

Figure 4-13.   Tool type filters dialog

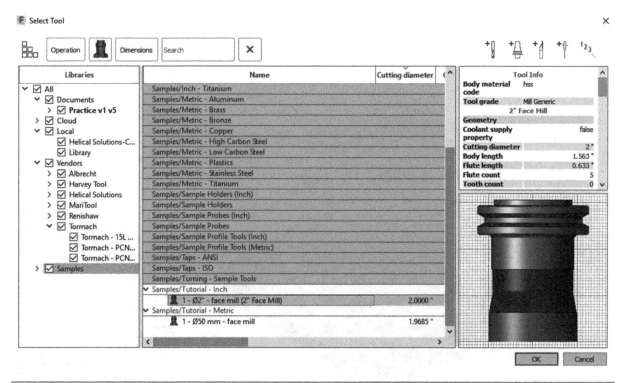

Figure 4-14.   Tool selection dialog

geometry. No changes are required to our example.

4. **Heights tab.** The setting in this dialog control the way the tool approaches and retracts from the material along the Z-axis. In this example, the default settings can be used. There are a multitude of options available from the drop-downs for Clearance, Retract, Feed, Top and

Bottom Heights, giving complete control to how the tool approaches the part. Be aware of fixture or part obstructions that may cause collisions; these settings can be useful for avoidance moves. The Heights tab is used in every type of tool path (Figure 4-15).

5. **Face, Passes tab.** In this example, the default settings can be used. Common adjustments are made to the following:

   a. **Pass Direction.** By default, the path is orthogonal to the stock; by using this parameter, the angle can be changed to any desired angular value.

   b. **Pass Extension.** This field can be used to extend the tool path beyond the part edge by a desired distance.

   c. **Stepover.** The amount of stepover is dependent on cutter characteristics and the geometry. As a general rule, a minimum of two-thirds of the diameter of the tool is a good place to start.

   d. **Direction.** This setting controls the pass direction and allows continuous feed travel when "Both ways" is used. "One way" will only allow cutting in one direction and Climb will force all passes to be climb cuts.

   e. **Multiple Depths.** If this box is checked, a new set of parameters are made available where multiple depth cuts can be defined.

   f. **Stock to Leave.** If this box is checked, an amount can be entered to allow stock to be left for a finishing operation.

   g. **Linking tab.** In this example, the default settings can be used. This dialog controls rapid traverse behaviors for the tool and the lead-in and lead-out transitions into and out of cutting conditions.

Figure 4-15. Face heights tab

6. **Press the OK button.** A back-plot of the tool path and a translucent image of the tool will be displayed on the Canvas area.

If adjustments are necessary, right-click on the [T1] Face1 path in the browser and select the Edit function (Figures 4-16 and 4-17).

Figure 4-16. Face passes tab

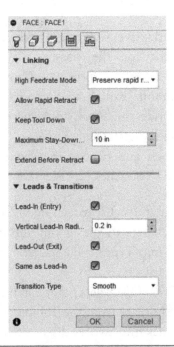

Figure 4-17 Face linking tab

Tool Path Verification and Simulation

A simulation should be completed for each path. To do this, select Simulate from the ACTIONS group in the Toolbar Ribbon. When you hover over the icon, a hint will be displayed (Figure 4-18).

Figure 4-18. Simulate flyout

Once the icon is selected, the SIMULATE dialog is displayed (Figure 4-19). Adjusting the setting in the SIMULATE dialog determines the graphical representation for the tool path during playback. The Playback buttons are located at the bottom center of the Canvas. Note the adjustments made for this example.

Display tab. The display tab offers adjustment to the Tool, Tool path and Stock appearance. The "Stop on collision" checkbox will stop the simulation if a collision of the tool occurs, when the checkbox is ticked.

Info tab. The Info tab merely displays information about the Position, Operation, Machine and Verification of the selected path.

Figure 4-19. Simulate dialog

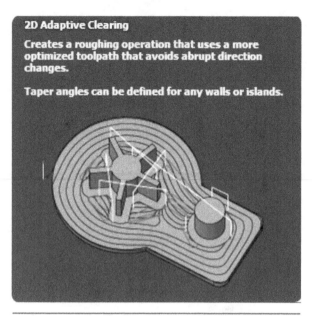

Figure 4-20. 2D Adaptive clearing flyout

Statistics tab. The Statistics tab displays information regarding machining time, linear machining distance, the number of operations and the number of tool changes. When the tool path has been simulated and verified correct, the work should be saved.

The next step is to follow the same basic sequence for all remaining features of the part. For this example, the next feature will be the step depth around the raised island portion of the part.

2D Adaptive Clearing

For this operation, the same tool can be used to remove the majority of material around the raised island. Only the items that require changes will be presented here (Figure 4-20).

2D Adaptive Clearing Geometry tab. Use the mouse to hover over the model until the base of the raised island contour is highlighted and left-click to select the chain. The red directional arrow should be to the outside of the island (if it is not, left-click on it to change it). No other changes are needed to this tab item.

2D Adaptive Clearing Heights tab. The only things that need to be changed in this tab is the Top Height and Bottom Height. For the Top Height, use the "From" drop-down and choose Model top from the list. Use the "From" drop-down and pick Selection from the list. Use the mouse to position the cursor over the face at the base of the raised island and left-click to select it.

2D Adaptive Clearing Passes tab. The changes for the passes tab are: tick the Multiple Depths checkbox; set the Maximum Roughing Stepdown to .125 inch; tick the Stock to Leave checkbox; change the Radial Stock to Leave to .06 inch and change the Axial Stock to Leave to 0 inches (see Figure 4-21).

2D Adaptive Clearing Linking tab. The only change to the linking tab is: change the Ramp Type to Plunge from the drop-down list.

Press OK. Select Setup1 in the browser and Simulate from the Toolbar Ribbon to playback the two paths created thus far. Make adjustment, if needed, by editing the path required.

Figure 4-21. 2D Adaptive clearing passes tab

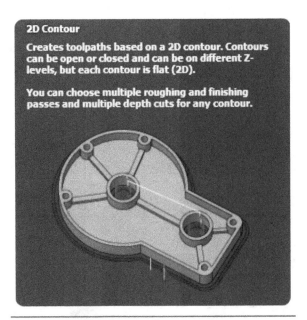

Figure 4-22. 2D Contour flyout

2D Contour

A little forethought is in order for the following operations. First, the raised island has .06 inch of material remaining on the radial contour that must be removed. Since the operation after Contour is the pocket, the idea is to use the same cutter for both—or as much as possible. A ½-inch diameter cutter is a functional choice. Select the drop-

down from the 2D icon on the Toolbar Ribbon and select 2D Contour from the list (Figure 4-22).

2D Contour Tool tab. Press the Select button to change the tool. From there, choose the Samples from Libraries, choose Type to activate filters and pick Flat. From the Samples/Inch Aluminum list, select the ½" – flat (½" Flat Endmill) (Figure 4-23).

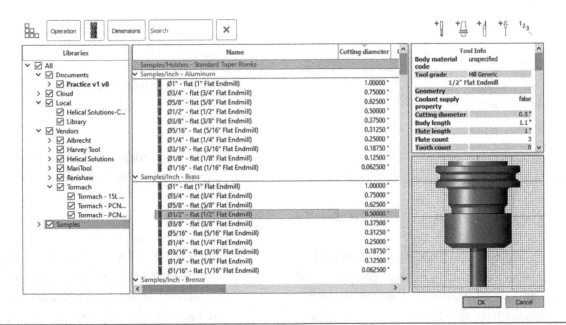

Figure 4-23. 2D Contour tool selection dialog

2D Contour Geometry tab. From the Geometry tab, select the chain for the raised island at the bottom boundary. Make sure the red direction arrow is on the outside of the contour (Figure 4-24).

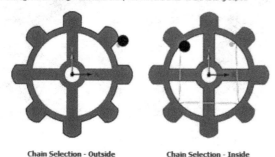

Contour Selection

Error: No contour selected to machine.

Select any Face, Edge or Sketch to define the machining boundary. Selecting a Face creates toolpaths on all the edges. Use Edge selection for areas with holes or pockets on the Face. Selecting the lower Edge will automatically set the reference for the cutting depth.

Chain Selection - Outside   Chain Selection - Inside

NOTE: Selecting a Face will detect all inside and outside boundaries.

Figure 4-24. 2D Contour selection flyout

2D Contour Heights tab. No changes are required in this tab.

2D Contour Passes tab. Make sure to change the Compensation Type to "In control". See Figure 4-25 for explanations for cutter compensation uses. There are no other changes required to this tab.

Compensation Type

Specifies the compensation type.

- **In computer** - Tool compensation is calculated automatically by the program, based on the selected tool diameter. The post-processed output contains the compensated path directly, instead of G41/G42 codes.
- **In control** - Tool compensation is *not* calculated, but rather G41/G42 codes are output to allow the operator to set the compensation amount and wear on the machine tool control.
- **Wear** - Works as if *In computer* was selected, but also outputs the G41/G42 codes. This lets the machine tool operator adjust for tool wear at the machine tool control by entering the difference in tool size as a negative number.
- **Inverse wear** - Identical to the *Wear* option, except that the wear adjustment is entered as a positive number.

Note that control compensation (including Wear and Inverse wear) is only done on finishing passes.

Figure 4-25. 2D Contour compensation type

2D Contour Linking tab. No changes are required in this tab.

Press OK to back-plot the tool path.

Select Setup1 in the Browser and Simulate to playback all of the paths finished up to this point. Make any adjustments needed and save the file.

2D Pocket

2D Pocket Tool tab. No changes are required in this tab because the same tool from the prior operation will be used.

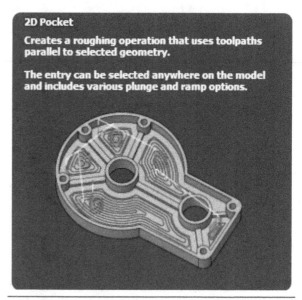

2D Pocket

Creates a roughing operation that uses toolpaths parallel to selected geometry.

The entry can be selected anywhere on the model and includes various plunge and ramp options.

Figure 4-26 2D Pocket flyout

2D Pocket Geometry tab. Use the mouse to select the bottom contour of the pocket in the center of the part. The red direction arrow should be on the inside of the pocket. If it is not, left-click it to change. No other changes are required in this tab.

2D Pocket Heights tab. No changes are required in this tab.

2D Pocket Passes tab. The only change to this tab is to untick the Stock to Leave checkbox.

2D Pocket Linking tab. Use the default values in all fields but take special note of the values in Leads & Transitions, and Ramp and adjust to specific needs in the future (see Figure 4-27).

Press OK to back-plot the tool path. Select Setup1 in the Browser and Simulate to playback all of the paths finished up to this point. Make any adjustments needed and save the file.

Figure 4-27. 2D Pocket linking tab

Slot

Because the slot width is ⁵⁄₁₆ inch in width, the tool will need to be the same size (Figure 4-28).

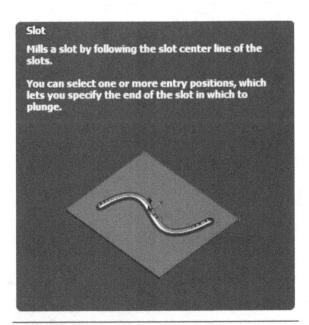

Figure 4-28. Slot flyout

Note: If the tolerance and surface finish for the slot requires, it may be better to program the slot as a contour operation.

Slot Tool tab. Use the methods described above to select a ⁵⁄₁₆-inch diameter flat end mill from the Samples Library and be sure the Coolant field is set to Flood.

Slot Geometry tab. Select the bottom edge inside of the slot for the geometry chain.

Slot Heights tab. Change the Top Height to the surface at the top of the slot. Use Selection from the drop-down list for the "From" value and pick the face of the stepped down surface.

Slot Passes tab. Tick the Multiple Depths checkbox and set the Maximum Roughing Stepdown to .125 inch.

Slot Linking tab. The settings may be left at the defaults as shown in Figure 4-29. Press OK to back-plot the tool path.

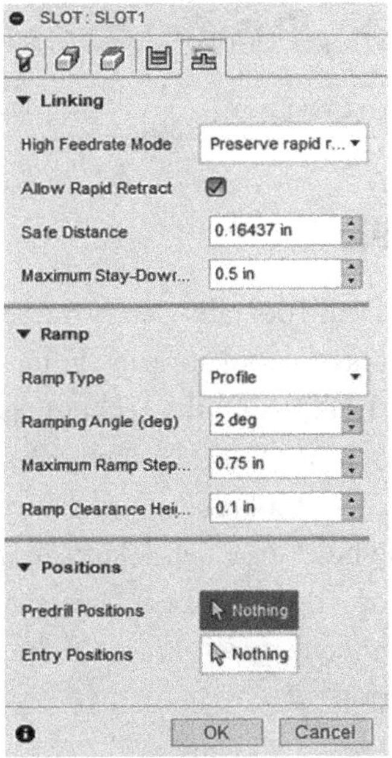

Figure 4-29 Slot linking tab

Select Setup1 in the Browser and Simulate to playback all of the paths finished up to this point. Make any adjustments needed and save the file.

Drilling

For the example, there are several holes that must be machined using drilling operations. The Drilling icon on the Toolbar Ribbon is used to initiate access to the different cycle variations, the first of which is Spot Drilling each hole. Press the Drilling icon to start the Drill dialog (Figure 4-30).

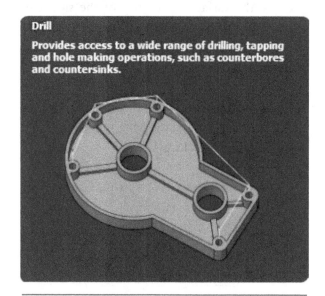

Figure 4-30.   Drill flyout

Spot Drill

Spot Drill Tool tab.   Press the Select button to access the tool libraries. Filter the tools to Spot Drill, using the techniques described earlier. Choose the ⅝-inch 90-degree Spot Drill. This tool is big enough to machine all of the holes to a depth sufficient to deburr the top edge yet not interfere with any of the other features. Press OK to accept the selection.

Spot Drill Geometry tab.   Use the default Hole Mode: Selected faces, to pick the chamfer geometry for each of the holes. The graphics in Figures 4-31 and 4-32 provide details and explain the remaining checkboxes in the Drill Geometry tab.

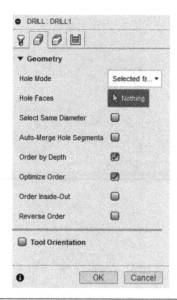

Figure 4-31.   Drill geometry tab

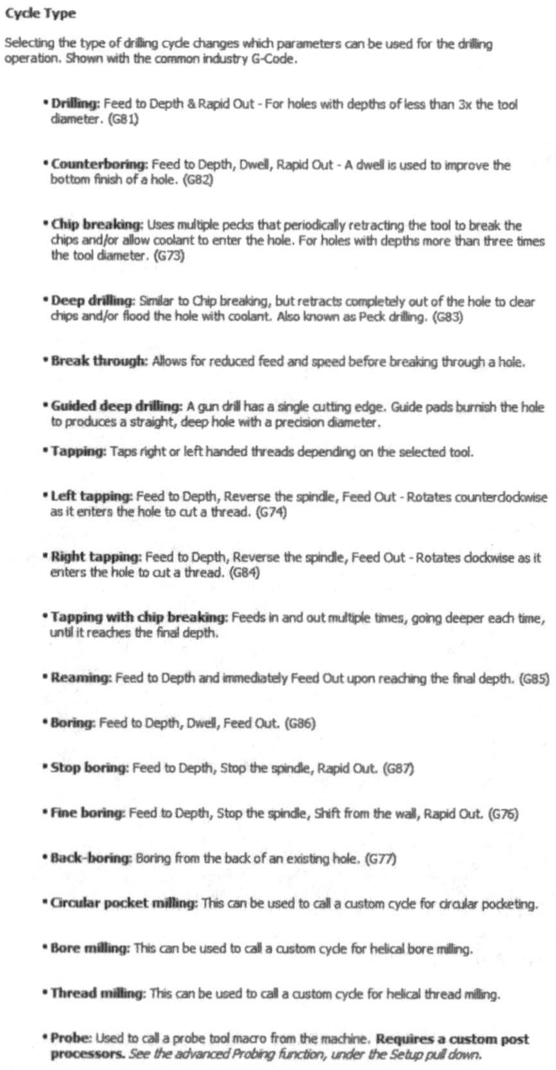

Cycle Type

Selecting the type of drilling cycle changes which parameters can be used for the drilling operation. Shown with the common industry G-Code.

- **Drilling:** Feed to Depth & Rapid Out - For holes with depths of less than 3x the tool diameter. (G81)

- **Counterboring:** Feed to Depth, Dwell, Rapid Out - A dwell is used to improve the bottom finish of a hole. (G82)

- **Chip breaking:** Uses multiple pecks that periodically retracting the tool to break the chips and/or allow coolant to enter the hole. For holes with depths more than three times the tool diameter. (G73)

- **Deep drilling:** Similar to Chip breaking, but retracts completely out of the hole to clear chips and/or flood the hole with coolant. Also known as Peck drilling. (G83)

- **Break through:** Allows for reduced feed and speed before breaking through a hole.

- **Guided deep drilling:** A gun drill has a single cutting edge. Guide pads burnish the hole to produces a straight, deep hole with a precision diameter.

- **Tapping:** Taps right or left handed threads depending on the selected tool.

- **Left tapping:** Feed to Depth, Reverse the spindle, Feed Out - Rotates counterclockwise as it enters the hole to cut a thread. (G74)

- **Right tapping:** Feed to Depth, Reverse the spindle, Feed Out - Rotates clockwise as it enters the hole to cut a thread. (G84)

- **Tapping with chip breaking:** Feeds in and out multiple times, going deeper each time, until it reaches the final depth.

- **Reaming:** Feed to Depth and immediately Feed Out upon reaching the final depth. (G85)

- **Boring:** Feed to Depth, Dwell, Feed Out. (G86)

- **Stop boring:** Feed to Depth, Stop the spindle, Rapid Out. (G87)

- **Fine boring:** Feed to Depth, Stop the spindle, Shift from the wall, Rapid Out. (G76)

- **Back-boring:** Boring from the back of an existing hole. (G77)

- **Circular pocket milling:** This can be used to call a custom cycle for circular pocketing.

- **Bore milling:** This can be used to call a custom cycle for helical bore milling.

- **Thread milling:** This can be used to call a custom cycle for helical thread milling.

- **Probe:** Used to call a probe tool macro from the machine. **Requires a custom post processors.** *See the advanced Probing function, under the Setup pull down.*

Figure 4-32.   Drill cycle type

The checkbox for Optimize Order, controls the paths between holes for the shortest distance in order to reduce cycle time (optimized). The checkbox for Reverse Order, changes the order from Inside-Out to Outside-In when the checkbox is ticked.

Spot Drill Heights tab. No changes are needed for this tab because of our selection method of the heights are preset accordingly.

Spot Drill Cycle tab. This tab controls the cycle type for the drilling. For the Spot Drill, the first in the list (default), setting of Drilling—rapid out, is correct. Figure 4-33 gives a detailed explanation for the different Drill Cycle Types.

Press OK to back-plot the tool path. Select Setup1 in the Browser and Simulate to playback all of the paths finished up to this point. Make any adjustments needed and save the file.

Hole Mode

Determines what you can select for your hole locations.

Selected Faces - *Model based selection.*

Cylinder Face Selection   **Chamfer Face Selection**

Selected Points - *Geometry based selection.*

Feature Edge Selection   **Sketch Point Selection**   **Sketch Circle Selection**

Diameter Range - *Size based selection.*

The system will evaluate the model based on the Minimum and Maximum Diameter values. Use this range to include or exclude hole sizes.
Example: Select all .250 - .2501 diameter holes for drilling and all .2505 - .2506 diameter holes for reaming.

Using Model based selections insures associativity to any future changes of the feature.

Note: Spot Drilling Options
* If you select the modeled chamfered face of a hole. The Bottom Height (depth) will automatically be determined to the theoretical tip.
* If you select by Diameter Range to include all holes. Set the Bottom Height from Chamfer Width and set the Chamfer Width value.

Auto-Merge Hole Segments

Check to merge multiple hole segments. When enable all hole segments are included to determine the starting height for drilling.

Example: If a hole was Spot Drilled or Counter Bored first, you may want to start drilling from a clearance above that machined area. Enabling Auto-Merge will start the drilling from above the highest hole segment.

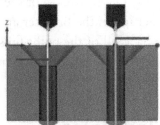

Left: Auto-Merge Disabled - Right: Auto-Merge Enabled
Blue line indicates the starting height for drilling.

Select Same Diameter

Check to select all holes with the same diameter as the currently select feature.

Example: If you activate this option, select a single 6mm hole and a single 12mm hole, every 6mm and 12mm hole on the part will automatically be selected.

A single selection can detect matching holes.

Note: This option is associative to the model. If additional holes with the same diameter are added to the model, regenerating the operation automatically includes the new holes.

Order by Depth

For holes with multiple Z start heights.

Changes the order from the highest to lowest, or lowest to highest. Unchecked, the order will start with the holes at the highest Z level and progressively move down. Check to reverse the order.

When Unchecked   **When Checked**
First hole is at the highest Z.   *First hole is at the lowest Z.*

Order Inside-Out

Check to change the order of the toolpath to start at the inner most position of the part. Unchecked and the toolpath will start from an outside edge and work it's way across the part.

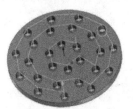

When Checked
First hole is at the inner most location.

Figure 4-33.   Drill geometry tab settings

Drill ⅜ Diameter Holes

There are three holes that are ⅜ inch diameter in the model, plus there are two more holes that can be predrilled at the same time on the opposite corners. All of these five holes can be set in this operation.

Drill ⅜ Diameter Tool tab. Press the Drilling icon to start the Drill dialog.

Press the Select button to access the tool Libraries. Filter the tools to Drill, using the techniques described earlier. Choose the ⅜" 118 degree Drill and press OK to accept it.

Drill ⅜ Diameter Geometry tab.

1. Change the Hole Mode to Diameter Range.
2. Set the Minimum Diameter to .375 inch.
3. Set the Maximum Diameter to .75 inch.
4. Tick the checkbox for Order by Depth.
5. Tick the checkbox for Optimize Order.
6. Ensure the settings match Figure 4-34, then proceed to the Geometry tab.

Figure 4-34. Drill diameter hole range

Drill ⅜ Diameter Heights tab. The only changes to this tab are:

1. Tick the checkbox for Drill Tip Through Bottom.
2. Set the Break-Through Depth to .03 inch.
3. Ensure the settings match Figure 4-35, then proceed to the Heights tab.

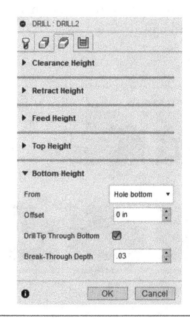

Figure 4-35. Drill bottom heights tab

Drill ⅜-inch Diameter Cycle tab.

1. Change the Cycle Type to Deep drilling—full retract.
2. Change the Pecking Depth to ⅔ of the tool diameter (.25 inch).

Ensure the settings match Figure 4-36, then press OK to back-plot the tool path. Select Setup1 in the Browser and Simulate to playback all of the paths finished up to this point. Make any adjustments needed and save the file.

Tap Drilled Holes

The tap drilled holes for the three 8-32 UNC threaded holes must be completed next.

Figure 4-36. Drill deep pecking cycle type

Drill #29 Tap Drill Tool tab.

1. Press the Drilling icon to start the Drill dialog.
2. Press the Select button to access the tool libraries.
3. Filter the tools to Drill, using the techniques described earlier.
4. Choose the .136-inch 118-degree (#29) Drill and press OK to accept it.

Drill #29 Tap Drill Geometry tab.

1. Use the default Hole Mode of Select faces.
2. Select each of the three threaded hole cylinder faces.

Drill #29 Tap Drill Heights tab. No changes are needed to this tab. The depth is controlled by the modeled feature in this case.

Drill #29 Tap Drill Cycle tab.

1. Change the Cycle Type to Deep drilling—full retract.
2. Change the Pecking Depth to ⅔ of the tool diameter (.09").
3. Press OK to back-plot the tool path.

Select Setup1 in the Browser and Simulate to playback all of the paths finished up to this point. Make any adjustments needed and save the file.

Tapped Holes

The three 8-32 UNF tapped holes are completed next.

Tap Drill Tool tab. Using the methods described above, select the Tap Right Hand to filter the tools in the tool Library.
Select the #8-32 UNC tap from the list and press OK.

Tap Geometry tab. Select the three tapped hole cylinder faces.

Tap Heights tab. Change the Bottom Height setting for "From" to Hole Top.
Change the Offset to -.312".

Tap Cycle tab.

1. Change the Cycle Type to Right tapping.
2. Press OK to back-plot the tool path.

Select Setup1 in the Browser and Simulate to playback all of the paths finished up to this point. Make any adjustments needed and save the file.

Counterbored Holes

The two counter bored holes are ½ inch in diameter so tool #2 can be used again (the tap drilled holes for the three 8-32 UNC threaded holes must be completed next). In this case, a drilling cycle will be used to plunge to the required depth.

Counterbore Drill Tool tab. Press the Drilling icon to start the Drill dialog.
Press the Select button and choose the ½" end mill that was used earlier (T2) and press OK.

Counterbore Drill Geometry tab. Use the default Hole Mode of Select faces. Select each of the two counterbored hole cylinder faces.

Counterbore Drill Heights tab. No changes are needed to this tab. The depth is controlled by the modeled feature in this case.

Counterbore Drill Cycle tab.

1. Change the Cycle Type to Counterboring – dwell rapid out (Figure 4-37).
2. Change the Dwelling Period to 2 seconds.
3. Press OK to back-plot the tool path.

Select Setup1 in the Browser and Simulate to playback all of the paths finished up to this point. Make any adjustments needed and save the file.

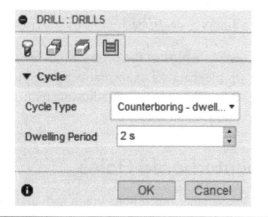

Figure 4-37 Counterbore cycle tab

Bored Holes

2D Bore Tool tab. Press the 2D drop-down from the 2D icon and select Bore to start the Bore dialog.

Press the Select button and choose the ½" end mill that was used earlier (T2) and press OK.

2D Bore Geometry tab. Use the Circular Face Selections Mode to select each of the two bored hole cylinder faces.

2D Bore Heights tab. Change the Bottom Height "From" Offset to –.03 inch.

2D Bore Passes tab. Tick the checkbox for Finishing Passes and leave the Stepover at .010 inch.

2D Bore Linking tab.

1. Change the Vertical Lead In/Out Radius to 0.
2. Press OK to back-plot the tool path.

Select Setup1 in the Browser and Simulate to playback all of the paths finished up to this point. Make any adjustments needed and save the file.

2D Chamfer

The final operation is chamfering the remaining sharp edges. To do so, select the 2D drop-down from the Toolbar Ribbon and choose the 2D Chamfer from the list (Figure 4-38).

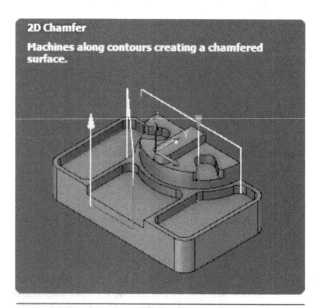

Figure 4-38 2D Chamfer flyout

2D Chamfer Tool tab. You will notice that the ⅝ in. x 90-degree Spot Drill (T4) is automatically selected. No further action is required.

2D Chamfer Geometry tab. Select the top edges for the outside step, island, pocket and slot (4 chains). No further changes are required for the Heights, Passes or Linking tabs. Press OK and the paths will be generated. Select Setup1 in the Browser and Simulate to playback all of the paths finished up to this point. Make any adjustments needed and save the file.

Post Processing

Now all of the paths are ready to be processed into CNC code (G-code) for the machine tool.

Press the Post Process icon in the ACTIONS section of the Toolbar Ribbon. The Post Process dialog will display (Figure 4-39).

Figure 4-39. Post process flyout

For this example, you should note:

- Under the Post Configuration, Milling has been selected from the drop-down list and

Haas Automation has been selected from the list of vendors (Figure 4-40).

- The specific post, Haas (preNGC), has been selected.
- The Output folder shown is the default. This is commonly changed to a location that meets company requirements.
- Under Program Settings, the program number and Program Comment determined at the beginning of the CAM process in the Setup are listed.
- In the window to the right is the Property list. These values can be changed by clicking on the item and changing the value as desired.
- The tick-box is checked for the Open NC file in editor so the file will open in the "Brackets" editor once the Post button is pressed and a file location is selected.
- The output will be displayed similar to what is shown below. This is the code that will be loaded into the machine to run. *Note:* The output may vary based on the specific post processor chosen.

```
%
O21519 (Mill Example)
(Using high feed G1 F500. instead of G0.)
(T1 D=2. CR=0. - ZMIN=-0.29 - face mill)
(T2 D=0.5 CR=0. - ZMIN=-1.07 - flat end
mill)
(T3 D=0.3125 CR=0. - ZMIN=-0.54 - flat
end mill)
(T4 D=0.625 CR=0. TAPER=90deg -
ZMIN=-0.586 - spot drill)
(T5 D=0.375 CR=0. TAPER=118deg -
ZMIN=-1.1827 - drill)
(T6 D=0.164 CR=0. - ZMIN=-0.477 - right
hand tap)
N10 G90 G94 G17
N15 G20
N20 G53 G0 Z0.
```

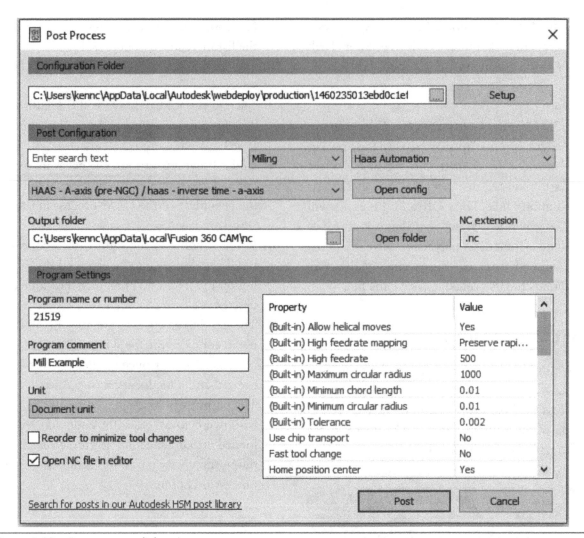

Figure 4-40.   Post process dialog

(Face1)

N30 T1 M6

N35 S6500 M3

N40 G54

N45 M11

N50 G0 A0.

N55 M10

N60 M8

N75 G0 X3.3 Y-1.6437

N80 G43 Z0.6 H1

N85 T2

N90 G0 Z0.2

N95 G1 Z0.16 F65.

N100 G18 G3 X3.1 Z-0.04 I-0.2 K0. F65.

N105 G1 X2. F65.

N110 X-2. F65.

N115 G17 G2 Y-0.3063 I0. J0.6687 F65.

N120 G1 X2. F65.

N125 G3 Y1.031 I0. J0.6687 F65.

N130 G1 X-2. F65.

N135 G18 G3 X-2.2 Z0.16 I0. K0.2 F65.

N140 G0 Z0.6

(2D Adaptive2)

N150 G1 X-2.8947 Y2.9052 F500.

N155 G0 Z0.56

N160 Z0.16

N165 Z0.035

N170 G1 X-2.8945 Y2.905 Z0.0238 F65.

lines N175 –N4655 that perform the 2D Adaptive milling are truncated here to save space.

N4655 G0 Z0.56

N4660 M5

N4665 G53 G0 Z0.

(2D Contour1)

N4675 M9

N4680 M1

N4685 T2 M6

N4690 S7640 M3

N4695 G54

N4700 M11

N4705 G0 A0.

N4710 M10

N4715 M8

N4730 G0 X1.145 Y-2.0509

N4735 G43 Z0.6 H2

N4740 T3

N4745 G0 Z0.2

N4750 G1 Z0.0394 F30.

N4755 Z-0.24 F30.

N4760 G19 G3 Y-2.0009 Z-0.29 J0.05 K0. F30.

N4765 G17

N4770 G1 G41 X1.42 Y-1.9509 D2 F92.

N4775 G3 X1.095 Y-1.6259 I-0.325 J0. F92.

N4780 G1 X-0.5042 F92.

N4785 X-1.375 Y-0.9615 F92.

N4790 Y1. F92.

N4795 G2 X-1.1639 Y1.3373 I0.375 J0. F92.

N4800 G1 X1.625 Y0.0596 F92.

N4805 Y-1.095 F92.

N4810 G3 X1.095 Y-1.6259 I0. J-0.53 F92.

N4815 X1.4205 Y-1.9503 I0.325 J0.0005 F92.

N4825 G1 G40 X1.4701 Y-1.6753 F92.

N4830 X1.4757 Z-0.2897 F92.

N4835 X1.4812 Y-1.6752 Z-0.2887 F92.

N4840 X1.4866 Z-0.2872 F92.

N4845 X1.4918 Z-0.285 F92.

N4850 X1.4967 Z-0.2823 F92.

N4855 X1.5013 Z-0.2791 F92.

N4860 X1.5054 Z-0.2754 F92.

N4865 X1.5092 Z-0.2712 F92.

N4870 X1.5124 Z-0.2666 F92.

N4875 X1.5151 Z-0.2617 F92.

N4880 X1.5173 Z-0.2565 F92.

N4885 X1.5188 Z-0.2511 F92.

N4890 X1.5198 Z-0.2456 F92.

N4895 X1.5201 Z-0.24 F92.

N4900 G0 Z0.6

(2D Pocket1)

N4910 G1 X0.2505 Y-0.16 F500.

N4915 G0 Z0.6

N4920 Z0.2

N4925 G1 Z0.15 F92.

N4930 G3 X0.2513 Y-0.1591 Z0.1391 I-0.1755 J0.16 F92.

N4935 X0.2537 Y-0.1565 Z0.1287 I-0.1763 J0.1591 F92.

N4940 X0.2574 Y-0.1521 Z0.1194 I-0.1787 J0.1565 F92.

N4945 X0.2621 Y-0.1462 Z0.1115 I-0.1824 J0.1521 F92.

N4950 X0.2676 Y-0.1389 Z0.1055 I-0.1871 J0.1462 F92.

N4955 X0.2734 Y-0.1305 Z0.1016 I-0.1926 J0.1389 F92.

N4960 X0.2792 Y-0.1213 Z0.1 I-0.1984 J0.1305 F92.

N4965 X0.2678 Y-0.1386 Z0.0486 I-0.2042 J0.1213 F92.

N4970 X0.255 Y-0.1549 Z-0.0028 I-0.1928 J0.1386 F92.

N4975 X0.2408 Y-0.17 Z-0.0542 I-0.18 J0.1549 F92.

N4980 X0.2254 Y-0.1838 Z-0.1055 I-0.1658 J0.17 F92.

N4985 X0.2088 Y-0.1962 Z-0.1569 I-0.1504 J0.1838 F92.

N4990 X0.3125 Y0. Z-0.165 I-0.1338 J0.1962 F92.

N4995 X0.1625 I-0.075 J0. F92.

N5000 X0.3125 I0.075 J0. F92.

N5005 X-0.3125 I-0.3125 J0. F92.

N5010 X0.3125 I0.3125 J0. F92.

N5015 X0.3113 Y0.011 Z-0.1637 I-0.05 J0. F92.

N5020 X0.3079 Y0.021 Z-0.16 I-0.0488 J-0.011 F92.

N5025 X0.3031 Y0.0292 Z-0.1541 I-0.0454 J-0.021 F92.

N5030 X0.298 Y0.0352 Z-0.1462 I-0.0406 J-0.0292 F92.

N5035 X0.2935 Y0.0392 Z-0.1367 I-0.0355 J-0.0352 F92.

N5040 X0.2906 Y0.0414 Z-0.1261 I-0.031 J-0.0392 F92.

N5045 X0.2895 Y0.0421 Z-0.115 I-0.0281 J-0.0414 F92.

N5050 G0 Z0.6

N5055 M5

N5060 G53 G0 Z0.

(Slot1)

N5070 M9

N5075 M1

N5080 T3 M6

N5085 S10000 M3

N5090 G54

N5095 M11

N5100 G0 A0.

N5105 M10

N5110 M8

N5125 G0 X-0.2031 Y1.5571

N5130 G43 Z0.6 H3

N5135 T4

N5140 G0 Z-0.09

N5145 G1 Z-0.19 F75.

N5150 X1.1606 Y0.9323 Z-0.235 F75.

N5155 X-0.2031 Y1.5571 Z-0.28 F75.

N5160 X1.1606 Y0.9323 Z-0.325 F75.

N5165 X-0.2031 Y1.5571 Z-0.37 F75.

N5170 X1.1606 Y0.9323 Z-0.415 F75.

N5175 X-0.2031 Y1.5571 F75.

N5180 X1.1606 Y0.9323 Z-0.4567 F75.

N5185 X-0.2031 Y1.5571 Z-0.4983 F75.

N5190 X1.1606 Y0.9323 Z-0.54 F75.

N5195 X-0.2031 Y1.5571 F75.

N5200 G0 Z0.6

N5205 M5

N5210 G53 G0 Z0.

(Drill1)

N5220 M9

N5225 M1

N5230 T4 M6

N5235 S6110 M3

N5240 G54

N5245 M11

N5250 G0 A0.

N5255 M10

N5260 M8

N5275 G0 X1.625 Y-1.625

N5280 G43 Z0.6 H4

N5285 T5

N5295 G0 Z0.2

N5300 G98 G81 X1.625 Y-1.625 Z-0.555 R-0.09 F30.

N5305 G80

N5315 G1 X-1.5 Y-1.5 Z0.2 F500.

N5320 G98 G81 X-1.5 Y-1.5 Z-0.586
R-0.09 F30.

N5325 G80

N5335 G1 X-1.625 Y1.625 Z0.2 F500.

N5340 G98 G81 X-1.625 Y1.625 Z-0.555
R-0.09 F30.

N5345 G80

N5355 G1 X1.5 Y1.5 Z0.2 F500.

N5360 G98 G81 X1.5 Y1.5 Z-0.586
R-0.09 F30.

N5365 G80

N5375 G1 X0. Y0.35 Z0.2 F500.

N5380 G98 G81 X0. Y0.35 Z-0.2472
R0.035 F30.

N5385 X0.3031 Y-0.175

N5390 G80

N5400 G1 X0. Y0. Z0.2 F500.

N5405 G98 G81 X0. Y0. Z-0.3675 R0.035
F30.

N5410 G80

N5420 G1 X-0.3031 Y-0.175 Z0.2 F500.

N5425 G98 G81 X-0.3031 Y-0.175
Z-0.2472 R0.035 F30.

N5430 G80

N5435 G0 Z0.6

N5440 M5

N5445 G53 G0 Z0.

(Drill2)

N5455 M9

N5460 M1

N5465 T5 M6

N5470 S3060 M3

N5475 G54

N5480 M11

N5485 G0 A0.

N5490 M10

N5495 M8

N5510 G0 X-1.625 Y1.625

N5515 G43 Z0.6 H5

N5520 T6

N5530 G0 Z0.2

N5535 G98 G83 X-1.625 Y1.625 Z-1.1827
R-0.105 Q0.25 F29.

N5540 X1.5 Y1.5

N5545 X1.625 Y-1.625

N5550 X-1.5 Y-1.5

N5555 G80

N5565 G1 X0. Y0. Z0.2 F500.

N5570 G98 G83 X0. Y0. Z-1.1827 R0.02
Q0.25 F29.

N5575 G80

N5580 G0 Z0.6

(Drill3)

N5585 S8430 M3

N5595 G1 X-0.3031 Y-0.175 F500.

N5600 G0 Z0.6

N5610 Z0.2

N5615 G98 G83 X-0.3031 Y-0.175 Z-0.665
R0.02 Q0.09 F29.

N5620 X0. Y0.35

N5625 X0.3031 Y-0.175

N5630 G80

N5635 G0 Z0.6

N5640 M5

N5645 G53 G0 Z0.

(Drill4)

N5655 M9

N5660 M1

N5665 T6 M6

N5670 S500 M3

N5675 G54

N5680 M11

N5685 G0 A0.

N5690 M10

N5695 M8

N5710 G0 X-0.3031 Y-0.175

N5715 G43 Z0.6 H6

N5720 T2

N5730 G0 Z0.2

N5735 G98 G84 X-0.3031 Y-0.175
Z-0.477 R0.035 F15.625

N5740 X0. Y0.35

N5745 X0.3031 Y-0.175

N5750 G80

N5755 G0 Z0.6

N5760 M5

N5765 G53 G0 Z0.

(Drill5)

N5775 M9

N5780 M1

N5785 T2 M6

N5790 S7640 M3

N5795 G54

N5800 M11

N5805 G0 A0.

N5810 M10

N5815 M8

N5830 G0 X-1.625 Y1.625

N5835 G43 Z0.6 H2

N5840 T4

N5850 G0 Z0.2

N5855 G98 G82 X-1.625 Y1.625 Z-0.665
R-0.105 P2000 F30.

N5860 X1.625 Y-1.625

N5865 G80

N5870 G0 Z0.6

(Bore1)

N5880 G1 X-1.501 Y-1.49 F500.

N5885 G0 Z0.6

N5890 Z-0.225

N5895 G1 Z-0.305 F40.

N5900 X-1.511 F40.

N5905 G3 X-1.521 Y-1.5 I0. J-0.01 F40.

lines N5910 –N6275, that perform the Bore1
operation are truncated here to save space.

N6280 G0 Z0.6

N6285 M5

N6290 G53 G0 Z0.

(2D Chamfer1)

N6300 M9

N6305 M1

N6310 T4 M6

N6315 S6110 M3

N6320 G54

N6325 M11

N6330 G0 A0.

N6335 M10

N6340 M8

N6355 G0 X0.475 Y0.

N6360 G43 Z0.6 H4

N6365 T1

N6370 G0 Z0.2

N6375 G1 Z0.08 F30.

N6380 Z-0.12 F30.

N6385 X0.5375 F40.

N6390 G3 X-0.5375 I-0.5375 J0. F40.

N6395 X0.5375 I0.5375 J0. F40.

N6400 G1 X0.475 F40.

N6405 G0 Z0.2

N6410 G1 X1.0802 Y-1.7134 F500.

N6415 Z0.08 F30.

N6420 Z-0.12 F30.

N6425 Y-1.6509 F40.

N6430 X-0.4991 F40.

N6435 G2 X-0.5234 Y-1.6427 I0. J0.04 F40.

N6440 G1 X-1.3843 Y-0.9859 F40.

N6445 G2 X-1.4 Y-0.9541 I0.0243 J0.0318 F40.

N6450 G1 Y1. F40.

N6455 G2 X-1.1818 Y1.3563 I0.4 J0. F40.

N6460 X-1.1469 Y1.357 I0.0182 J-0.0356 F40.

N6465 G1 X1.6267 Y0.0864 F40.

N6470 G2 X1.65 Y0.05 I-0.0167 J-0.0364 F40.

N6475 G1 Y-1.0802 F40.

N6480 G2 X1.6111 Y-1.1202 I-0.04 J0. F40.

N6485 G3 X1.1202 Y-1.6119 I0.0139 J-0.5048 F40.

N6490 G2 X1.0802 Y-1.6509 I-0.04 J0.001 F40.

N6495 G1 Y-1.7134 F40.

N6500 G0 Z0.6

N6505 G1 X-0.1817 Y1.6225 F500.

N6510 G0 Z0.2

N6515 G1 Z0.08 F30.

N6520 Z-0.37 F30.

N6525 X-0.1623 Y1.6819 F40.

N6530 G3 X-0.2577 Y1.4378 I-0.0408 J-0.1248 F40.

N6535 G1 X1.106 Y0.813 F40.

N6540 G3 X1.2153 Y1.0517 I0.0547 J0.1193 F40.

N6545 G1 X-0.1484 Y1.6764 F40.

N6550 G3 X-0.1623 Y1.6819 I-0.0547 J-0.1193 F40.

N6555 G1 X-0.1817 Y1.6225 F40.

N6560 G0 Z0.2

N6565 G1 X0. Y2.0875 F500.

N6570 Z0.08 F30.

N6575 Z-0.37 F30.

N6580 Y2.025 F40.

N6585 X1.985 F40.

N6590 G2 X2.025 Y1.985 I0. J-0.04 F40.

N6595 G1 Y-1.985 F40.

N6600 G2 X1.985 Y-2.025 I-0.04 J0. F40.

N6605 G1 X-1.985 F40.

N6610 G2 X-2.025 Y-1.985 I0. J0.04 F40.

N6615 G1 Y1.985 F40.

N6620 G2 X-1.985 Y2.025 I0.04 J0. F40.

N6625 G1 X0. F40.

N6630 Y2.0875 F40.

N6635 G0 Z0.6

N6640 M5

N6645 M9

N6650 G53 G0 Z0.

N6655 M11

N6660 G0 A0.

N6665 M10

N6675 X0.

N6680 G53 Y0.

N6685 M30

%

Setup Documentation

Part of process planning documentation is the setup sheet for the CNC Mill program. This gives the operator a go-to source for tooling and operation related information (Charts 4-1a and 4-1b).

1. Press the Setup Sheet icon in the ACTIONS section of the Toolbar Ribbon.
2. Select the destination folder.
3. The setup sheet will be generated in .html format.
4. To save the file, right-click in the white space and choose Save as . . . to save it to a location of your choosing.

Setup

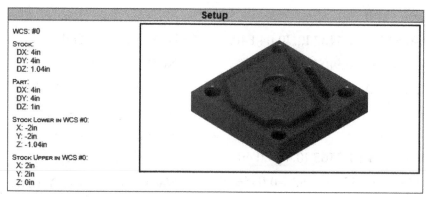

WCS: #0

Stock:
DX: 4in
DY: 4in
DZ: 1.04in

Part:
DX: 4in
DY: 4in
DZ: 1in

Stock Lower in WCS #0:
X: -2in
Y: -2in
Z: -1.04in

Stock Upper in WCS #0:
X: 2in
Y: 2in
Z: 0in

Total

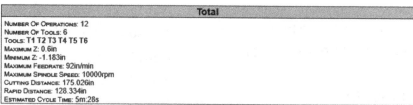

Number Of Operations: 12
Number Of Tools: 6
Tools: T1 T2 T3 T4 T5 T6
Maximum Z: 0.6in
Minimum Z: -1.183in
Maximum Feedrate: 92in/min
Maximum Spindle Speed: 10000rpm
Cutting Distance: 175.026in
Rapid Distance: 128.334in
Estimated Cycle Time: 5m:28s

Tools

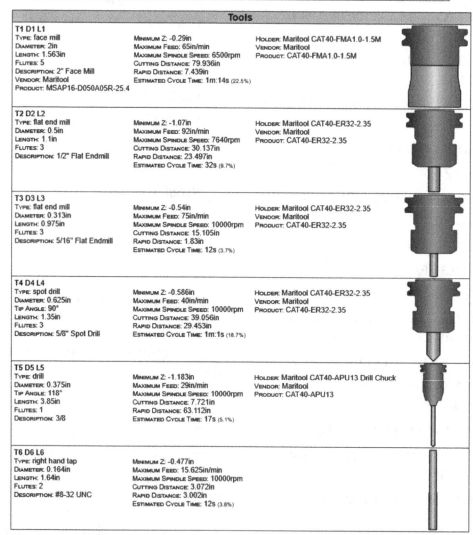

T1 D1 L1
Type: face mill
Diameter: 2in
Length: 1.563in
Flutes: 5
Description: 2" Face Mill
Vendor: Maritool
Product: MSAP16-D050A05R-25.4

Minimum Z: -0.29in
Maximum Feed: 65in/min
Maximum Spindle Speed: 6500rpm
Cutting Distance: 79.936in
Rapid Distance: 7.439in
Estimated Cycle Time: 1m:14s (22.5%)

Holder: Maritool CAT40-FMA1.0-1.5M
Vendor: Maritool
Product: CAT40-FMA1.0-1.5M

T2 D2 L2
Type: flat end mill
Diameter: 0.5in
Length: 1.1in
Flutes: 3
Description: 1/2" Flat Endmill

Minimum Z: -1.07in
Maximum Feed: 92in/min
Maximum Spindle Speed: 7640rpm
Cutting Distance: 30.137in
Rapid Distance: 23.497in
Estimated Cycle Time: 32s (9.7%)

Holder: Maritool CAT40-ER32-2.35
Vendor: Maritool
Product: CAT40-ER32-2.35

T3 D3 L3
Type: flat end mill
Diameter: 0.313in
Length: 0.975in
Flutes: 3
Description: 5/16" Flat Endmill

Minimum Z: -0.54in
Maximum Feed: 75in/min
Maximum Spindle Speed: 10000rpm
Cutting Distance: 15.105in
Rapid Distance: 1.83in
Estimated Cycle Time: 12s (3.7%)

Holder: Maritool CAT40-ER32-2.35
Vendor: Maritool
Product: CAT40-ER32-2.35

T4 D4 L4
Type: spot drill
Diameter: 0.625in
Tip Angle: 90°
Length: 1.35in
Flutes: 3
Description: 5/8" Spot Drill

Minimum Z: -0.586in
Maximum Feed: 40in/min
Maximum Spindle Speed: 10000rpm
Cutting Distance: 39.056in
Rapid Distance: 29.453in
Estimated Cycle Time: 1m:1s (18.7%)

Holder: Maritool CAT40-ER32-2.35
Vendor: Maritool
Product: CAT40-ER32-2.35

T5 D5 L5
Type: drill
Diameter: 0.375in
Tip Angle: 118°
Length: 3.85in
Flutes: 1
Description: 3/8

Minimum Z: -1.183in
Maximum Feed: 29in/min
Maximum Spindle Speed: 10000rpm
Cutting Distance: 7.721in
Rapid Distance: 63.112in
Estimated Cycle Time: 17s (5.1%)

Holder: Maritool CAT40-APU13 Drill Chuck
Vendor: Maritool
Product: CAT40-APU13

T6 D6 L6
Type: right hand tap
Diameter: 0.164in
Length: 1.64in
Flutes: 2
Description: #8-32 UNC

Minimum Z: -0.477in
Maximum Feed: 15.625in/min
Maximum Spindle Speed: 10000rpm
Cutting Distance: 3.072in
Rapid Distance: 3.002in
Estimated Cycle Time: 12s (3.6%)

Chart 4-1a. CNC Mill Example Setup Sheet page 1

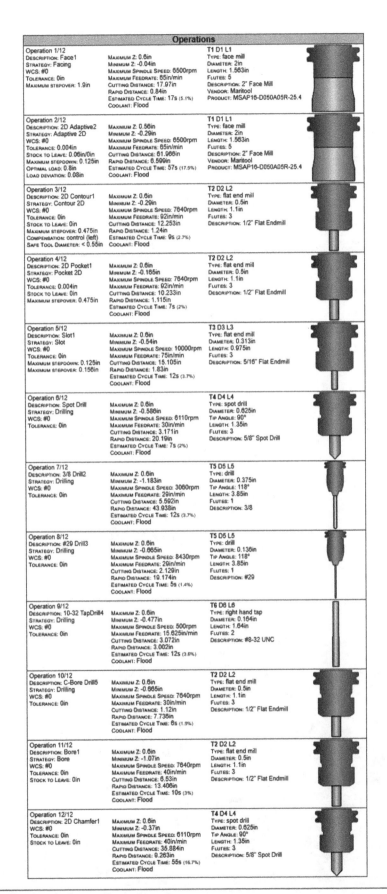

Operations

Operation 1/12
DESCRIPTION: Face1
STRATEGY: Facing
WCS: #0
TOLERANCE: 0in
MAXIMUM STEPOVER: 1.9in

MAXIMUM Z: 0.6in
MINIMUM Z: -0.04in
MAXIMUM SPINDLE SPEED: 6500rpm
MAXIMUM FEEDRATE: 65in/min
CUTTING DISTANCE: 17.97in
RAPID DISTANCE: 0.84in
ESTIMATED CYCLE TIME: 17s (5.1%)
COOLANT: Flood

T1 D1 L1
TYPE: face mill
DIAMETER: 2in
LENGTH: 1.563in
FLUTES: 5
DESCRIPTION: 2" Face Mill
VENDOR: Maritool
PRODUCT: MSAP16-D050A05R-25.4

Operation 2/12
DESCRIPTION: 2D Adaptive2
STRATEGY: Adaptive 2D
WCS: #0
TOLERANCE: 0.004in
STOCK TO LEAVE: 0.06in/0in
MAXIMUM STEPDOWN: 0.125in
OPTIMAL LOAD: 0.8in
LOAD DEVIATION: 0.08in

MAXIMUM Z: 0.56in
MINIMUM Z: -0.29in
MAXIMUM SPINDLE SPEED: 6500rpm
MAXIMUM FEEDRATE: 65in/min
CUTTING DISTANCE: 61.966in
RAPID DISTANCE: 6.599in
ESTIMATED CYCLE TIME: 57s (17.5%)
COOLANT: Flood

T1 D1 L1
TYPE: face mill
DIAMETER: 2in
LENGTH: 1.563in
FLUTES: 5
DESCRIPTION: 2" Face Mill
VENDOR: Maritool
PRODUCT: MSAP16-D050A05R-25.4

Operation 3/12
DESCRIPTION: 2D Contour1
STRATEGY: Contour 2D
WCS: #0
TOLERANCE: 0in
STOCK TO LEAVE: 0in
MAXIMUM STEPOVER: 0.475in
COMPENSATION: control (left)
SAFE TOOL DIAMETER: < 0.55in

MAXIMUM Z: 0.6in
MINIMUM Z: -0.29in
MAXIMUM SPINDLE SPEED: 7640rpm
MAXIMUM FEEDRATE: 92in/min
CUTTING DISTANCE: 12.253in
RAPID DISTANCE: 1.24in
ESTIMATED CYCLE TIME: 9s (2.7%)
COOLANT: Flood

T2 D2 L2
TYPE: flat end mill
DIAMETER: 0.5in
LENGTH: 1.1in
FLUTES: 3
DESCRIPTION: 1/2" Flat Endmill

Operation 4/12
DESCRIPTION: 2D Pocket1
STRATEGY: Pocket 2D
WCS: #0
TOLERANCE: 0.004in
STOCK TO LEAVE: 0in
MAXIMUM STEPOVER: 0.475in

MAXIMUM Z: 0.6in
MINIMUM Z: -0.165in
MAXIMUM SPINDLE SPEED: 7640rpm
MAXIMUM FEEDRATE: 92in/min
CUTTING DISTANCE: 10.233in
RAPID DISTANCE: 1.115in
ESTIMATED CYCLE TIME: 7s (2%)
COOLANT: Flood

T2 D2 L2
TYPE: flat end mill
DIAMETER: 0.5in
LENGTH: 1.1in
FLUTES: 3
DESCRIPTION: 1/2" Flat Endmill

Operation 5/12
DESCRIPTION: Slot1
STRATEGY: Slot
WCS: #0
TOLERANCE: 0in
MAXIMUM STEPDOWN: 0.125in
MAXIMUM STEPOVER: 0.156in

MAXIMUM Z: 0.6in
MINIMUM Z: -0.54in
MAXIMUM SPINDLE SPEED: 10000rpm
MAXIMUM FEEDRATE: 75in/min
CUTTING DISTANCE: 15.105in
RAPID DISTANCE: 1.83in
ESTIMATED CYCLE TIME: 12s (3.7%)
COOLANT: Flood

T3 D3 L3
TYPE: flat end mill
DIAMETER: 0.313in
LENGTH: 0.975in
FLUTES: 3
DESCRIPTION: 5/16" Flat Endmill

Operation 6/12
DESCRIPTION: Spot Drill
STRATEGY: Drilling
WCS: #0
TOLERANCE: 0in

MAXIMUM Z: 0.6in
MINIMUM Z: -0.586in
MAXIMUM SPINDLE SPEED: 6110rpm
MAXIMUM FEEDRATE: 30in/min
CUTTING DISTANCE: 3.171in
RAPID DISTANCE: 20.19in
ESTIMATED CYCLE TIME: 7s (2%)
COOLANT: Flood

T4 D4 L4
TYPE: spot drill
DIAMETER: 0.625in
TIP ANGLE: 90°
LENGTH: 1.35in
FLUTES: 3
DESCRIPTION: 5/8" Spot Drill

Operation 7/12
DESCRIPTION: 3/8 Drill2
STRATEGY: Drilling
WCS: #0
TOLERANCE: 0in

MAXIMUM Z: 0.6in
MINIMUM Z: -1.183in
MAXIMUM SPINDLE SPEED: 3060rpm
MAXIMUM FEEDRATE: 29in/min
CUTTING DISTANCE: 5.592in
RAPID DISTANCE: 43.938in
ESTIMATED CYCLE TIME: 12s (3.7%)
COOLANT: Flood

T5 D5 L5
TYPE: drill
DIAMETER: 0.375in
TIP ANGLE: 118°
LENGTH: 3.85in
FLUTES: 1
DESCRIPTION: 3/8

Operation 8/12
DESCRIPTION: #29 Drill3
STRATEGY: Drilling
WCS: #0
TOLERANCE: 0in

MAXIMUM Z: 0.6in
MINIMUM Z: -0.665in
MAXIMUM SPINDLE SPEED: 8430rpm
MAXIMUM FEEDRATE: 29in/min
CUTTING DISTANCE: 2.129in
RAPID DISTANCE: 19.174in
ESTIMATED CYCLE TIME: 5s (1.4%)
COOLANT: Flood

T5 D5 L5
TYPE: drill
DIAMETER: 0.136in
TIP ANGLE: 118°
LENGTH: 3.85in
FLUTES: 1
DESCRIPTION: #29

Operation 9/12
DESCRIPTION: 10-32 TapDrill4
STRATEGY: Drilling
WCS: #0
TOLERANCE: 0in

MAXIMUM Z: 0.6in
MINIMUM Z: -0.477in
MAXIMUM SPINDLE SPEED: 500rpm
MAXIMUM FEEDRATE: 15.625in/min
CUTTING DISTANCE: 3.072in
RAPID DISTANCE: 3.002in
ESTIMATED CYCLE TIME: 12s (3.6%)
COOLANT: Flood

T6 D6 L6
TYPE: right hand tap
DIAMETER: 0.164in
LENGTH: 1.64in
FLUTES: 2
DESCRIPTION: #8-32 UNC

Operation 10/12
DESCRIPTION: C-Bore Drill5
STRATEGY: Drilling
WCS: #0
TOLERANCE: 0in

MAXIMUM Z: 0.6in
MINIMUM Z: -0.665in
MAXIMUM SPINDLE SPEED: 7640rpm
MAXIMUM FEEDRATE: 30in/min
CUTTING DISTANCE: 1.12in
RAPID DISTANCE: 7.736in
ESTIMATED CYCLE TIME: 6s (1.9%)
COOLANT: Flood

T2 D2 L2
TYPE: flat end mill
DIAMETER: 0.5in
LENGTH: 1.1in
FLUTES: 3
DESCRIPTION: 1/2" Flat Endmill

Operation 11/12
DESCRIPTION: Bore1
STRATEGY: Bore
WCS: #0
TOLERANCE: 0in
STOCK TO LEAVE: 0in

MAXIMUM Z: 0.6in
MINIMUM Z: -1.07in
MAXIMUM SPINDLE SPEED: 7640rpm
MAXIMUM FEEDRATE: 40in/min
CUTTING DISTANCE: 6.53in
RAPID DISTANCE: 13.406in
ESTIMATED CYCLE TIME: 10s (3%)
COOLANT: Flood

T2 D2 L2
TYPE: flat end mill
DIAMETER: 0.5in
LENGTH: 1.1in
FLUTES: 3
DESCRIPTION: 1/2" Flat Endmill

Operation 12/12
DESCRIPTION: 2D Chamfer1
WCS: #0
TOLERANCE: 0in
STOCK TO LEAVE: 0in

MAXIMUM Z: 0.6in
MINIMUM Z: -0.37in
MAXIMUM SPINDLE SPEED: 6110rpm
MAXIMUM FEEDRATE: 40in/min
CUTTING DISTANCE: 35.884in
RAPID DISTANCE: 9.263in
ESTIMATED CYCLE TIME: 55s (16.7%)
COOLANT: Flood

T4 D4 L4
TYPE: spot drill
DIAMETER: 0.625in
TIP ANGLE: 90°
LENGTH: 1.35in
FLUTES: 3
DESCRIPTION: 5/8" Spot Drill

Chart 4-1b. CNC Mill Example Setup Sheet page 2

Introduction to CAM for CNC Turning

Lathe Example

Use the steps described in the first section of this chapter to startup Autodesk® Fusion 360. Open the file by using the Open from My Computer or from projects that have been saved to the cloud via the Data Panel (Figure 4-41).

The user interface will be similar to what was described in the first section until changes specific to the lathe are made. For this example, the frontside operation only will be completed.

Manufacture Setup for an Existing Lathe Model

1. Open the model file in Autodesk® Fusion 360.
2. Save the file to the cloud with the name: Lathe Example.
3. Use the drop-down for User Account, select Preferences and set the following:

a. Under General, select Manufacture and ensure that the checkbox for Enable Cloud Libraries is checked.
b. For Material, change the Physical Material Category to Metal and the Physical Material name to steel AISI 4340 242 HR. The Appearance Category should also be changed to Metal/Steel and Material Name to Steel/Satin. These settings ensure proper calculation of feeds and speeds based on the cutting tool selected.
c. For Default Units, verify that Manufacture is set to inches, press Apply and then OK.

4. Change the Workspace to Manufacture.
5. In the Browser tree directly under CAM Root, check the Units to confirm that inch is selected. Change if necessary.

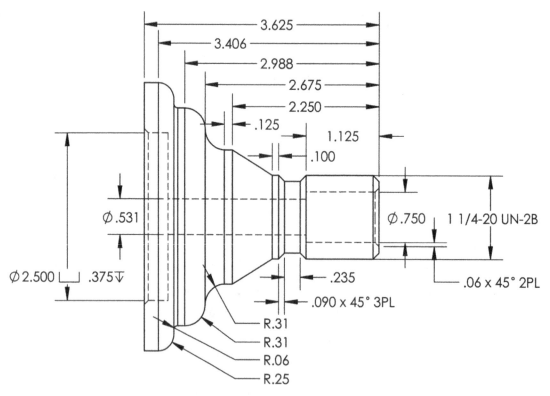

Material = 4340 Alloy Steel, 4.0 Diameter Bar Stock

Figure 4-41.   Design model for CAM lathe example

6. Press the Setup button. This step automatically generates stock data based on the maximum part geometry boundaries and the Setup dialog is displayed. Set the view to Front.

7. Change the default settings to match what is shown in the Figure 4-42. Each of the dialog items should be changed to suit specific needs. The Operation Type is changed to Turning or mill/turn for this example.

Lathe Example—Work Coordinate System (WCS)

1. **Z-Axis (Rotary Axis): Face.** If the Z-axis Gnomon is pointing toward the right, this is correct. If it is not, check the flip Z-axis checkbox.

2. **X-Axis.** If the X-axis Gnomon is pointing upward, this is correct. If it is not, check the flip X-axis checkbox.

3. **Origin.** Use the drop-down and change to Model front. This will change the location of the Gnomon to the finished face of the model. What this means is that the part zero (G54) is set to the center of the part for the X and Z-axes and at the finished face in the Z-axis.

> **Lathe Example Stock tab.** Change the default settings to match the values shown in Figure 4-43. Change the Model Position to Offset from front and set the Offset to .04 inch.

> **Lathe Example Post Process tab.** Change the default settings to match the values shown in Figure 4-44. Set the Program Name/Number to 2819 and the Program Comment to Lathe Example. Remember that the WCS value of (0) will output a G54 for the Work Offset. Press OK to accept the Setup1 settings.

Figure 4-42. Lathe example setup tab

Figure 4-43. Lathe example stock tab

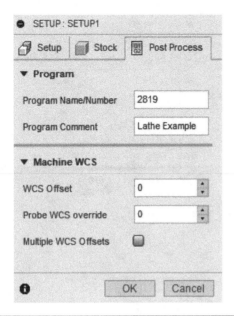

Figure 4-44. Lathe example stock tab

Loading a CAM Tool Library

For the example presented here, we will use the tools in the Tormach – 15L Slant-PRO – Starter. To install the tool library to the cloud, follow these steps:

1. Use your browser and search to: https://cam. autodesk.com/hsmtools, Tools for Fusion 360 and Autodesk HSM.
2. Filter the Vendor to Tormach.
3. Left-click on "For Fusion 360", to download the file to the Download folder.
4. Open the Data Panel and use the Leave Data Details (back arrow)—if necessary, scroll to Libraries and double-click to open Assets.
5. Select the CAM Tools folder.
6. Press the Upload button.
7. Press the Select button, go to the Downloads folder and select the Tormach-Tormach – 15L Slant-PRO – Starter.json file you just downloaded and press Open.
8. Press the Upload button.
9. Close the Data Panel.

Now the tools will be available via the cloud library for use wherever you log-in.

Turning

The first operation for this example will be to remove the excess (.040 inch) material from the face of the stock. Select the drop-down from the Turning icon and choose the Turning Face option from the menu (Figures 4-45 and 4-46).

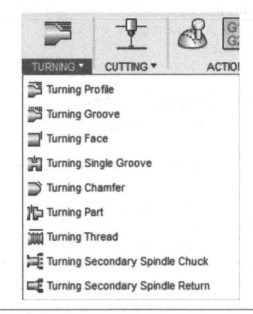

Figure 4-45. Turning menu

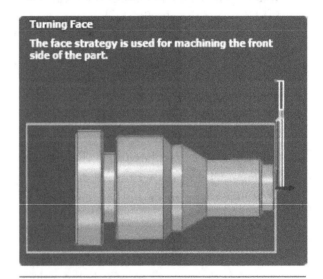

Figure 4-46. Turning face flyout

Lathe Example Face Tool tab. For Tool, press the Select button and the Select Tool dialog will display (Figure 4-47). Check the tick-box for Cloud Libraries and the Tormach-Tormach – 15L Slant-PRO – Starter.

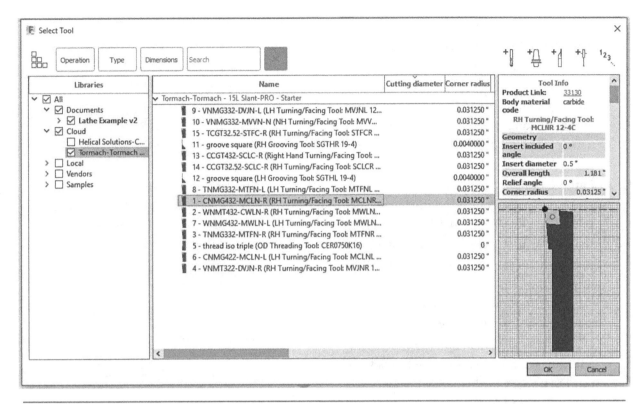

Figure 4-47. Select tool library dialog

From the list of tools, highlight the CNMG432-MCLN-R (RH Turning/Facing Tool MCLNR and press OK.

Observe the settings shown in Figure 4-48. Set the Coolant to Flood, check the tick-box to Use Constant Surface Speed and also to Use Feed per Revolution.

Lathe Example Face Stock tab. In this case, no changes are needed to the dialog (Figure 4-49).

Figure 4-48. Face tool tab

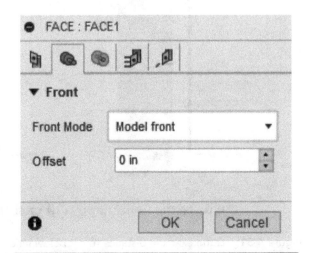

Figure 4-49. Lathe example face stock tab dialog

Lathe Example Face Radii tab. The only change needed to this dialog is the Inner Radius section (see Figures 4-50 and 4-51).

Lathe Example Face Passes tab. There are no changes required to the Passes dialog (Figure 4-52).

Lathe Example Face Linking tab. There are no changes required to the Lathe Face Linking dialog (Figure 4-53). Press OK to accept the tool path. A back-plot of the tool path and an image of the tool will be displayed on the Canvas area.

Use the Simulate icon to access the Simulate dialog. Playback the simulation and observe for satisfactory conditions. If adjustments are necessary, right-click on the [T1] Face1 path in the browser and select the Edit function.

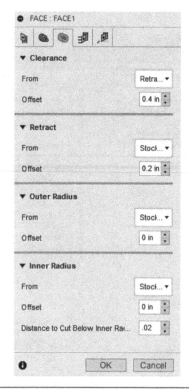

Figure 4-50. Lathe example face radii tab dialog

Distance to Cut Below Inner Radius

This is an adjustment to a Face or Part cut to position the tool nose past the Inner Radius position. Use this to cut past the Centerline of the part.

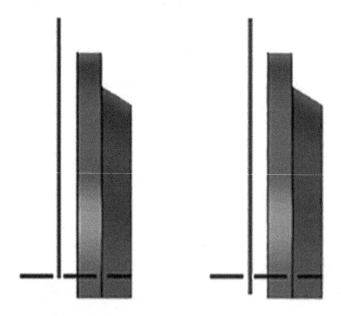

Cut up to the CenterLine   Cut past the CenterLine

Figure 4-51. Lathe example distance to cut below inner radius

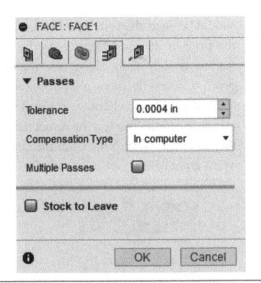

Figure 4-52. Lathe example face passes dialog

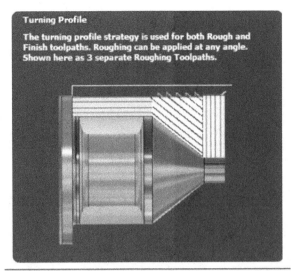

Figure 4-54. Turning profile flyout

used as for the facing operation just completed. See the settings in Figure 4-55 and don't forget to observe the helpful visual hints when you hover the mouse over each field. Change the Grooving to "Don't allow grooving" by accessing the drop-down in order to force the tool to avoid the grooved section of the profile (see Figure 4-56 for functional descriptions).

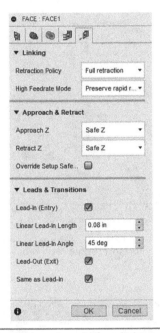

Figure 4-53. Lathe example face linking dialog

Turning Profile Roughing

The second operation for this example will be to remove stock material along the profile of the finished model. Select the drop-down from the Turning icon and choose the Turning Profile option from the menu (Figure 4-45). The flyout will be displayed as shown in Figure 4-54.

Lathe Example Profile Tool tab. For this roughing operation the same tool can be

Figure 4-55. Turning profile tool tab

Grooving

Use this to allow or restrict undercut toolpath motion. Can be used to keep the tool from dipping into channels along the diameter, face or end of the part.

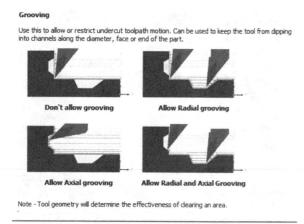

Note - Tool geometry will determine the effectiveness of clearing an area.

Figure 4-56. Turning profile grooving field

Lathe Example Profile Geometry tab. The model geometry is automatically recognized. In the dialog Front section, use the drop-down to change the setting to Model front and, for the Back section, change the setting to Model back and set the Offset value to –.2 inches (Figure 4-57). This will cut material in order to allow room for the parting tool starting.

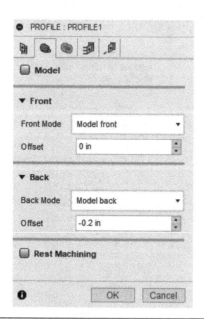

Figure 4-57. Turning profile geometry tab

Lathe Example Profile Radii tab. All of the settings are default except in the Inner Radius section of the dialog. In this case, the starting point for the profile cutting

should begin at the vertex of the chamfer at the front of the model profile. In the Inner Radius section of the dialog use the drop-down to access the menu of options (Figures 4-58 and 4-59) and choose Selection. Use the mouse to manually select the vertex between the face of the part model and the chamfer at the front of the part (Figure 4-60).

Figure 4-58. Turning profile radii tab

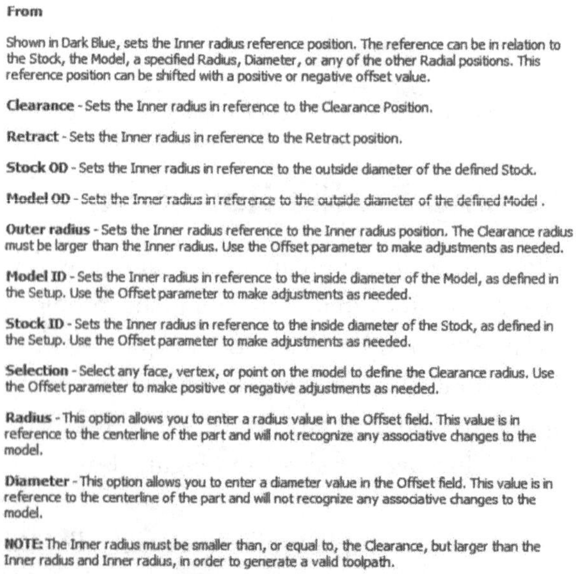

From

Shown in Dark Blue, sets the Inner radius reference position. The reference can be in relation to the Stock, the Model, a specified Radius, Diameter, or any of the other Radial positions. This reference position can be shifted with a positive or negative offset value.

Clearance - Sets the Inner radius in reference to the Clearance Position.

Retract - Sets the Inner radius in reference to the Retract position.

Stock OD - Sets the Inner radius in reference to the outside diameter of the defined Stock.

Model OD - Sets the Inner radius in reference to the outside diameter of the defined Model.

Outer radius - Sets the Inner radius reference to the Inner radius position. The Clearance radius must be larger than the Inner radius. Use the Offset parameter to make adjustments as needed.

Model ID - Sets the Inner radius in reference to the inside diameter of the Model, as defined in the Setup. Use the Offset parameter to make adjustments as needed.

Stock ID - Sets the Inner radius in reference to the inside diameter of the Stock, as defined in the Setup. Use the Offset parameter to make adjustments as needed.

Selection - Select any face, vertex, or point on the model to define the Clearance radius. Use the Offset parameter to make positive or negative adjustments as needed.

Radius - This option allows you to enter a radius value in the Offset field. This value is in reference to the centerline of the part and will not recognize any associative changes to the model.

Diameter - This option allows you to enter a diameter value in the Offset field. This value is in reference to the centerline of the part and will not recognize any associative changes to the model.

NOTE: The Inner radius must be smaller than, or equal to, the Clearance, but larger than the Inner radius and Inner radius, in order to generate a valid toolpath.

Figure 4-59. Turning profile inner radius tab

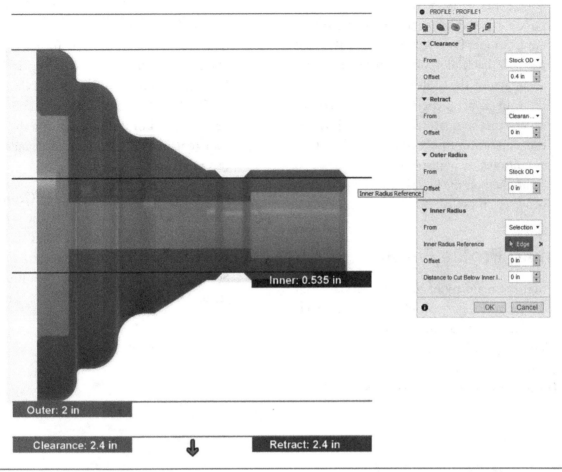

Figure 4-60. Turning profile inner radius reference selection

This forces the turning profile to start at this point; otherwise, the profile will begin at X-axis zero by default.

Lathe Example Profile Passes tab. The Passes tab controls how the material is removed from the part profile (Figure 4-61). For this example, check the tick-box for Roughing Passes and change the Maximum Roughing Stepdown to .08 in. This value should always fall within the insert manufacturers recommendations based on the material type and can be found on the insert container box labeling. Check the tick-box for Stock to Leave and ensure the Radial Stock to Leave and Axial Stock to Leave are set to .004 in.

Lathe Example Profile Linking tab. No changes are required for the Linking tab

Figure 4-61. Turning profile passes tab dialog

dialog. Take a moment to study all of the possible options in each section for future applications. Press OK to accept the tool path. A back-plot of the tool path and an image of the tool will be displayed on the Canvas area. Use the Simulate icon to access the Simulate dialog and playback the simulation to observe for satisfactory conditions. If adjustments are necessary, right-click on the [T1] Profile1 path in the browser and select the Edit function.

Turning Profile Finish

Now that the majority of the material has been removed from the part profile, a finishing pass can be added to make the final cut. A different finishing tool will be used (refer back to Figure 4-49). Activate the Turning Profile again using the Turning drop-down (see Figure 4-62).

Figure 4-62.   Turning profile linking tab dialog

Lathe Example Turning Profile Tool tab. On the Tool section, press the Select button and make sure the Tormach-Tormach – 15L Slant-PRO – Starter tool library is highlighted. From the list, select the VNMT322-DVJN-R (RH Turning/Facing Tool and press OK. All of the settings from the prior tool remain and need not be changed on this tab (Figure 4-55).

Lathe Example Turning Profile Geometry tab. This tab is setup exactly the same as for the roughing tool except that the tick-box for Rest Machining is checked and the Source is "From previous operation(s)" (Figure 4-63).

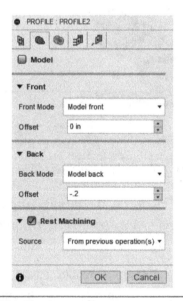

Figure 4-63.   Turning profile geometry tab dialog

Lathe Example Turning Profile Radii tab. Once again, this tab should be set the same as for the roughing tool operation (Figure 4-58 and 4-60).

Lathe Example Turning Passes tab. For this tab, change the Compensation Type to "In control" and un-tick the checkboxes for Roughing Passes and Stock to Leave (refer to Figure 4-61).

Lathe Example Turning Profile Linking tab. No changes are needed for this tab dialog. Press OK to accept the settings for the tool path. A back-plot of the tool path and an image of the tool will be displayed on the Canvas area.

Highlight Setup1 in the browser and playback the simulation to observe all tool paths for satisfactory conditions. If adjustments are necessary, right-click on the [T4] Profile2 path in the browser and select the Edit function.

Turning Groove

In this section, the clearance groove at the back of the threaded portion is completed. From the Turning icon drop-down choose Turning Groove and the flyout (Figure 4-64) will be displayed from the menu (Figure 4-45).

Lathe Example Turning Groove Tool tab. From the Tool tab press the Select button and choose the grooving tool as shown in Figure 4-65 and press OK. The remaining settings on the Tool tab are appropriate.

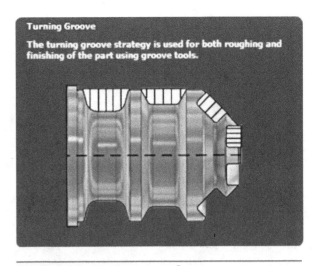

Turning Groove

The turning groove strategy is used for both roughing and finishing of the part using groove tools.

Figure 4-64. Turning groove flyout

Lathe Example Turning Groove Geometry tab. For the Geometry tab Front and Back Mode settings must be adjusted to capture the start and end of the groove geometry as shown in Figure 4-66. For the Front Mode, use the drop-down and pick Selection; then, use the cursor to select the front edge of the groove. For the

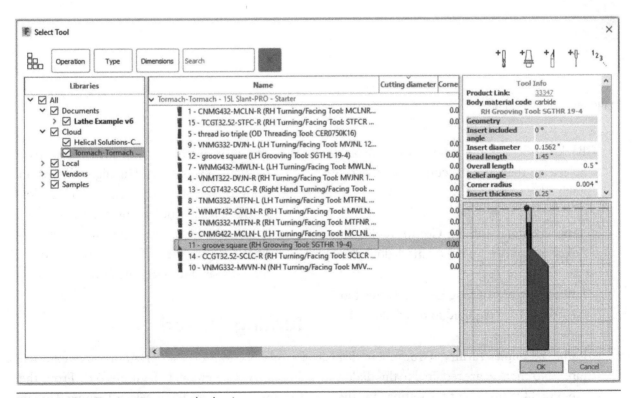

Figure 4-65. Turning groove tool selection

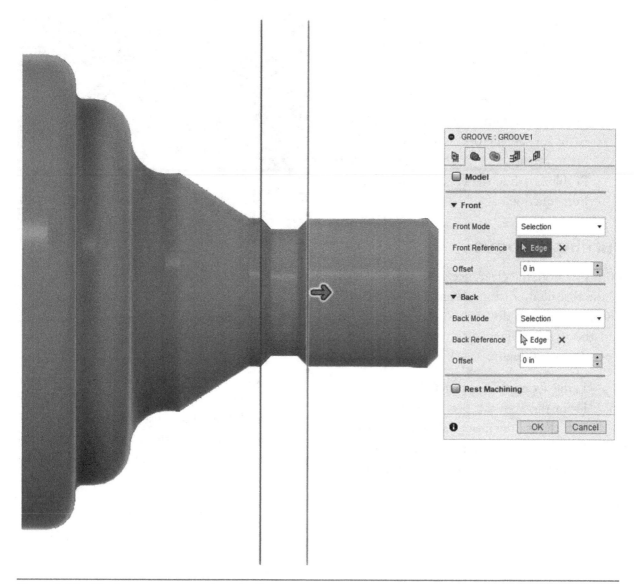

Figure 4-66. Turning groove geometry selection

Back Mode, use the drop-down and pick Selection; then, use the cursor to select the back edge of the groove.

Lathe Example Turning Groove Radii tab. No changes are required to this dialog tab.

Lathe Example Turning Groove Passes tab. No changes are required to this dialog tab.

Lathe Example Turning Groove Linking tab. No changes are required to this dialog tab. Press OK to accept the settings for the tool path. A back-plot of the tool path

and an image of the tool will be displayed on the Canvas area. Highlight Setup1 in the browser and playback the simulation to observe all tool paths for satisfactory conditions. If adjustments are necessary, right-click on the [T11] Groove1 path in the browser and select the Edit function.

Turning Thread

In this section of the program, the external threads will be created (Figure 4-67). From the Turning icon drop-down select Turning Thread from the menu (Figure 4-45).

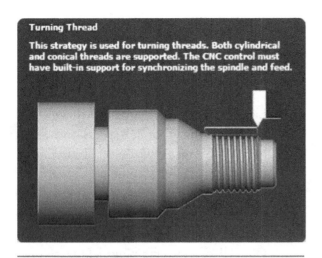

Figure 4-67. Turning thread flyout

Lathe Example Turning Thread Tool tab. From the Tool tab press the Select button and choose tool 5 – thread iso triple (OD Threading Tool) from the same library we have been using. Ensure the settings match those in Figure 4-68 and press OK.

Figure 4-68. Turning thread tool tab dialog

Lathe Example Turning Thread Geometry tab. For the Geometry tab, use the cursor to select the face of the diameter to be threaded. Leave the Frontside Stock Offset at the default value of 0.25 in. and change

the Backside Stock Offset to .125 to allow the tool to travel into the escape groove sufficiently.

Lathe Example Turning Thread Radii tab. Change the "Clearance From" to Selection and choose the thread diameter start at the top of the chamfer. Leave the Offset value at the default value of 0.4 in.

Lathe Example Turning Thread Passes tab. All of the settings on this tab are based on the geometry face selected and the thread pitch listed on the print. Hover over each field to get a detailed description for each setting. Especially note the changes for Infeed Mode to Reduced infeed, and Infeed Angle to 29.5. Check the tick-boxes for Spring Pass and Use cycle (see Figure 4-69).

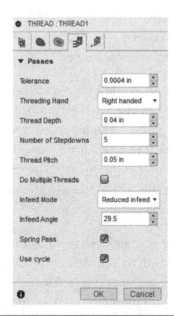

Figure 4-69. Turning thread passes tab dialog

Lathe Example Turning Thread Linking tab. The default settings for the Linking tab are acceptable. Press OK to accept the settings for the tool path. A back-plot of the tool path and an image of the tool will be displayed on the Canvas area. Highlight

Setup1 in the browser and playback the simulation to observe all tool paths for satisfactory conditions. If adjustments are necessary right-click on the [T5] Thread1 path in the browser and select the Edit function.

Lathe Example Center Drilling

For the internal operations the feature must first be Center Drilled for the subsequent drilling operation.

Press the Drilling icon on the Toolbar Ribbon.

Lathe Example Center Drilling Tool tab. On the Tool tab press the Select button to open the Select Tool dialog. From the Libraries column highlight the Samples. Filter the tool Type to Center Drills and press OK. From the list, choose the ½" 118 degrees 60 degrees – center drill (#6 Center Drill) and press OK. All of the remaining default settings are not changed.

Lathe Example Center Drilling Geometry tab. On the Geometry tab, the Selected faces is the default setting and this does not need to be changed. No other changes are required on this tab. Use the cursor to select the face of the chamfered .750-inch diameter hole.

Lathe Example Center Drilling Heights tab. All of the default settings on this dialog are fine, except Change the Bottom Height, From: to Hole Top and the Offset to –.375 inch.

Lathe Example Center Drilling Cycle tab. In this case, the default value for Cycle Type is Drilling – rapid out; press OK to accept the settings.

Follow common procedures for Simulation the results and make adjustments if needed.

Lathe Example Drilling

Press the Drilling icon on the Toolbar Ribbon to start the tool path.

Lathe Example Center Drilling Tool tab. On the Tool tab press the Select button to open the Select Tool dialog. From the Libraries column highlight the Samples. Filter the tool Type to: Drills and press OK. From the list, choose the $^{17}\!/_{32}$-inch 118-degree – drill ($^{17}\!/_{32}$ in.) and press OK. All of the remaining default settings are not changed.

Lathe Example Center Drilling Geometry tab. On the Geometry tab, use the cursor to select the cylindrical face of the .531-inch diameter through hole. There are no other changes required to this tab.

Lathe Example Center Drilling Heights tab. For the Heights tab: Change the Top Height, From: to Model Top. Change Bottom Height, From: to the Model Bottom. Tick the check-box for Drill Tip Through Bottom. Set the Break-Through Depth to .1 inch (Figure 4-70).

Figure 4-70. Lathe example drilling heights tab dialog

Lathe Example Center Drilling Cycle tab. In this case, change the Cycle Type to Deep drilling—full retract—and set the Pecking Depth to .35 inch. Press OK to accept the settings. Follow common procedures for Simulation the results and make adjustments if needed.

Lathe Boring Tool Creation

For this example, a tool must be built from scratch that matches the following Sandvik part #: E08-SCLPR 2-R. Follow these steps:

1. Press the Tool Folder icon from the Toolbar Ribbon Manage group. This will display the CAM Tool Library dialog.
2. In the Libraries column, check the tick-boxes for Local and Library and highlight Library. A blank screen should be displayed unless you have loaded other tools to this location.
3. In the upper-right corner of the dialog, press the plus sign for New Turning Tool. This will open the tool creation dialog.
   a. For the General tab, input the Description, Vendor, Product I.D. and Product Link (if known) for the tool given above.
   b. For the Insert tab, change the Type to Turning boring and, under Geometry, change the Corner radius to 1 = $^1/_{64}$ in.
   c. For the Holder tab, see Figure 4-71.
   d. No changes are required for the Setup tab.
   e. No changes are required for the Feed & Speed tab.
   f. For the Post Process tab, change the Number and Compensation offset to 7.
   g. Press OK to accept the new tool and close the CAM Tool Library.

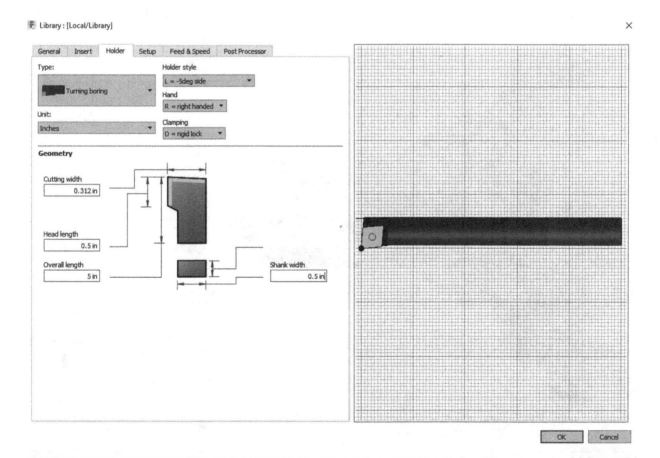

Figure 4-71. Lathe example boring tool creation dialog

Lathe Example Boring

The counterbored diameter at the front of the part can be bored now that we have created an appropriate tool. Start the process by pressing the Turning icon on the Toolbar Ribbon and selecting Turning Profile just like when doing outside diameters.

Lathe Example Bore Turning Profile Tool tab. For the Tool, press the Select button and highlight the Local Library. Choose the 7 – CNMT?2.51-DCLN-R (E08-SCLPR 2-R) tool. In the Mode & Direction section change the Turning Mode to Inside profiling.

Lathe Example Bore Turning Profile Geometry tab. On the Geometry tab Front section, change the Front Mode to Model front. On the Geometry tab Back section, change the Back Mode to Model front and set the Offset to –1.125 inches.

Lathe Example Bore Turning Profile Radii tab. On the Radii tab Outer Radius section, change the From to Selection and use the cursor to select the outer diameter of the ID Chamfer at the front of the part (.435). For the Inner Radius section, change the From setting to Selection and use the cursor to select the inner drilled diameter edge (.2655). The Retract section does not need any changes. For the Clearance section, change the From setting to Selection and use the cursor to select the inner drilled diameter edge (.2655) and change the Offset to –.01 in. (Figure 4-72).

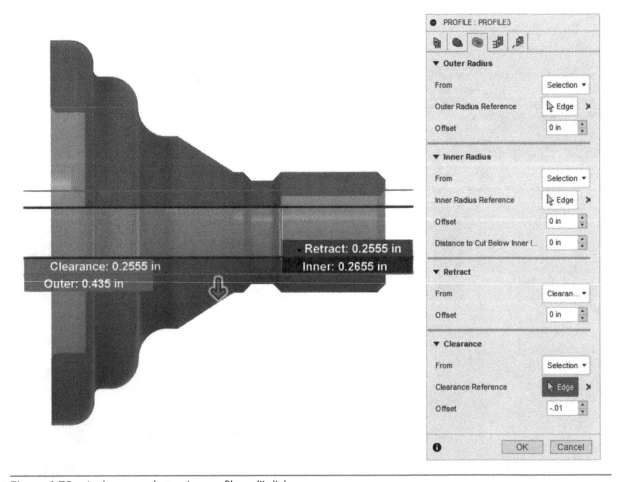

Figure 4-72. Lathe example turning profile radii dialog

Lathe Example Bore Turning Profile Passes tab. On the Passes tab, check the tick-box for Finishing Passes and change the Stepover amount to .04 in.

Check the tick-box for Repeat Finishing Pass. Check the tick-box for Roughing Passes and set the Maximum Roughing Stepdown to .04 in.

Lathe Example Bore Turning Profile Linking tab. In the Linking section, change the Safe Distance to .04 in.

In the Leads & Transitions section, change the Linear Lead-In-Length to .04 in.

Lathe Example Part-Off

The final operation to finish the project is to part-off. The existing grooving tool will not have enough length to perform the operation, so a copy will be made and edited for use.

Lathe Part-Off Tool Editing

1. Press the Tool Folder icon from the Toolbar Ribbon Manage group. This will display the CAM Tool Library dialog.
2. In the Libraries column, highlight the active Lathe Example list. Select the Grooving Tool used earlier in the program, right-click and select Copy Tool.
3. Highlight the Local Library and right-click in the white space below the existing tool, then press the Paste tool in the menu.
4. Double-click on the newly inserted tool to open it for editing.
5. Go to the Holder tab and change the Head Length to 2.1 in.
6. On the Post Processor tab, change the Number and Compensation offset to 8, and press OK to accept the changes.

Lathe Example Part-Off Tool tab. For the Tool tab, press the Select button and choose the grooving tool just created from the Local Library.

In the Feed & Speed section, change the Maximum Spindle Revolution to 1500.

Lathe Example Part-Off Geometry tab. There are no changes required to this dialog.

Lathe Example Bore Turning Profile Radii tab. There are no changes required to this dialog.

Lathe Example Bore Turning Profile Passes tab. On the Passes tab, check the tick-box for Use Pecking and set the Pecking Depth to .4 in and the Pecking Retract to .08 in.

Lathe Example Bore Turning Profile Linking tab. There are no changes required to this dialog.

Lathe Example Post Processing

Now that all of the tool paths have been generated and simulated, it is time to post process the CNC code for the machine. Press the Post Process icon on the Actions portion of the Toolbar Ribbon.

Change the settings in the Post Process dialog as shown below in Figure 4-73 and press the POST button when ready.

The Brackets Editor will open and display the code specific to the machine chosen. In this case, the actual CNC code (422 lines) will not be listed here to save space.

Lathe Example Setup Documentation

To generate a CNC Setup Sheet that lists all of the tools and their operations, press the Setup Sheet icon in the Toolbar Ribbon and select the folder location for the file to be saved. This will produce a document similar to the one depicted in Charts 3-2a, 2b and 2c in Chapter 3.

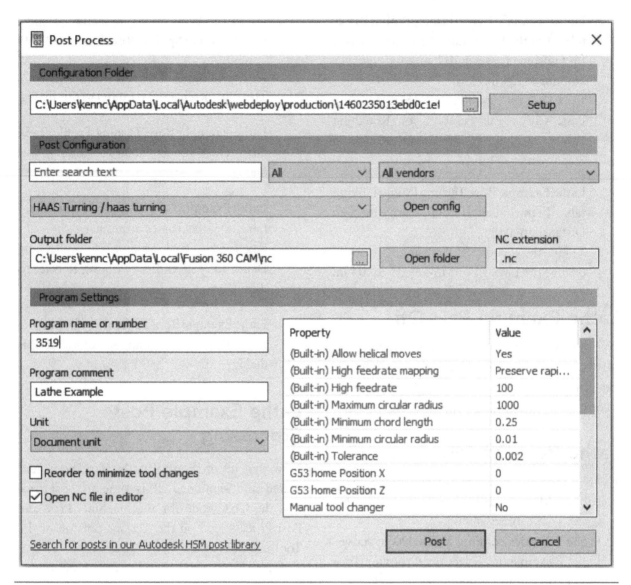

Figure 4-73.   Lathe example post processor dialog

Summary

In this section, the most prevalent method for CNC mill and lathe programming using CAM was introduced. Autodesk® Fusion 360 is widely used in manufacturing and machining environments and is one of many software programs available to meet the CNC programming task.

The skills gained using one CAM program is easily applied to similar programs. Solid models and electronic data are commonly transferred from the customer to the shop to aid in the process. Apply the knowledge you have learned within the workplace and you will be rewarded with the output of successful CNC programs and quality parts more quickly.

Introduction to CAM for CNC Milling, Study Questions

1. The icon on the Toolbar Ribbon used to access CAM in Autodesk® Fusion 360 is:
   a. Sketch
   b. Model drop-down and then Manufacture
   c. Model drop-down and then Simulation
   d. Model drop-down and then Drawing

2. The large display area where the part is shown is called?
   a. Graphical User Interface
   b. Display Screen
   c. Canvas
   d. View Cube

3. Where is the part material information for the project input?
   a. From the Setup icon
   b. In User Preferences
   c. In each tool path
   d. All of the above

4. Where is the stock size established for your model?
   a. From the Setup icon
   b. In the User Preferences
   c. In each tool path
   d. All of the above

5. The tool path tree is located under the:
   a. Canvas
   b. Data Panel
   c. Browser
   d. Toolbar Ribbon

6. The View Cube allows:
   a. Access to Top, Front and Right views
   b. Access to Bottom, Left and Back views
   c. View rotation
   d. All of the above

7. When using 2D Contour, what is the meaning of Compensation in computer?
   a. The value in the tools machine Wear Offset controls the amount the tool is compensated.
   b. The value in the tools machine Geometry Offset controls the amount the tool is compensated.
   c. The value in the machine Work Offset controls the amount the tool is compensated.
   d. The value of the tool radius amount of the tool is compensated within the program code.

8. When selecting tool path geometry, the red arrow that is displayed indicates?
   a. The tool path direction
   b. The tool offset side
   c. Both a and b
   d. The path starting point

9. For a drilled hole that is greater than two times its diameter in depth, which drilling cycle type should be used?
   a. Tapping
   b. Counterboring – dwell and rapid out
   c. Deep drilling – full retract
   d. Circular pocket milling

10. The action that creates the final CNC "G-code" for the machine tool is called?
    a. Simulation
    b. Post Processing
    c. Setup Sheet
    d. Generate

Introduction to CAM for CNC Turning, Study Questions

1. Cutting tools used for creation of CNC Turning program can be:
   a. Selected from Cloud Libraries
   b. Created from scratch
   c. Selected from Samples Libraries
   d. All of the above

2. Where is the information related to raw material stock size found?
   a. Post Process icon
   b. Setup icon, Stock tab
   c. Preferences, Material
   d. In each tool path

3. Where is the program number input?
   a. Either b or c
   b. In the Post Process Dialog
   c. Setup icon, Post Process tab
   d. Preferences

4. What is the most common final operation when CNC Turning?
   a. Facing
   b. Part-Off
   c. Profile
   d. Grooving

5. Which path type is used to remove the majority of materials from internal or external shapes?
   a. Part-Off
   b. Grooving
   c. Turning Profile
   d. Both b and c

6. When drilling on a CNC Lathe in hard materials, which cycle type should be used to reach the final depth for deep holes?
   a. Drilling rapid out
   b. Tapping
   c. Deep drilling – full retract
   d. Chip breaking – partial retract

7. Constant Surface Speed control is suitable for use for when:
   a. Drilling
   b. Grooving
   c. Part-Off
   d. None of the above

8. After each tool path is completed, an on-screen back plot is displayed for that path. What is the next best practice activity?
   a. Complete the next tool path
   b. Post Process the CNC code
   c. Simulate the individual operation
   d. Simulate all operations

9. Where are manufacturing units determined?
   a. Preferences, Manufacturing
   b. In the Browser, Change Active Units
   c. Either a or b
   d. During creation of the Model

10. Which of the following is not a function of the Setup Sheet?
    a. Provides tool data, stock and work piece position
    b. Generates the CNC code for the machine
    c. Generates an overview of the CNC program for the CNC operator.
    d. Provides and machining statistics

5

Preparing for Certification Exam

Exam Preparations

The goal of the *CNC Machining Exam Guide: for CNC Setup, Operation and Programming* is to prepare aspiring CNC machinists for written certification or credential testing that is offered through technical schools where CNC machine tool technology or similar courses are taught. The material can also be used for company-sponsored apprenticeship programs along with the practical application of the skills on relevant machine tools present on-the-job. The practice exam that follows contains questions based on the body of knowledge required for a CNC machinist to perform duties on-the-job that are similar to those on official exams presented at the schools or at the workplace. Study questions at the end of each section are based on the subject of the section, whereas the practice exam is a collection of all subject areas in a mixed format.

You should set a time limit that closely matches the requirement for official testing (typically 1.5–2 hours). The practical application portion of the certification will be completed at the school or workplace with the appropriate machine tools, inspection and gauging equipment. A certificate or credential is not a guarantee of employment, but it is a powerful ingredient for a persuasive resume that could open doors to new opportunities. These skills are verified by having proven proficiency (certification) and are widely sought after all over the country. Specifically, National Institute for Metalworking Skills (NIMS) credentialing is recognized and accepted nationally. Certificates from local technical schools and colleges are also a valued asset recognized around the country and especially in the surrounding region.

Exam Preparations

By reading this exam guide and completing the chapter study and practice exam questions, you will become familiar with the content on the written certification exam. The actual test questions will be different from, yet similar to, what is presented here. The types of questions presented in the practice exam will include true or false, multiple choice, matching and word answers. The actual certification exam may differ.

Each chapter in this exam guide covers specific skill areas that make up the body of knowledge for the exam questions. While the subject matter may be separated into sections in the actual certification exam, the subject areas are scrambled here in the practice exam to further solidify your understanding.

Common Sense Exam Preparation

Take the time to review all of the study questions at the end of each chapter (and review the chapter, if necessary) where a weakness is present. Consult with your instructor prior to the test to review any questions that remain.

Arrive at the test site early and well-rested with your attention focused on the test. Familiarize yourself with resources that are provided and make sure you bring reading glasses with you if you need them.

Test-Taking Techniques

1. Read and follow any special instructions for the test before starting. Just because there are many questions with a time limit, don't let that rush you.
2. Take the time to carefully read each question and all answers thoroughly. Answer every question; remember, you have studied the material, so even a guess will be an educated one.
3. As you proceed with the questions, pay attention to the amount of time. Avoid dwelling on one question for a long period; if you can't answer it, mark it and come back to it (remember to return and answer, though).
4. Watch carefully for disqualifying words in the question (such as "not") and be aware

that, in some cases, there is more than one correct answer.

5. If calculations are necessary, double-check your answer at least once to mitigate the chance of data entry errors.

6. In every case, you should go with your first instinct, don't second-guess yourself.

7. For multiple choice questions, read the question and then each of the answers. Eliminate the obvious incorrect answer first. From the remaining choices, pick the best fit answer.

Resources Allowed

The following resources items will be allowed during testing: *Machinery's Handbook*, scientific calculator, pencil, and blank scratch paper (you may be required to turn in your calculations and notes).

At the Testing Site

At the testing site, the instructor/proctor will go over important rules, special instructions and the time allowance with you. The instructor/proctor will indicate when you can begin the test. Follow these instructions precisely. Best of luck!

Practice Exam

The questions are multiple choice, true/false and matching. Enter the answer you feel most accurately answers the question by circling or entering the answer in the space provided.

Return the completed exam to the instructor/proctor at the end of the testing period. You may use a calculator and *Machinery's Handbook*. No other reference materials can be used unless provided by your instructor/proctor. You may use the space on this sheet or separate scratch paper for calculations as needed. You must turn in your calculations to the instructor/proctor.

1. Safety glasses should be worn:

   a. most of the time
   b. the entire time you are in the shop
   c. only when working on machines
   d. all of the above

2. Programming is a method of defining tool movements through the application of numbers and corresponding coded letter symbols.

   T or F

3. A CNC lathe has the following axes:

   a. X, Y and Z
   b. X and Y only
   c. X and Z only
   d. Y and Z only

4. The tool used to locate part zero on a CNC milling machine is a(n) _____.

   a. wiggler
   b. edge finder
   c. centering scope
   d. all of the above

5. Program coordinates that are based on a fixed origin are called:

   a. incremental
   b. absolute
   c. relative
   d. polar

6. When referring to the polar coordinate system, the clockwise rotation direction has a positive value.

   T or F

7. On a two-axis turning center, the diameter controlling axis is:

   a. B
   b. A
   c. X
   d. Z

8. You should avoid using compressed air to clean chips from CNC machine tools because?

   a. flying chips may cause eye injuries
   b. it could create a dangerous mist that might be injurious to your health
   c. it can create a dangerous environmental situation
   d. all of the above

9. The inch-based Vernier Micrometer is capable of measuring accurately to:

   a. 0.0001" and 0.00001"
   b. 0.001" and 0.0001"
   c. 0.01" and 0.001"
   d. none of the above

10. A symbol on a blueprint that looks similar to a check mark and is accompanied by a number such as 63 is a _____ requirement.

    a. surface lay
    b. surface texture
    c. dimensional
    d. none of the above

11. A program block is a single line of code followed by an end-of-block character.

    T or F

12. Dial indicators can be used for:

    a. visual inspection
    b. centering and aligning work on machine tools
    c. checking for eccentricity
    d. all of the above

13. When using carbide inserted tools, cutting fluids _____.

    a. are sometimes not needed
    b. are used only when copious amounts may be applied evenly
    c. aid in the removal of chips
    d. all of the above

14. Each CNC program block contains one or more program words.

    T or F

15. When finish turning an aluminum alloy bar that is 2.3125 inches in diameter with an HSS turning tool, if the r/min (RPM) is 991, what is the cutting speed (SFPM)?

    a. 600
    b. 400
    c. 500
    d. 750

16. Which button is used to activate automatic operation of a CNC program?

    a. Emergency Stop
    b. CYCLE STOP
    c. CYCLE START
    d. AUTO

17. A surface comparison gauge can be used to _____.

    a. check whether a part's surface is parallel with given surface
    b. check angular surfaces
    c. verify machined part surface finish requirements
    d. none of the above

18. When referring to the polar coordinate system, the clockwise rotation direction has a(n) _____ value.

    a. positive
    b. incremental
    c. absolute
    d. negative

19. The shape of a CNMG432 carbide insert is:

    a. round
    b. 80-degree diamond
    c. square
    d. 55-degree diamond

20. A program block is _____.

    a. a single block switch on the control
    b. a single line of code followed by an end-of-block character
    c. a program word that skips the block
    d. none of the above

21. A bolt circle is to be programmed for 6 equally spaced holes on a 5.0-inch diameter. The first hole is at 0 degrees. Identify the rectangular coordinate for the fifth hole in the Y-axis (rotation is positive).

    a. .500
    b. 2.1651
    c. 2.1662
    d. 1.250

22. The counterclockwise direction of rotation is always a negative axis movement when referring to the HANDLE (pulse generator).

    T or F

23. What is the MMC size of a .500 +.003/−.000 diameter pin?

    a. .500
    b. .497
    c. .503
    d. .5015

24. In which quadrant are both X and Y coordinate values positive?

    a. one
    b. two
    c. three
    d. four

25. The Position tolerance zone is:

    a. circular
    b. cylindrical
    c. square
    d. linear

26. In a feature control frame, datums are listed in precedence of primary, secondary and _____.
    a. position
    b. tertiary
    c. basic
    d. MMC

27. If a 2.625-inch diameter bolt circle is to be programmed and the first hole starts at 15 degrees, identify rectangular coordinate for the X-axis.
    a. 1.2678
    b. .3397
    c. .6563
    d. .6794

28. CNC programming is a method of defining tool movements through the application of _____.
    a. coded letters
    b. symbols
    c. coordinates
    d. all of the above

29. Which would be considered a feature of size?
    a. profile
    b. hole
    c. flat surface
    d. the location of a hole

30. Calculate the amount of tool travel necessary to allow for the drill point, plus another .090", to drill through a .875" thick plate using a .453" diameter drill and with a point angle of 135 degrees.
    a. 1.0588
    b. .9650
    c. .9688
    d. 1.000

31. What is the MMC size of a .375 +.005/-.000 diameter hole?
    a. .375
    b. .380
    c. .3725
    d. .370

32. Which display screen lists the CNC program?
    a. POSITION page
    b. OFFSET page
    c. CURRENT COMMANDS
    d. PROGRAM page

33. The acronym CAD/CAM stands for Computer-Aided Design and Computer-Aided Manufacturing.
    T or F

34. If the Reset button is pressed during automatic operation, then spindle rotations, feed, and coolant will stop.
    T or F

35. When programming a tool path in CAM, why would you set the Compensation Type to "Control" in the parameters?
    a. so that an undersized or worn cutter can be compensated and used
    b. so that the control offsets the tool path for only one set size of cutter
    c. to set the value in the Wear Offset column of the Offset register
    d. because the "In Computer" setting requires calculation for the tool radius

36. Match the following definitions with the proper CNC milling M-code.

a.	Program Stop	M30	___
b.	Optional Stop	M06	___
c.	End of Program	M03	___
d.	Spindle on Clockwise	M01	___
e.	Spindle on Counter Clockwise	M00	___
f.	Spindle Off	M19	___
g.	Flood Coolant On	M05	___
h.	Flood Cooloant Off	M04	___
i.	Spindle Oreienation	M09	___
j.	Tool Change	M08	___

37. Match the following CNC milling definitions with the proper G-code.

a.	Linear Interpolation	G54	___
b.	Circular Interpolation Clockwise	G81	___
c.	Rapid Traverse	G28	___
d.	Dwell	G40	___
e.	Absolute Coordinate System	G00	___
f.	Incremental Coordinate System	G41	___
g.	Canned Cycle Cancellation	G20	___
h.	Peck Drilling Cycle	G84	___
i.	Drilling Cycle	G21	___
j.	Cutter Compensation Left	G80	___
k.	Cutter Compensation Right	G83	___
l.	Cutter Compensation Cancel	G04	___
m.	Zero Return	G91	___
n.	Inch Programming	G42	___
o.	Millimeter Programming	G43	___
p.	R.H. Tapping Cycle	G01	___
q.	Fixture Offset Command	G03	___
r.	Postive Tool Length Compensation	G90	___
s.	Circular Interpolation Counter Clockwise	G02	___

38. Match the following CNC milling definitions with the proper letter address.

a.	Program Number	I	___
b.	Sequence Number	D	___
c.	Preparatory Function	T	___
d.	Miscellaneous Function	R	___
e.	Feed Rate	H	___
f.	Spindle Function	N	___
g.	Tool Function	J	___
h.	Tool Length Compensation	M	___
i.	Diameter Compensation	S	___
j.	Dwell in Seconds	Q	___
k.	Arc Center X-axis	P	___
l.	Arc Center Y-axis	G	___
m.	Arc Radius	F	___
n.	Peck Amount in Canned Drilling Cycle	O	___

39. A basic dimension is _____.
   a. a numerical value denoting the exact size, profile, orientation, or location of a feature or datum
   b. a dimension used for information only
   c. the measured size of a part after manufacture
   d. none of the above

40. During setup, the mode button used to allow for manual movement of the machine axes is:
   a. AUTO
   b. MDI
   c. EDIT
   d. HANDLE/JOG

41. Of the following choices, which is the best method for compensating for dimensional inaccuracies caused by tool deflection or wear when turning?
   a. Geometry Offsets
   b. Wear Offsets
   c. Tool Length Offsets
   d. Absolute Position Register

42. Plug gauges are used to check whether _____.
   a. hole diameters are within specified tolerances
   b. the hole was located correctly
   c. the micrometer measurement was correct
   d. none of the above

43. The default (at machine start-up) feed rate on lathes is typically measured in:
   a. cutting speed
   b. inches per revolution in/rev
   c. inches per minute in/min
   d. constant surface speed

44. Calculate the r/min (RPM) best suited for drilling a 9/16-inch diameter hole through a plate that is 1-1/4 inches thick. The material is 303 stainless steel.

    a. 85
    b. 577
    c. 679
    d. 55

45. The advantage of using G96 Constant Cutting Speed in turning is that as the diameter changes (position of the tool changes in relation to the centerline), the r/min increases or decreases to accomplish the programmed cutting speed.

    T or F

46. The cutting edge-angle of a DNMG332 carbide insert is?

    a. 80 degrees
    b. 90 degrees
    c. 55 degrees
    d. 60 degrees

47. When Position tolerance on an MMC basis is applied to a feature, the tolerance on the actually produced feature _____ as the size departs from MMC size.

    a. stays the same
    b. increases
    c. decreases
    d. none of the above

48. Which operation selection button allows for the execution of a single CNC block command?

    a. DRY RUN
    b. SINGLE BLOCK
    c. BLOCK SKIP
    d. OPTIONAL STOP

49. When programming arc and angular cuts using tool nose radius compensation G41 and G42, which is used for facing and turning?

    a. G41 facing, G42 turning
    b. G42 facing, G41 turning
    c. neither, the tool nose radius amount must be calculated and programmed to compensate
    d. G41 for facing if contours or angles are involved, G42 for turning if contours or angles are involved

50. Tool tip orientation needs to be identified in the controller when TNRC is used to program functions G41 or G42. How is this information input?

    a. The number is input by R into the program
    b. The number is input in the Tip Direction column of the offset page
    c. The number is input by R into the offset page
    d. The number is input by T into the program

51. Cutter diameter compensation G41 and G42 offset the cutter to the left or the right. Which command is used for climb (down) milling?

    a. G40
    b. G41
    c. G42
    d. G43

52. Which mode switch/button enables the operator to make changes to the program?

    a. EDIT
    b. MDI
    c. AUTO
    d. JOG

53. A datum is _____.
    a. a general term applied to a physical portion of a part
    b. the actual feature of a part
    c. an exact point, axis, or plane
    d. all of the above

54. Sequence (N) numbers in programs may be omitted entirely and the program will execute without any problem.

    T or F

55. When chaining part geometry, the direction the chaining arrow points determines whether the tool will travel on the left or right of the selected line.

    T or F

56. When incremental programming is required in turning diameters and facing, which letters identify the axis movements respectively?
    a. I and J
    b. U and W
    c. I and K
    d. U and V

57. A block of codes at the beginning of the program are used to cancel modal commands and are called the "safety block." They are:
    a. G90 G54 G00
    b. G20 G90 G00
    c. G40 G80 G49
    d. G91 G28 G00

58. What other program word is necessary when programming G01 linear interpolation?
    a. S
    b. F
    c. T
    d. H

59. The letter address, O, is used to identify the program number and has no other use in programming.

    T or F

60. What is the angular value for the 3 o'clock position in the polar coordinate system?
    a. 0 degrees
    b. 90 degrees
    c. 180 degrees
    d. 270 degrees

Further Reading

American Society of Mechanical Engineers. *Dimensioning and Tolerancing: ANSI Y14.5m-1982 (American National Standard Engineering Drawings and Related Documentation Practices).* New York: American Society of Mechanical Engineers; 1983.

American Society of Mechanical Engineers. *Dimensioning and Tolerancing: ASME Y14.5-1994 (Engineering Drawing and Related Documentation Practices).* New York: American Society of Mechanical Engineers; 1995.

American Society of Mechanical Engineers. *Dimensioning and Tolerancing: ASME Y14.5-2009 (Engineering Drawing and Related Documentation Practices).* New York: American Society of Mechanical Engineers; 2009.

American Society of Mechanical Engineers. *Engineering Drawing Practices: ASME Y14.100 (Engineering Product Definition and Related Documentation Practices).* New York: American Society of Mechanical Engineers; 2017.

Evans, K. *Student Workbook for Programming of CNC Machines* (4th ed.). South Norwalk, CT: Industrial Press; 2016.

Gillis, C., Hammer, W. (deceased). *Hammer's Blueprint Reading Basics* (4th ed.). South Norwalk, CT: Industrial Press; 2018. http://new.industrialpress.com/hammer-s-blueprint-reading-basics-4th-edition.html.

Jackson, D. *Statistics for Quality Control.* South Norwalk, CT: Industrial Press; 2015. https://books.industrialpress.com/statistics-for-quality-control.html.

Oberg, E. *Machinery's Handbook.* South Norwalk, CT: Industrial Press; 2016.

References

Evans, K. *Programming of CNC Machines* (4th ed). South Norwalk, CT: Industrial Press; 2016. (excerpt in Chapter 1, p. 62–71)

Harvey Performance Co. *Helical Machining Guidebook: Quick Reference eBook for CNC Milling Practices and Techniques*. Gorham, ME: Helical Solutions; 2016. http://www.helicaltool.com/secure/Content/Documents/Helical_MachiningGuidebook.pdf (excerpt in Chapter 1, pp. 38–47)

L.S. Starrett Co. *Tools and Rules for Precision Measuring* (bulletin #1211). Athol, MA: Starrett; 2011. (excerpt in Chapter 1, pp. 194–217)

National Fire Protection Association. *NFPA 10: Standard for Portable Fire Extinguishers*. Washington, DC: NFPA; 2018.*

Puncochar, D., Evans, K. *Interpretation of Geometric Dimensioning and Tolerancing, Based on ASME Y14.5-2009*. South Norwalk, CT: Industrial Press; 2011. http://new.industrialpress.com/interpretation-of-geometric-dimensioning-and-tolerancing.html (excerpt in Chapter 1, pp. 91–184)

Waters, T.R., Putz-Anderson, V., Garg, A. *Applications Manual for the Revised NIOSH Lifting Equation* (No. 94-110). Springfield, VA: US Department of Commerce; 2014. https://www.cdc.gov/niosh/docs/94-110/

---

* Reprinted with permission from NFPA 704-2017, System for the Identification of the Hazards of Materials for Emergency Response, Copyright © 2016, National Fire Protection Association. This reprinted material is not the complete and official position of the NFPA on the referenced subject, which is represented solely by the standard in its entirety. The classification of any particular material within this system is the sole responsibility of the user and not the NFPA. NFPA bears no responsibility for any determinations of any values for any particular material classified or represented using this system.

Answer Key for Chapter Study Questions

Chapter 1, CNC Machining Fundamentals

Machine Safety

1. d 2. e 3. d 4. d 5. d 6. c 7. a 8. c 9. b 10. d 11. d

Types of CNC Machines

1. d 2. a 3. c 4. a 5. d

Machine Maintenance and Workspace Efficiency

1. d 2. c 3. d 4. c 5. d 6. d 7. a 8. d 9. b

Cutting Tool Selection

1. b 2. c 3. c 4. d 5. c 6. a 7. a 8. b 9. d 10. d

Machining Mathematics

1. c 2. d 3. a 4. c 5. a 6. a 7. a 8. b 9. d

Understanding Engineering Drawings

1. a 2. c 3. b 4. d 5. a 6. a 7. c 8. b 9. a 10. d

Basic Geometric Dimensioning and Tolerancing

Basic Geometric Dimensioning and Tolerancing, Study Questions

1. communications 2. symbols 3. American Society of Mechanical Engineers (ASME)
4. clarity 5. replace 6. total 7. form 8. size 9. function 10. tolerances, interchangeability
11. tolerance 12. plus/minus 13. size

Part 1, GD&T Symbols and Abbreviations, Study Questions

1. Q 2. B 3. E 4. T 5. S 6. F 7. E 8. A 9. V 10. X 11. G 12. C 13. H
14. U 15. K 16. N 17. W 18. L 19. P 20. I 21. M 22. J 23. D 24. O 25. R
26. Y 27. Z

Part 2, GD&T Datums, Study Questions

1. specified 2. dimension, orientation 3. surface (planes), axis, center lines, edges (lines), points,
areas 4. no 5. three 6. 90 7. circle 8. size 9. lines, points areas 10. tolerance
11. simulated 12. modifiers 13. convey 14. rock 15. flat

Part 3, GD&T Feature Control Frames, Study Questions

1. control 2. combined 3. surface 4. geometric characteristic 5. tolerance 6. left, right
7. primary 8. line 9. compartments 10. no 11. I and O

Part 4, Form, Orientation, Profile and Runout Tolerances, Study Questions

1. directions 2. radially 3. element 4. boundaries 5. entire 6. error 7. tolerance
8. diameter 9. attachment 10. implied 11. assumed 12. size 13. axis 14. distance
15. width 16. RFS 17. bilateral 18. line 19. profile 20. revolution 21. datums

Part 5, GD&T Tolerances of Location, Study Questions

1. position, concentricity 2. variation 3. RFS 4. cylindrical, non-cylindrical 5. basic
6. specified 7. increases 8. shift/rotation 9. axis 10. length, depth 11. virtual condition
12. MMC 13. MMC 14. size 15. non-cumulative

Deburring Tools and Techniques

1. a 2. File, Rout-A-Burr and Countersinks 3. c 4. c 5. c

Deburring Tools and Techniques, Practical Exercises

Demonstrate to your instructor as parts are completed.

Machined Part Inspection

1. 199 2. .344 3. .3238 4. b 5. d 6. d 7. c 8. d 9. b 10. d 11a. 4.0, 1.0, .2, .119,
.1007 (5) 11b. 2.0, .7, .108, .1001 (4)

Basic CNC Program Structure and Format

1. d 2. d 3. b 4. c 5. a

Chapter 2, Setup, Operation and Programming of CNC Mills

CNC Mill Setup and Operation

1. c 2. T 3. c 4. d 5. b 6. a 7. c 8. b 9. T 10. d

Introduction to CNC Mill Programming

1. b 2. d 3. c 4. b 5. d 6. c 7. d 8. a 9. b 10. b 11. a 12. c 13. d 14. a 15. c

Chapter 3, Setup, Operation and Programming of CNC Lathes

CNC Lathe Setup and Operation

1. d 2. a 3. c 4. a 5. d 6. d 7. c 8. d 9. c 10. c

Introduction to CNC Lathe Programming

1. b 2. b 3. d 4. b 5. T 6. a 7. b 8. b 9. d 10. b 11. d

Chapter 4, Introduction to Computer Aided Manufacturing

Introduction to CAM for CNC Milling

1. b 2. c 3. b 4. a 5. c 6. d 7. d 8. c 9. c 10. b

Introduction to CAM for Turning

1. d 2. b 3. a 4. b 5. c 6. d 7. d 8. d 9. c 10. b

Chapter 5, Preparing for the Certification Exam

Practice Exam

1. b 2. T 3. c 4. b 5. b 6. F 7. c 8. a 9. b 10. b 11. T 12. b 13. d 14. T
15. a 16. c 17. c 18. d 19. b 20. b 21. b 22. T 23. c 24. a 25. b 26. b
27. a 28. d 29. b 30. a 31. a 32. d 33. T 34. T 35. a

36.

a.	Program Stop	M30	c.
b.	Optional Stop	M06	j.
c.	End of Program	M03	d.
d.	Spindle on Clockwise	M01	b.
e.	Spindle on Counter Clockwise	M00	a.
f.	Spindle Off	M19	i.
g.	Flood Coolant On	M05	f.
h.	Flood Cooloant Off	M04	e.
i.	Spindle Oreienation	M09	h.
j.	Tool Change	M08	g.

37.

a.	Linear Interpolation	G54	q.
b.	Circular Interpolation Clockwise	G81	i.
c.	Rapid Traverse	G28	m.
d.	Dwell	G40	l.
e.	Absolute Coordinate System	G00	c.
f.	Incremental Coordinate System	G41	j.
g.	Canned Cycle Cancellation	G20	n.
h.	Peck Drilling Cycle	G84	p.
i.	Drilling Cycle	G21	o.
j.	Cutter Compensation Left	G80	g.
k.	Cutter Compensation Right	G83	h.
l.	Cutter Compensation Cancel	G04	d.
m.	Zero Return	G91	f.
n.	Inch Programming	G42	k.
o.	Millimeter Programming	G43	r.
p.	R.H. Tapping Cycle	G01	a.
q.	Fixture Offset Command	G03	s.
r.	Postive Tool Length Compensation	G90	e.
s.	Circular Interpolation Counter Clockwise	G02	b.

38.

a.	Program Number	I	k.
b.	Sequence Number	D	i.
c.	Preparatory Function	T	g.
d.	Miscellaneous Function	R	m.
e.	Feed Rate	H	h.
f.	Spindle Function	N	b.
g.	Tool Function	J	l.
h.	Tool Length Compensation	M	d.
i.	Diameter Compensation	S	f.
j.	Dwell in Seconds	Q	n.
k.	Arc Center X-axis	P	j.
l.	Arc Center Y-axis	G	c.
m.	Arc Radius	F	e.
n.	Peck Amount in Canned Drilling Cycle	O	a.

39. a   40. d   41. b   42. a   43. b   44. b   45. T   46. c   47. b   48. b   49. d   50. b   51. b   52. a
53. c   54. T   55. T   56. b   57. c   58. b   59. T   60. a

Index

About the Author

Ken Evans has enjoyed a career of more than 40 years working in the aerospace and manufacturing industry as an Innovation Center Director, Prototype Lab Manager, CNC Programmer, Author, Sales Engineer, Instructor, Quality Control Inspector, Prototype Machinist, Tool & Die and Mold-Maker. He is recognized for building partnerships, working within teams and passionate about continuous improvement to produce exceptional results.

Ken is currently the Director at the USTAR Innovation Center where member–partners design projects that are prototyped using CAD/CAM software, CNC machine tools, waterjet, laser, 3D printing, assembly, and testing.

Ken was a CNC Machine Tool Technology instructor at the Davis Applied Technology College in Kaysville, Utah, where he dedicated 20 years of service. At the college, he taught foundational through advanced-level courses in the machining curriculum, including CNC Programming, CAD/CAM, and GD&T classes for students, educators, and private industry.

Ken began his teaching career in 1984 at the T.H. Pickens Technical Center in Colorado, while he worked full-time as a CNC machinist and Quality Control Inspector for a local job shop. He completed his formal machinist OJT in 1976 at Cessna Aircraft in Wichita, Kansas.

Ken loves the outdoors; he enjoys gardening, golf, and high-performance autos.